Dopamine in the Gut

Jin-Xia Zhu
Editor

Dopamine in the Gut

Springer

Editor
Jin-Xia Zhu
Physiology and Pathophysiology
Capital Medical University
Beijing, Beijing, China

ISBN 978-981-33-6585-8 ISBN 978-981-33-6586-5 (eBook)
https://doi.org/10.1007/978-981-33-6586-5

This Springer imprint is published by the registered company Springer Nature Singapore Pte Ltd.
The registered company address is: 152 Beach Road, #21-01/04 Gateway East, Singapore 189721, Singapore

Contents

Chapter 1
Introduction

Sumei Liu and Jin-Xia Zhu

Abstract The digestive system is essential to life. It breaks food into small molecules and absorbs nutrients into the blood stream so then they can be used for energy, growth, and repair. Functions of the digestive system are controlled by the autonomic nervous system, gastrointestinal hormones, local paracrine messengers, immune factors, and microbiota in the gut. Dopamine is a type of catecholamine that is found not only in the brain but also in the gut, including the enteric nervous system, epithelial cells, endocrine cells, and immune cells. A great mount of dopamine is also detected in the feces. Dopamine receptors and the enzymes involved in dopamine synthesis and metabolism are widely distributed in the gut. Therefore, dopamine in the gut has attracted increasing attention in recent years. Degeneration of dopaminergic neurons in the substantia nigra of the midbrain has been found in patients with Parkinson's disease (PD). Paradoxically, enzymes for dopamine synthesis and dopamine transporter levels are higher in the gut of PD animal models. Patients with PD often have impairment in gastrointestinal function such as gastroparesis or constipation, which usually appears many years before motor symptoms are diagnosed. Dopamine has been found to influence gastrointestinal motility, secretion, and mucosal barrier function. Recently, dopamine has also been reported to have anti-inflammation and anti-tumor functions. This chapter provides a brief overview of the dopaminergic system and the latest advances in dopamine receptor signaling. The role of dopamine in the regulation of gut functions will be discussed in detail in the following chapters.

Keywords Digestive system · Catecholamine · Dopamine · D1-like dopamine receptors · D2-like dopamine receptors

---

S. Liu
Department of Biology, College of Science and Health, University of Wisconsin-La Crosse, La Crosse, WI, USA
e-mail: sliu@uwlax.edu

J.-X. Zhu (✉)
Department of Physiology and Pathophysiology, School of Basic Medical Science, Capital Medical University, Beijing, China
e-mail: zhu_jx@ccmu.edu.cn

J.-X. Zhu (ed.), *Dopamine in the Gut*, https://doi.org/10.1007/978-981-33-6586-5_1

Abbreviations

3-MT	3-Methoxytyramine
Akt	Protein kinase B
ALDH	Aldehyde dehydrogenase
ALR/AR	Aldehyde/aldose reductase
AMPA	α-Amino-3-hydroxy-5-methyl-4-isoxazolepropionic acid
CaMKII	Calcium/calmodulin-dependent protein kinase II
cAMP	Cyclic adenosine monophosphate
CDK5	Cyclin-dependent kinase 5
CNS	Central nervous system
COMT	Catechol-o-methyltransferase
CREB	cAMP response element-binding protein
D_1R	Dopamine D_1 receptor
D_{2L}	Dopamine D_2 receptor-long
D_2R	Dopamine D_2 receptor
D_{2S}	Dopamine D_2 receptor-short
D_3R	Dopamine D_3 receptor
D_4R	Dopamine D_4 receptor
D_5R	Dopamine D_5 receptor
DA	Dopamine
DAG	Diacylglycerol
DARPP-32	Dopamine and cAMP-regulated neuronal phosphoprotein
DAT	Dopamine transporter
DOPAC	3,4-Dihydroxyphenylacetic acid
DOPAL	3,4-Dihydroxyphenylacetaldehyde
DOPET	3,4-Dihydroxyphenylethanol
ENS	Enteric nervous system
$GABA_A$	Gamma-aminobutyric acid A receptor
GIRKs	Inwardly rectifying potassium channels
GPCR	G protein-coupled receptor
GSK-3	Glycogen synthase kinase 3
HVA	Homovanillic acid
IP_3	Inositol trisphosphate
L-DOPA	3,4-Dihydroxyphenylalanine
MAO	Monoamine oxidase
MAPK	Mitogen-activated protein kinase
MB-COMT	Membrane-bound COMT
NMDA	*N*-Methyl-D-aspartic acid
PD	Parkinson's disease
PIP_2	Phosphatidylinositol-4,5-bisphosphate
PKA	Protein kinase A
PKC	Protein kinase C
PLA_2	Phospholipase A_2

PLC	Phospholipase C
PLD	Phospholipase D
PP1	Protein phosphatase 1
PP2A	Protein phosphatase 2A
PP2B	Protein phosphatase calcineurin/protein phosphatase 2B
PPP1R1B	Protein phosphatase 1 regulatory subunit 1B
S-COMT	Soluble COMT
VMAT	Vesicular monoamine transporter

1.1 Physiological Function and Regulation of the Digestive System

The digestive system consists of the gastrointestinal tract and accessory organs. The gastrointestinal tract is a continuous hollow, twisting tube extending from the mouth to the anus. The accessory organs include the liver, pancreas, and gallbladder which add secretions to the lumen of the gastrointestinal tract. The primary functions of the digestive system are to take in food, break it down into simple nutrient molecules, absorb nutrients into the bloodstream, and eliminate the indigestible matter from the body. In addition, the digestive system works closely with other organ systems to maintain homeostasis of the body.

Four major processes occur in the digestive system: digestion, absorption, motility, and secretion. Digestion involves a series of catabolic chemical reactions in which complex food molecules are broken down into smaller, absorbable units by the digestive enzymes. Absorption refers to the passage of digested end products from the lumen of the gastrointestinal tract to the blood or lymph. Motility is generated by smooth muscle contractions, of which there are two basic types: peristalsis and segmentation. Peristalsis is a type of propulsive movement which propels the luminal contents forward through the gastrointestinal tract. Segmentation predominantly happens in the small intestine during the digestive state where it mixes food with digestive juices and facilitates absorption by increasing physical contact between the intestinal epithelial surfaces to the digested nutrients. Various digestive juices are secreted by exocrine glands into the lumen of the gastrointestinal tract. These secretions contain water, electrolytes, mucus, and other organic constituents such as enzymes and bile salts that are essential for chemical digestion and absorption.

Regulation of the digestive system functions is primarily targeted on motility and secretion to ensure maximal digestion and absorption of ingested food. The autonomic nervous system, gastrointestinal endocrine and paracrine, enteric amine, and immune factors are all involved in the regulation of digestive system functions. The autonomic nervous system consists of three divisions, the sympathetic, parasympathetic, and enteric nervous system (ENS). The ENS is considered the intrinsic control system of the gut and is referred to as "the-brain-in-the-gut" (Wood 2016). It is located within the gut wall and has two neuronal networks, the myenteric and submucosal plexus. The ENS controls gut motility, secretion, and local blood flow moment-by-moment and can function independently from the central nervous system (CNS). The extrinsic control system consists of both parasympathetic and sympathetic nerves and regulates gut motility and secretion via the ENS. In some instances, extrinsic nerve fibers directly innervate gastrointestinal smooth muscle and glands. For example, sympathetic nerve fibers directly innervate arterioles in the wall of the gastrointestinal tract and cause vasoconstriction.

The gastrointestinal tract can be considered as the largest endocrine/paracrine organ due to large number of endocrine/paracrine cells interspersing throughout the mucosa of the entire gastrointestinal tract. Under appropriate stimulation, the endocrine cells release hormones into the blood, which carries the hormones to other areas of the gut to regulate gastrointestinal functions. In contrast to endocrine signaling, paracrine signaling functions at relatively short range when secreted chemical messengers move limited distances by passive diffusion in the extracellular fluid and act on neighboring cells. The serotonin- and histamine-containing enterochromaffin-like cells in the lamina propria are examples of cells using paracrine signaling to regulate gut functions. Overall, gut hormones and paracrine agents seem to be involved in the control of various digestive processes, such as gastric acid secretion, bicarbonate secretion, enzyme secretion from the pancreas and gut, gastrointestinal motility, and local blood flow. In addition, they may have indirect effects on these processes by activating enteric neurons and/or other endocrine cells.

The gastrointestinal tract is also the largest immune organ in the body. The gut immune system comprises scattered immune cells in the epithelium and lamina propria, as well as various organized lymphoid structures, including the Peyer's patches of the small intestine, the appendix, the solitary lymphoid follicles of the large intestine, and the mesenteric lymph nodes (Yap and Marino 2018). The gut is constantly exposed to a vast array of food antigens, commensal microbiota, and pathogenic microorganisms. As such, the gut immune system has evolved mechanisms to avoid deleterious responses from food antigens, to harness the beneficial effects of commensal microbiota, and to detect and eliminate invading pathogenic organisms gaining entry to the body through the gut (Janeway et al. 2001). Disruption of the gut immune homeostasis may lead to diseases such as food allergy, inflammatory bowel disease, infection, or autoimmune diseases.

The lumen of the gastrointestinal tract contains a large number of bacteria, archaea, and eukarya, which are collectively called the "gut microbiota." The gut microbiota is an integral component of the human body and is involved in a range of physiological functions such as strengthening gut mucosal barrier, protecting against pathogens, modulating host immunity, and harvesting energy from luminal contents. Recently, the functions of gut microbiota have drawn a lot of attention. Maintaining the right balance of microorganisms that live in the gut is vital for physical and mental health, immunity, and more. Altered gut microbiota has been related to many gastrointestinal diseases, such as functional gastrointestinal disorders and inflammatory bowel diseases (Putignani et al. 2016), extraintestinal diseases, such as obesity, diabetes, and cardiovascular disease (Marzullo et al. 2020; Sekirov et al. 2010), and neurodegenerative disorders, such as Alzheimer's disease and Parkinson's disease (Ceppa et al. 2020). A healthy gut communicates with the brain through afferent nerves, hormones, immune factors, etc., which helps maintain overall body health and well-being.

1.2 Definitions of Monoamine, Biogenic Amine, Catecholamine, and Enteric Amine

Monoamines (also known as *biogenic amines*) refer to a group of small molecules that have only one amine group. All monoamines are produced by the enzymatic decarboxylation of aromatic amino acids such as phenylalanine, tyrosine, tryptophan, and histidine. Monoamines are classified into three types based on their amino acid origins: catecholamines, indolamines, and histamine. The *catecholamines* (including dopamine, norepinephrine, and epinephrine) are named on the basis of the presence of a catechol group. They are synthesized from tyrosine through a common pathway. Phenylalanine can be converted into tyrosine, which is then used to generate catecholamines. Indolamines (including serotonin and melatonin) are derivatives of tryptophan and share a common molecular structure tryptamine. Histamine is synthesized from histidine.

Monoamines serve as neurotransmitters in both the CNS and the peripheral nervous system. Monoamine neurotransmitters in the brain play important roles in motion, emotion, learning and memory, sleep and wakefulness, appetite, etc. Norepinephrine is also a main neurotransmitter used by the sympathetic nervous system. Many non-neuronal cells are able to synthesize monoamines. Melatonin is a hormone that is produced by the pineal gland and regulates sleep and wakefulness. Epinephrine is predominantly produced by the chromaffin cells in the adrenal medulla and serves as a fight-or-flight hormone. Histamine is produced by mast cells and basophils in the connective tissues where it is involved in local immune responses. Serotonin, histamine, and dopamine are the primary monoamines found in the gut (Hakanson et al. 1969); and they are collectively called *enteric amines*. Serotonin and histamine are involved in a variety of functions of the gut, and they have been reviewed extensively in several excellent review articles (Galligan 2004; Gershon 2013; Wood 2004, 2006). In this book, we focus on the recent progress toward understanding the role of dopamine in the regulation of gut functions (Fig. 1.1) and the underlying mechanisms. We hope the information could provide novel targets and approaches for body health, disease prevention, and pharmacological intervention in dopamine-related gastrointestinal disorders.

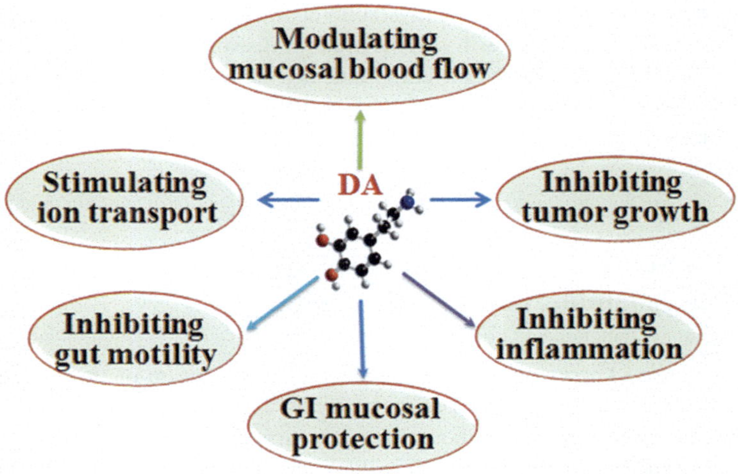

Fig. 1.1 Functions of dopamine in the gut. Dopamine (DA) is involved in the regulation of gut motility, epithelial ion transport, mucosal barrier, mucosal blood flow, inflammation, and tumor growth

1.3 Sources, Synthesis, and Metabolism of Dopamine

1.3.1 Sources of Dopamine

The major sources of dopamine in the brain are localized in the substantia nigra and ventral tegmental area of the midbrain. Dopamine neurons in the midbrain project to three forebrain areas, including the corpus striatum, the limbic areas, and the prefrontal cortex (Iversen and Iversen 2007; Klein et al. 2019; Wang et al. 2020). *The nigrostriatal pathway* carries signals from the substantia nigra pars compacta to the caudate and putamen. This pathway plays an essential role in the coordination of body movements. Degeneration of dopaminergic neurons in the substantia nigra causes Parkinson's disease with a characteristic motor dysfunction. *The mesolimbic pathway* originates in the ventral tegmental area and projects to the limbic areas, including the amygdala, pyriform cortex, lateral septal nuclei, and nucleus accumbens. In this pathway, dopamine mediates reward, reinforcement, and motivation functions. Many drugs of abuse work by affecting dopaminergic synapses in the mesolimbic pathway. *The mesocortical pathway* originates in the ventral tegmental area and projects to the prefrontal cortex. Mesocortical dopamine mediates emotion and cognitive behavior. Levels of dopamine in the prefrontal cortex help in improved working memory and attentions (Iversen and Iversen 2007; Klein et al. 2019; Wang et al. 2020). Besides the midbrain, a distinct group of neurons in the arcuate nucleus of the hypothalamus also produce dopamine. These tuberoinfundibular dopamine neurons project to the median eminence, release dopamine into the hypothalamo-hypophyseal portal system, and inhibit prolactin release from the anterior pituitary (Fitzgerald and Dinan 2008). More than 50% dopamine in the body come from multiple sources in the periphery, including sympathetic nerves, some primary sensory nerves, adrenal medulla, kidney, pancreas, retina, and gut (Rubi and Maechler 2010; Vieira-Coelho and Soares-da-Silva 1993). Some immune cells and gut microbiota can also produce dopamine (Asano et al. 2012; Rubi and Maechler 2010; Thomas Broome et al. 2020). Sources, synthesis, and metabolism of dopamine in the gut are discussed in Chap. 2 of this book.

1.3.2 Synthesis of Dopamine

All catecholamines are derived from a common precursor, the amino acid tyrosine (Klein et al. 2019). Phenylalanine can be used to synthesize tyrosine by the enzyme phenylalanine hydroxylase. The specific catecholamine that a cell can produce depends on the presence or absence of specific enzymes of the metabolic pathway (Fig. 1.2):

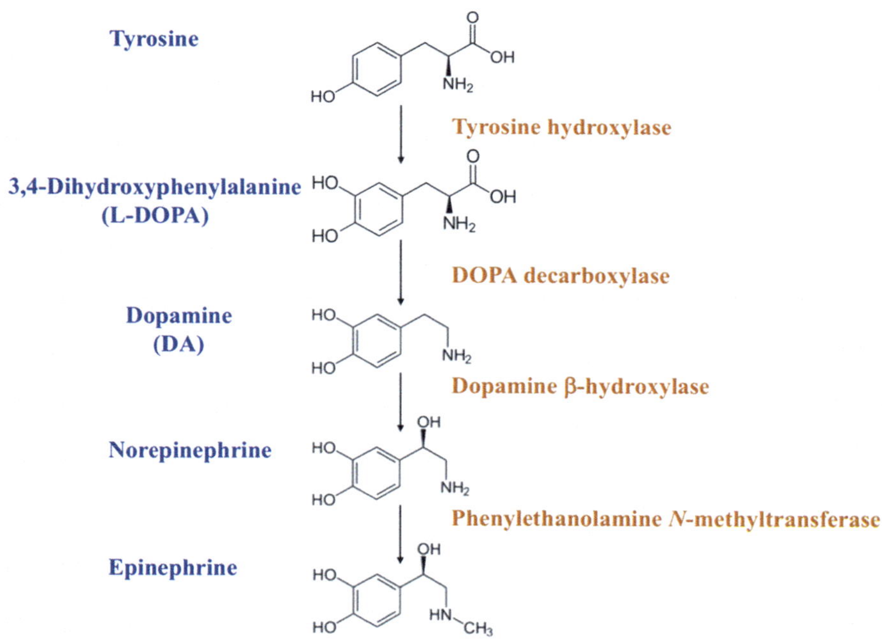

Fig. 1.2 Synthesis of catecholamines. The major catecholamine synthesis pathway starts at tyrosine. Tyrosine is hydroxylated to form 3,4-dihydroxyphenylalanine (L-DOPA) by tyrosine hydroxylase. Decarboxylation of L-DOPA by DOPA decarboxylase leads to the production of dopamine (DA). Besides acting as a neurotransmitter, DA also serves as the precursor of norepinephrine and epinephrine. Dopamine β-hydroxylase and phenylethanolamine *N*-methyltransferase catalyze the depicted reactions in a cell type-dependent manner

1. Tyrosine is hydroxylated into 3,4-dihydroxyphenylalanine (L-DOPA), catalyzed by the enzyme tyrosine hydroxylase. This is a rate-limiting step for the synthesis of all catecholamines.
2. L-DOPA is decarboxylated into dopamine, catalyzed by the enzyme aromatic amino acid decarboxylase (or DOPA decarboxylase). In dopamine-producing cells, chemical reactions stop at this step.
3. Dopamine is converted into norepinephrine by β-oxidation, catalyzed by the enzyme dopamine β-hydroxylase. This reaction happens in the noradrenergic neurons in the locus coeruleus of the brain stem, sympathetic ganglia, and chromaffin cells of the adrenal medulla.
4. Norepinephrine is converted into epinephrine. The enzyme that synthesizes epinephrine, phenylethanolamine *N*-methyltransferase, is present only in a restricted group of epinephrine-secreting neurons in the brain stem and in chromaffin cells of the adrenal medulla.

In dopamine-producing cells, following its synthesis in the cytoplasm, dopamine is transported into intracellular secretory vesicles via a vesicular monoamine transporter (VMAT). Two isoforms of VMAT (VMAT1 and VMAT2) have been identified in mammals (Erickson et al. 1996). VMAT2 is the exclusive isoform found in monoaminergic neurons (Erickson et al. 1996; Klein et al. 2019; Weihe and Eiden 2000). VMAT1 is the dominant isoform in endocrine cells, although both VMAT1 and VMAT2 are found to be co-expressed in some endocrine cells, such as the adrenomedullary chromaffin cells in human and rhesus monkey (Weihe and Eiden 2000). VMAT2 is also found in the pancreatic islets (Hong et al. 2014), as well as the platelets, basophils, mast cells, dendritic cells, etc. (Weihe and Eiden 2000). It is worth to mention that dopamine, like other catecholamines, constantly leaks out of the vesicular storage to the cytoplasm. Fortunately, VMAT avidly sequesters about 90% of the leaked catecholamines back into the vesicular storage. About 10% of the catecholamines escapes sequestration and is degraded in the cytoplasm (Eisenhofer et al. 2004). Therefore, the vesicular stores of catecholamines are not static. Instead, there is a dynamic equilibrium between passive outward leakage of catecholamines and active inward transport through VMAT.

1.3.3 Metabolism of Dopamine

Under physiological conditions, dopamine can be degraded mainly by three enzymes: monoamine oxidase (MAO), aldehyde dehydrogenase (ALDH), and catechol-*o*-methyltransferase (COMT). The degradation of dopamine mostly takes place in the same cells where it is produced, although the surrounding cells may also help dopamine degradation. In the brain, dopamine inactivation is mainly done by reuptake to presynaptic nerve terminals via dopamine transporter (DAT) and followed by sequestration into the synaptic storage vesicles by VMAT2 (Wang et al. 2020). Due to constant leakage from synaptic vesicles, DA accumulates in the

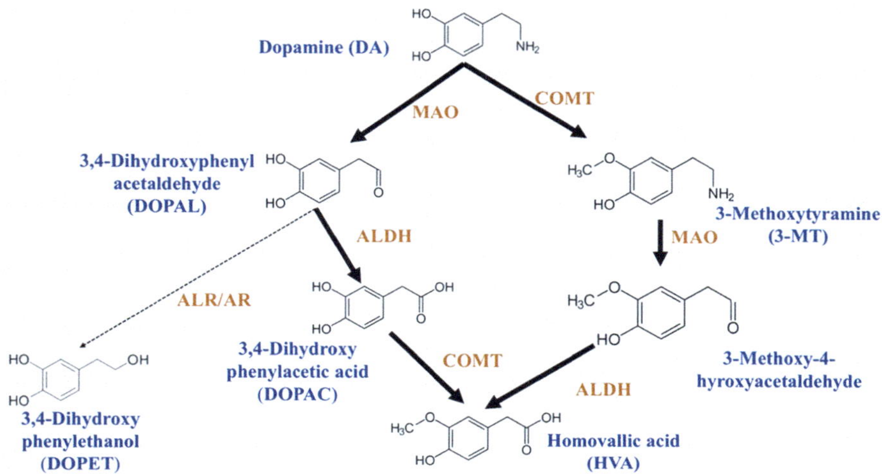

Fig. 1.3 Degradation of dopamine. DA degradation starts with deamination by MAO to generate DOPAL, which is then converted to DOPAC by ALDH. A small fraction of DOPAL is converted to DOPET by ALR/AR (dash arrow). In cells that express COMT, DOPAC is *O*-methylated by COMT to produce HVA. COMT can also directly convert DA to 3-MT, which is further metabolized by MAO and ALDH to produce HVA. DOPAC and HVA are the major end products of DA metabolism. *ALDH* aldehyde dehydrogenase, *ALR* aldehyde reductase, *AR* aldose reductase, *COMT* catechol-*O*-methyltransferase, *DA* dopamine, *DOPAC* 3,4-dihydroxyphenylacetic acid, *DOPAL* 3,4-dihydroxyphenylacetaldehyde, *DOPET* 3,4-dihydroxyphenylethanol, *HVA* homovanillic acid; *MAO* monoamine oxidase, *3-MT* 3-methoxytyramine

cytoplasm and is subsequently degraded by MAO (Craddock et al. 2006; Eisenhofer et al. 2004; Wang et al. 2020). MAO is located in the outer membrane of the mitochondria. Two isoforms of MAO, MAO-A and MAO-B, encoded by two separate genes, have been found, and both isoforms are involved in DA metabolism. In the CNS, both MAO-A and MAO-B are expressed in dopaminergic neurons, while MAO-B is mainly expressed in astrocytes (Agid et al. 1973; Masato et al. 2019). In dopaminergic neurons, oxidative deamination of dopamine by MAO results in the production of 3,4-dihydroxyphenylacetaldehyde (DOPAL), which is further metabolized to the carboxylic acid 3,4-dihydroxyphenylacetic acid (DOPAC) by ALDH. A smaller fraction of DOPAL is converted to 3,4-dihydroxyphenylethanol (DOPET) by aldehyde/aldose reductase (ALR/AR) (Masato et al. 2019; Meiser et al. 2013; Klein et al. 2019) (Fig. 1.3).

Besides dopaminergic neurons, surrounding glial cells can also take up DA from the synaptic cleft and degrade it. Within glial cells, MAO and ALDH convert DA to DOPAC. COMT catalyzes the transfer of methyl groups from *S*-adenosylmethionine to DOPAC, leading to the production of homovanillic acid (HVA), the primary end-product of DA metabolism in human (Klein et al. 2019; Meiser et al. 2013) (Fig. 1.3). Alternatively, COMT can also convert DA into 3-methoxytyramine (3-MT), which is further metabolized by MAO and ALDH to form HVA (Meiser et al. 2013) (Fig. 1.3). Two isoforms of COMT, the soluble (S-COMT) and membrane-bound (MB-COMT) isoforms, have been found, and they are coded by one single gene. MB-COMT is thought to be mainly responsible for the O-methylation of catecholamine neurotransmitters, whereas S-COMT is believed to be more important in the elimination of biologically active or toxic catechols and hydroxylated metabolites, mostly from exogenous sources (Myohanen et al. 2010). Both COMT isoforms are found in microglial cells and astrocytes in the CNS. COMT is also present in some neurons in the brain, such as the pyramidal neurons, cerebellar Purkinje and granular cells, and striatal spinal neurons, but not the dopaminergic nigro-striatal neurons (Myohanen et al. 2010).

MAO and COMT are not only expressed in the brain but also widely expressed in the peripheral tissues, such as adrenal medulla, liver, kidney, and gastrointestinal tract (Liu et al. 2018; Myohanen et al. 2010). MAO and COMT in the gut are further discussed in Chap. 2. Both MAO and COMT are responsible for the breakdown of norepinephrine and epinephrine. MAO is also involved in the breakdown of serotonin (Wang et al. 2020).

Although the main processes of DA degradation are based on oxidation reactions, DA and its metabolites can further undergo conjugation reactions before excretion, namely sulfation and glucuronidation. These reactions are known as phase II reactions of DA metabolism and can occur in both CNS and periphery (Abrantes Dias et al. 2020; Meiser et al. 2013).

1.4 Dopamine Receptor Classification, Signaling, and Function

Dopamine exerts its physiological functions by binding to five distinct but closely related dopamine receptors (D_1R-D_5R), which belong to the rhodopsin-like class A, seven-transmembrane domain, G protein-coupled receptor (GPCR) superfamily. Dopamine receptors are broadly expressed in the brain and in the periphery. Based on their structural, biochemical, and pharmacological properties, the five dopamine receptors are subdivided into two major groups, the D1-like (including D_1R and D_5R) and D2-like (including D_2R, D_3R, and D_4R) dopamine receptors (Beaulieu et al. 2015; Beaulieu and Gainetdinov 2011; Martel and Gatti McArthur 2020; Vallone et al. 2000).

1.4.1 D1-Like Dopamine Receptors

The two D1-like dopamine receptors, D_1R and D_5R, share a very high homology in their transmembrane domain regions (Sunahara et al. 1991). The genes for the D_1R and D_5R are intronless in the coding regions, but pseudogenes exist for the human D_5R and encode non-functional receptors (Beaulieu and Gainetdinov 2011). The

D_1R and D_5R demonstrate differential distribution patterns. In the brain, D_1R is the most widespread and is expressed at higher levels than any other dopamine receptors. D_1R is heavily expressed in the nigrostriatal, mesocortical, and mesolimbic dopaminergic pathways, such as the caudate, putamen, nucleus accumbens, substantia nigra, olfactory bulb, amygdala, and frontal cortex. Lower level of D_1R expression is detected in the hippocampus, thalamus, hypothalamus, cerebellum, and brain stem (Beaulieu and Gainetdinov 2011; Cai et al. 2013; Ran et al. 2019; Wang et al. 2016; Yang et al. 2019; Zhou et al. 2014, 2019). D_5R is found at low levels in various areas of the brain, including the prefrontal cortex, premotor cortex, cingulated cortex, entorhinal cortex, hippocampus, dentate gyrus, caudate, nucleus accumbens, substantia nigra, thalamus, and hypothalamus (Beaulieu and Gainetdinov 2011; Missale et al. 1998). It is generally accepted that D_1R and D_5R are expressed exclusively on postsynaptic neurons receiving dopaminergic innervation (Beaulieu and Gainetdinov 2011). In the periphery, D_1R and D_5R are found in the kidneys, heart, blood vessels, adrenal glands, digestive tract, pancreatic islets, and immune cells (Amenta et al. 1995; Aperia 2000; Cavallotti et al. 2010; Chen et al. 2014; Feng et al. 2013; Li et al. 2006, 2019; Missale et al. 1998; Wang et al. 2012; Zhang et al. 2012, 2015, 2017). Distribution of the D1-like dopamine receptors in the gut is further discussed in Chap. 3 of this book.

The D1-like dopamine receptors are coupled to the $G\alpha_{s/olf}$ family of G proteins. Activation of either D_1R or D_5R increases intracellular cyclic adenosine monophosphate (cAMP) production by stimulating adenylate cyclase, which results in the activation of the protein kinase A (PKA) (Beaulieu and Gainetdinov 2011; Vallone et al. 2000). PKA phosphorylates cytoplasmic and nuclear target proteins and elicits cellular responses. Examples of PKA targets affected by D1-like dopamine receptor stimulation include ionotropic glutamate receptors (AMPA and NMDA), gamma-aminobutyric acid A ($GABA_A$) receptors, certain ion channels and membrane transporters, and cAMP response element-binding protein (CREB) (Greengard 2001) (Fig. 1.4).

Over the past 30 years, a major advance in the understanding of the downstream effectors of dopamine receptor signaling pathways is the discovery of a 32-kDa dopamine and cAMP-regulated neuronal phosphoprotein (DARPP-32), also known as protein phosphatase 1 regulatory subunit 1B (PPP1R1B) (Fig. 1.4). PKA phosphorylates DARPP-32 at Thr^{34} (DARPP-$32_{(p\text{-}Th34)}$) and converts DARPP-32 into an inhibitor of protein phosphatase 1 (PP1). PP1 dephosphorylates substrates of PKA and other protein kinases. Inhibition of PP1 by DARPP-$32_{(p\text{-}Th34)}$ amplifies PKA-mediated cellular responses (Svenningsson et al. 2004). DARPP-32 could also be phosphorylated at Thr^{75} (DARPP-$32_{(p\text{-}Th75)}$) by cyclin-dependent kinase 5 (CDK5) in response to the increase of protein kinase C (PKC) levels. DARPP-$32_{(p\text{-}Th75)}$ inhibits PKA and suppresses PKA-mediated cellular responses (Svenningsson et al. 2004). Depending on the phosphorylation sites, DARPP-32 serves as an inhibitor of either PP1 or PKA. By modulating the activity of PP1 and PKA, DARPP-32 plays an integrative role in the biological actions of dopamine, as well as many other neurotransmitters and neuromodulators through the $G\alpha_s$/cAMP/PKA signaling pathway (Fig. 1.4).

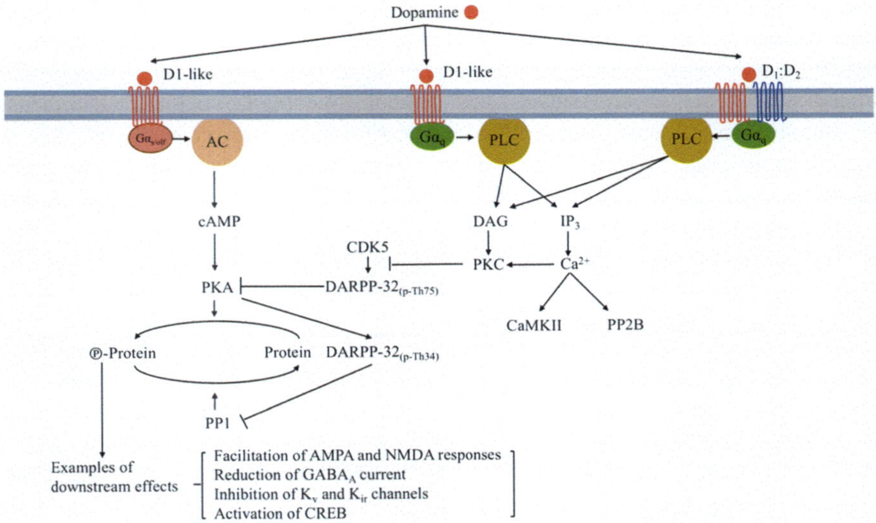

Fig. 1.4 Dopamine D1-like receptor family signaling pathways. The D1-like (D_1R and D_5R) dopamine receptors are generally coupled to the $G\alpha_{s/olf}$/cAMP/PKA signaling pathway. PKA phosphorylates DARPP-32 at Th^{34}, which results in inhibition of PP1 and amplification of PKA-mediated signaling. The D1-like dopamine receptors and the D_1:D_2 dopamine receptor heterodimers are also coupled to the $G\alpha_q$/PLC signaling pathway. Arrows indicate activation. T-arrows indicate inhibition. *AC* adenylate cyclase, *AMPA* α-amino-3-hydroxy-5-methyl-4-isoxazolepropionic acid receptor, *CaMKII* calmodulin kinase II, *CDK5* cyclin-dependent kinase 5, *CREB* cAMP response element-binding protein, *DAG* diacylglycerol, *DARPP-32* dopamine and cAMP-regulated neuronal phosphoprotein, *$GABA_A$* gamma-aminobutyric acid A receptor, *IP_3* inositol trisphosphate, *K_V* voltage-gated potassium channel, *K_{ir}* inward rectifier potassium channel, *NMDA* *N*-methyl-D-aspartate receptor, *PKA* protein kinase A, *PKC* protein kinase C, *PLC* phospholipase C, *PP1*, protein phosphatase 1, *PP2B* protein phosphatase 2B

In addition to coupling to $G\alpha_{s/olf}$ and stimulation of cAMP production, both D_1R and D_5R can activate $G\alpha_q$ to regulate phospholipase C (PLC) (Fig. 1.4). PLC cleaves phosphatidylinositol-4,5-bisphosphate (PIP_2) into diacylglycerol (DAG) and inositol trisphosphate (IP_3). DAG activates PKC. IP_3 induces intracellular calcium release from the endoplasmic reticulum, which leads to the activation of calcium-dependent PKC variants as well as calcium-regulated enzymes, such as the calcium/calmodulin-dependent protein kinase II (CaMKII) and the protein phosphatase calcineurin/protein phosphatase 2B (PP2B) (Beaulieu and Gainetdinov 2011; Beaulieu et al. 2015). Interestingly, a number of studies have shown that D_1R and D_2R can form a D_1-D_2 receptor heterodimer that regulates DAG and IP_3 signaling by activating $G\alpha_q$ in transfected cells as well as in the striatum of younger mice (Lee et al. 2004; Rashid et al. 2007) (Fig. 1.4). Co-localization of D_1 and D_2 receptors has been found in a subset of medium spiny neurons of the nucleus accumbens, globus pallidus, and caudate putamen (Perreault et al. 2010; Rashid et al. 2007). Thus, the D_1–D_2 heterodimer may regulate calcium-dependent cell signaling in these neuronal populations.

Under normal conditions, activation of D_1R has a stimulatory effect on locomotion (Beaulieu and Gainetdinov 2011; Vallone et al. 2000). D_1R is also involved in reward, reinforcement, learning, and memory (Beaulieu and Gainetdinov 2011; Vallone et al. 2000). The physiological roles played by D_5R in the brain are less well understood. However, a recent study using D_5R knockout mice suggests a role for D_5R in spatial and recognition memory, but not in working memory (Moraga-Amaro et al. 2016). In the periphery, D1-like receptors modulate cardiovascular, renal, immune/inflammatory, and gastrointestinal functions as well as gastrointestinal motility, secretion, epithelial barrier function, hormone secretion, and immune/inflammatory functions (Beaulieu and Gainetdinov 2011; Feng et al. 2013, 2017, 2020; Li et al. 2019; Missale et al. 1998; Zhang et al. 2012, 2017). The roles of the D1-like dopamine receptors in gastrointestinal motility, secretion, and epithelial barrier function are further discussed in detail in Chaps. 4 and 5 of this book.

1.4.2 D2-Like Dopamine Receptors

The family of the D2-like dopamine receptors is composed of three different subtypes, D_2R, D_3R, and D_4R. The D_2R is 75% homologous with D_3R and 53% homologous with D_4R in the transmembrane domains (Beaulieu and Gainetdinov 2011). The genes that encode the D2-like receptors have several introns, with six introns found in the D_2R gene, five in the D_3R gene, and three in the D_4R gene (Beaulieu and Gainetdinov 2011; Missale et al. 1998). The D_2R has two main splice variants, D_2-short (D_{2S}) and D_2-long (D_{2L}). The D_{2S} variant is mostly expressed on the presynaptic membrane of dopaminergic neurons and likely serves as an autoreceptor, whereas the D_{2L} variant is predominantly expressed on the postsynaptic membrane. Splice variants of the D_3R have also been identified, but they all encode nonfunctional proteins. Several polymorphic variants have been described for the D_4R; however, functions of these variants remain to be elucidated (Beaulieu and Gainetdinov 2011; Missale et al. 1998).

The D2-like dopamine receptors are expressed both postsynaptically on dopamine target cells and presynaptically on dopaminergic neurons (Beaulieu and Gainetdinov 2011). The highest levels of D_2R are found in the striatum, olfactory tubercle, and nucleus accumbens. D_2R is also expressed in the cortical areas, septum, amygdala, hippocampus, hypothalamus, substantia nigra pars compacta, ventral tegmental area, dorsal motor nucleus of the vagus, and hypoglossal nucleus (Beaulieu and Gainetdinov 2011; Cai et al. 2013; Missale et al. 1998; Ran et al. 2019; Yang et al. 2019; Zhou et al. 2014, 2019). Although both D_1R and D_2R are found in the striatum and nucleus accumbens, they are expressed on different subgroups of the medium spiny neurons that have distinct projections. Specifically, the medium spiny neurons that project to the medial globus pallidus and substantia nigra pars reticulata selectively express D_1R and comprise a direct striatonigral pathway. The medium spiny neurons that project to the lateral globus pallidus selectively express D_2R and form the indirect striatopallidal pathway. Interestingly, co-expression of D_1R and D_2R was observed in a small population of the medium spiny neurons in the dorsal striatum and pyramidal neurons in the prefrontal cortex of mice (Beaulieu and Gainetdinov 2011; Missale et al. 1998). D_3R is mainly expressed in limbic regions, such as the shell of the nucleus accumbens, olfactory tubercle, and islands of Calleja. D_3R is also detectable at significantly lower levels in the striatum, substantia nigra pars compacta, ventral tegmental area, hippocampus, septum, and various cortical areas (Beaulieu and Gainetdinov 2011; Missale et al. 1998). D_4R has the lowest level of expression in the brain, with documented expression in the frontal cortex, amygdala, hippocampus, hypothalamus, globus pallidus, substantia nigra pars reticulata, and thalamus (Beaulieu and Gainetdinov 2011; Missale et al. 1998). In the periphery, D_2R, D_3R, and D_4R have been detected at varying levels in the kidneys, heart, blood vessels, adrenal glands, gastrointestinal tract, and immune/inflammatory cells (Aperia 2000; Cavallotti et al. 2010; Chen et al. 2014; Feng et al. 2020; Li et al. 2006; Missale et al. 1998; Wang et al. 2012; Zhang et al. 2017). In particular, high levels of D_2R have been found in the anterior pituitary, where it mediates the inhibitory action of dopamine on prolactin release (Missale et al. 1998). Distribution of the D2-like dopamine receptors in the gut is further discussed in Chap. 3 of this book.

It is well known that the D2-like dopamine receptors are coupled to the $G\alpha_{i/o}$ family of G proteins (Fig. 1.5). Activation of D2-like dopamine receptors inhibits adenylate cyclase and decreases intracellular cAMP production, which results in a reduction of PKA activity (Beaulieu and Gainetdinov 2011; Beaulieu et al. 2015). Activation of D2-like dopamine receptors also reduces the phosphorylation of DARPP-32 at threonine 34, presumably due to the reduction in PKA activation (Beaulieu and Gainetdinov 2011; Beaulieu et al. 2015; Missale et al. 1998). Some of the responses generated by D2-like dopamine receptor activation are mediated by the $G_{\beta\gamma}$ subunits, which dissociated from the G_α subunit after receptor activation (Fig. 1.5). D2-like dopamine receptor-regulated $G_{\beta\gamma}$ subunits mediate the activation of PLC, leading to the production of DAG and IP_3, and triggering their corresponding downstream signaling events (Beaulieu and Gainetdinov 2011; Beaulieu et al. 2015). In addition, $G_{\beta\gamma}$ subunits directly open the G protein-coupled inwardly rectifying potassium channels (GIRKs) and inhibit the L/N-type Ca^{2+} channels (Beaulieu and Gainetdinov 2011; Beaulieu et al. 2015).

Several lines of evidence support the involvement of G protein-independent pathways in D2-like dopamine receptors signaling, which is represented by the β-arrestin 2-mediated mechanisms (Fig. 1.5). β-arrestin 2 is a multifunctional intracellular scaffolding protein that is involved in the desensitization of several GPCRs, including D2-like dopamine receptors. Besides shutting down D2-like receptor signaling, β-arrestin 2 can promote a new wave of signaling events that are G protein-independent. β-Arrestin 2 can serve as an adaptor protein that induces the formation of a protein complex composed of β-arrestin 2, protein phosphatase 2A (PP2A), and protein kinase B (also known as Akt). Formation of this complex results in the deactivation of Akt by PP2A and the subsequent stimulation of the glycogen synthase kinase 3 (GSK-3)-mediated signaling, which results in slower onset but

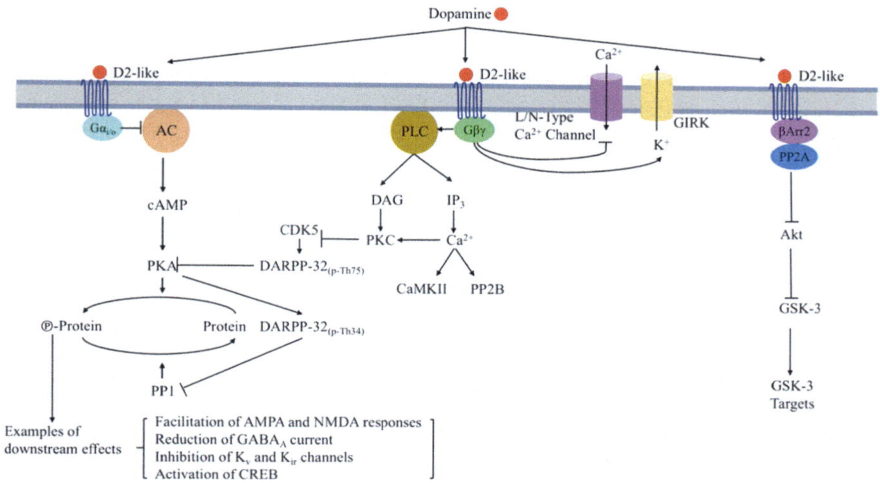

Fig. 1.5 Dopamine D2-like receptor family signaling pathways. The D2-like dopamine receptors (D_2R, D_3R, and D_4R) are coupled to the $G\alpha_{i/o}$ proteins. Activation of $G\alpha_{i/o}$ results in decrease in intracellular cAMP level and reduction in PKA activity. The $G_{\beta\gamma}$ subunits are also involved in D2-like dopamine receptor signaling. The G protein-independent pathways in D2-like dopamine receptor signaling are mediated by the intracellular scaffolding protein βArr2 and the associated proteins such as PP2A. Arrows indicate activation. T-arrows indicate inhibition. *AC* adenylate cyclase, *Akt* protein kinase B, *AMPA* α-amino-3-hydroxy-5-methyl-4-isoxazolepropionic acid receptor, *βArr2* β-arrestin 2, *CaMKII* calmodulin kinase II, *CDK5* cyclin-dependent kinase 5, *CREB* cAMP response element-binding protein, *DAG* diacylglycerol, *DARPP-32* dopamine and cAMP-regulated neuronal phosphoprotein, *$GABA_A$* gamma-aminobutyric acid A receptor, *GIRK* G protein-coupled inwardly rectifying potassium channels, *GSK-3* glycogen synthase kinase 3, *IP_3* inositol trisphosphate, *K_v* voltage-gated potassium channel, *K_{ir}* inward rectifier potassium channel, *NMDA*, *N*-methyl-D-aspartate receptor, *PKA* protein kinase A, *PKC* protein kinase C, *PLC* phospholipase C, *PP1* protein phosphatase 1, *PP2A* protein phosphatase 2A, *PP2B* protein phosphatase 2B

longer lasting responses comparing with the G protein-coupled signaling events (Beaulieu and Gainetdinov 2011; Beaulieu et al. 2015).

The functional roles of the D2-like dopamine receptors are complicated due to the fact that D2-like receptors have both presynaptic and postsynaptic locations. In general, activation of presynaptic D2-like receptors (i.e., D_{2S} and D_3R) by a lower concentration of dopamine agonists causes a decrease in dopamine release and a reduction in locomotor activity, whereas activation of postsynaptic receptors (i.e., D_{2L}) by a higher dose of dopamine agonists slightly stimulates locomotion (Beaulieu and Gainetdinov 2011; Missale et al. 1998). The role of D_4R on locomotion is minimal (Beaulieu and Gainetdinov 2011). Besides locomotion, D_3R is highly involved in reward and reinforcement (Beaulieu and Gainetdinov 2011). D_2R is critical for learning and memory. D_3R and D_4R have minor roles on some aspects of cognitive functions (Beaulieu and Gainetdinov 2011). In the anterior pituitary, activation of D_2R by dopamine inhibits prolactin release. A recent study has shown that both D_{2S} and D_{2L} isoforms are present in the anterior pituitary, and activation of each individual isoform is sufficient to inhibit prolactin release and maintain prolactin homeostasis (Radl et al. 2013). D2-like receptors are also involved in the modulation of renal, cardiovascular, gastrointestinal, metabolic, endocrine, immune, and defense function (Beaulieu and Gainetdinov 2011; Beaulieu et al. 2015; Feng et al. 2013, 2020; Li et al. 2006, 2019; Missale et al. 1998; Zhang et al. 2015, 2017; Zheng et al. 2014). Effects of D2-like receptors on gastrointestinal motility and secretion are further discussed in Chaps. 4 and 5 of this book.

1.5 Adrenoceptors, Functions, and Crosstalk with Dopamine in the Gut

Similar to the dopamine receptors, the adrenoceptors also belong to the rhodopsin-like class A, seven-transmembrane domain, GPCR superfamily. They are localized to almost all peripheral tissues and many neuronal populations within the CNS and are activated by the endogenous catecholamines norepinephrine and epinephrine. There are three major types of adrenoceptors (α_1, α_2, and β), each of which is further divided into at least three subtypes (Bylund et al. 1994; Hieble et al. 1995). The α_1 adrenoceptors (including α_{1A}, α_{1B}, and α_{1D}) are coupled to the $G_{q/11}$/PLC signaling pathway. In addition to mobilizing intracellular Ca^{2+} stores, the α_1 adrenoceptors have also been shown to activate Ca^{2+} influx via voltage-dependent and -independent Ca^{2+} channels. The α_1 adrenoceptors are also linked to other signaling pathways, including activation of phospholipase A_2 (PLA_2), phospholipase D (PLD), and mitogen-activated protein kinase (MAPK). Activation of the α_1 adrenoceptors primarily causes vasoconstriction and elevates resting blood pressure (Bylund et al. 1994; Hieble et al. 1995; Piascik and Perez 2001). The α_2 adrenoceptors (including α_{2A}, α_{2B}, and α_{2C}) are coupled to the $G_{i/o}$ proteins. Activation of the α_2 adrenoceptors causes inhibition of adenylyl cyclase and decreases intracellular cAMP concentration; however, other signaling mechanisms are also involved in α_2 adrenoceptor action (Bylund et al. 1994). The α_2 adrenoceptors are expressed both presynaptically and postsynaptically. The presynaptic α_2 adrenoceptors (mainly α_{2A} and α_{2C}) inhibit norepinephrine release and decrease sympathetic outflow, resulting in hypotension. The postsynaptic α_2 adrenoceptors mediate the sedative and analgesic effects. The α_{2B} receptors are also essential for placental angiogenesis (Philipp and Hein 2004). The β adrenoceptors (including β_1, β_2, and β_3) are linked to adenylyl cyclase activation through the G_s proteins, although β_2 also couples to $G_{i/o}$. The β_1 adrenoceptors are predominately expressed in the heart, where they mediate the positive inotropic and chronotropic effects. The β_2 adrenoceptors are widely distributed in several cell types, including cardiomyocytes, vascular smooth muscles, and pulmonary smooth muscles, where they mediate inotropic and chronotropic effects in the heart, vasodilation, and bronchodilation, respectively. The β_3 adrenoceptors are mainly expressed in adipose tissue, where they regulate lipolysis

(Bylund et al. 1994; Philipp and Hein 2004; Taylor and Bristow 2004). Interestingly, the distribution of β_3 adrenoceptors in the colonic muscularis externa has been reported recently. Pretreatment with SR59230A, a selective antagonist of β_3 adrenoceptor, significantly blocked norepinephrine-induced inhibition of colonic motility (Zhang et al. 2015).

Dopamine is very similar in structure to norepinephrine and epinephrine. Considerable evidence supports a crosstalk between dopamine and different subtypes of adrenoceptors in the CNS and peripheral tissues (Cornil et al. 2002; Lee et al. 1998; Malenka and Nicoll 1986; Ruffolo et al. 1984; Segawa et al. 1998; Zhang et al. 1999, 2004). The gut is well endowed with adrenoceptors and dopamine receptors. It is generally accepted that lower concentrations of dopamine activate the dopamine receptors, whereas medium to high concentrations of dopamine are needed to activate the adrenoceptors (Tsai and Cheng 1992; Walker et al. 2000; Zizzo et al. 2010).

Several pharmacological and functional studies have revealed that both dopamine receptors and adrenoceptors mediate the effect of dopamine on gastrointestinal smooth muscle contraction. Dopamine produces dose-dependent contractions in isolated longitudinal strips of chicken esophagus; these contractions are partially suppressed by the α_2-adrenergic receptor antagonist yohimbine, which suggests the participation of α_2-adrenergic receptors in dopamine-induced contraction in chicken esophagus (Sanchez et al. 1990). Kurosawa et al. have demonstrated that dopamine induces circular muscle contraction of the guinea pig stomach by activating the α adrenoceptors ($\alpha_1 > \alpha_2$) and longitudinal muscle relaxation by activating the D1-like dopamine receptors (Kurosawa et al. 1991). In the small and large intestines, exogenously administered dopamine is known to activate both α and β adrenoceptors to inhibit motility (Aguilar et al. 2005; Grivegnee et al. 1984; Kirschstein et al. 2009; Lucchelli et al. 1990; Tsai and Cheng 1992; Zhang et al. 2015), although D1- and D2-like dopamine receptors are also involved in this inhibition (Tsai and Cheng 1992; Zizzo et al. 2010; Walker et al. 2000; Zhang et al. 2015). Endogenously released dopamine has been demonstrated to inhibit colonic motility by activating both D1- and D2-like dopamine receptors, an effect that is potentiated in the DAT knockout mice (Walker et al. 2000). Mice lacking the D_2R have faster total gastrointestinal transit and regional colonic transit, suggesting that endogenous dopamine may act via enteric D_2R to inhibit propulsive gastrointestinal motor activity (Li et al. 2006). Dopamine is also found to have biphasic effects (an early contraction mediated by dopamine receptors and late relaxation mediated by adrenoceptors) in the rat small intestine (Kirschstein et al. 2009). DA-induced biphasic effects are common in the duodenum, less frequent in the jejunum, and rare in the ileum (Kirschstein et al. 2009).

Both adrenoceptors and dopamine receptors are involved in dopaminergic regulation of intestinal epithelial ion transport. In the rat duodenum, dopamine-induced K^+ secretion is mediated by both α adrenoceptors and D1-like dopamine receptors (Feng et al. 2013), and dopamine-induced HCO_3^- secretion is mediated by the apical D_2R on the duodenal epithelial cells (Feng et al. 2020). In fact, the α adrenoceptor antagonist can block the majority of the dopamine-induced changes of K^+ secretion in the rat duodenum compared with the D1-like receptor antagonist (Feng et al. 2013). In the canine and rabbit ileum, dopamine stimulates water and electrolyte absorption via α_1, α_2, and D2-like receptors (Barry et al. 1995; Donowitz et al. 1982). In the rat distal colon, dopamine increases Cl^- absorption and HCO_3^- secretion by activating the β (mainly β_2) adrenoceptors, but not dopamine receptors (Zhang et al. 2008, 2010). Taken together, these studies may have important implications for our understanding of the enteric dopaminergic system and its physiological and pathophysiological roles in the gastrointestinal function.

References

Abrantes Dias AS, Amaral Pinto JC, Magalhaes M, Mendes VM, Manadas B (2020) Analytical methods to monitor dopamine metabolism in plasma: moving forward with improved diagnosis and treatment of neurological disorders. J Pharm Biomed Anal 187:113323. https://doi.org/10.1016/j.jpba.2020.113323

Agid Y, Javoy F, Youdim MB (1973) Monoamine oxidase and aldehyde dehydrogenase activity in the striatum of rats after 6-hydroxydopamine lesion of the nigrostriatal pathway. Br J Pharmacol 48(1):175–178. https://doi.org/10.1111/j.1476-5381.1973.tb08238.x

Aguilar MJ, Estan L, Martinez-Mir I, Martinez-Abad M, Rubio E, Morales-Olivas FJ (2005) Effects of dopamine in isolated rat colon strips. Can J Physiol Pharmacol 83(6):447–452. https://doi.org/10.1139/y05-031

Amenta F, Ferrante F, Ricci A (1995) Pharmacological characterisation and autoradiographic localisation of dopamine receptor subtypes in the cardiovascular system and in the kidney. Hypertens Res 18(Suppl 1):S23–S27. https://doi.org/10.1291/hypres.18.supplementi_s23

Aperia AC (2000) Intrarenal dopamine: a key signal in the interactive regulation of sodium metabolism. Annu Rev Physiol 62:621–647. https://doi.org/10.1146/annurev.physiol.62.1.621

Asano Y, Hiramoto T, Nishino R, Aiba Y, Kimura T, Yoshihara K, Koga Y, Sudo N (2012) Critical role of gut microbiota in the production of biologically active, free catecholamines in the gut lumen of mice. Am J Physiol Gastrointest Liver Physiol 303(11):G1288–G1295. https://doi.org/10.1152/ajpgi.00341.2012

Barry MK, Maher MM, Gontarek JD, Jimenez RE, Yeo CJ (1995) Luminal dopamine modulates canine ileal water and electrolyte transport. Dig Dis Sci 40(8):1738–1743. https://doi.org/10.1007/BF02212695

Beaulieu JM, Gainetdinov RR (2011) The physiology, signaling, and pharmacology of dopamine receptors. Pharmacol Rev 63(1):182–217. https://doi.org/10.1124/pr.110.002642

Beaulieu JM, Espinoza S, Gainetdinov RR (2015) Dopamine receptors - IUPHAR review 13. Br J Pharmacol 172(1):1–23. https://doi.org/10.1111/bph.12906

Bylund DB, Eikenberg DC, Hieble JP, Langer SZ, Lefkowitz RJ, Minneman KP, Molinoff PB, Ruffolo RR Jr, Trendelenburg U (1994) International union of pharmacology nomenclature of adrenoceptors. Pharmacol Rev 46(2):121–136

Cai QQ, Zheng LF, Fan RF, Lian H, Zhou L, Song HY, Tang YY, Feng XY, Guo ZK, Wang ZY, Zhu JX (2013) Distribution of dopamine receptors D1- and D2-immunoreactive neurons in the dorsal motor nucleus of vagus in rats. Auton Neurosci 176(1–2):48–53. https://doi.org/10.1016/j.autneu.2013.01.007

Cavallotti C, Mancone M, Bruzzone P, Sabbatini M, Mignini F (2010) Dopamine receptor subtypes in the native human heart. Heart Vessel 25(5):432–437. https://doi.org/10.1007/s00380-009-1224-4

Ceppa FA, Izzo L, Sardelli L, Raimondi I, Tunesi M, Albani D, Giordano C (2020) Human gut-microbiota interaction in neurodegenerative disorders and current engineered tools for its modeling. Front Cell Infect Microbiol 10:297. https://doi.org/10.3389/fcimb.2020.00297

Chen Y, Hong F, Chen H, Fan RF, Zhang XL, Zhang Y, Zhu JX (2014) Distinctive expression and cellular distribution of dopamine receptors in the pancreatic islets of rats. Cell Tissue Res 357 (3):597–606. https://doi.org/10.1007/s00441-014-1894-9

Cornil CA, Balthazart J, Motte P, Massotte L, Seutin V (2002) Dopamine activates noradrenergic receptors in the preoptic area. J Neurosci 22(21):9320–9330

Craddock N, Owen MJ, O'Donovan MC (2006) The catechol-O-methyl transferase (COMT) gene as a candidate for psychiatric phenotypes: evidence and lessons. Mol Psychiatry 11(5):446–458. https://doi.org/10.1038/sj.mp.4001808

Donowitz M, Cusolito S, Battisti L, Fogel R, Sharp GW (1982) Dopamine stimulation of active Na and cl absorption in rabbit ileum: interaction with alpha 2-adrenergic and specific dopamine receptors. J Clin Invest 69(4):1008–1016. https://doi.org/10.1172/jci110504

Eisenhofer G, Kopin IJ, Goldstein DS (2004) Catecholamine metabolism: a contemporary view with implications for physiology and medicine. Pharmacol Rev 56(3):331–349. https://doi.org/10.1124/pr.56.3.1

Erickson JD, Schafer MK, Bonner TI, Eiden LE, Weihe E (1996) Distinct pharmacological properties and distribution in neurons and endocrine cells of two isoforms of the human vesicular monoamine transporter. Proc Natl Acad Sci U S A 93(10):5166–5171. https://doi.org/10.1073/pnas.93.10.5166

Feng XY, Li Y, Li LS, Li XF, Zheng LF, Zhang XL, Fan RF, Song J, Hong F, Zhang Y, Zhu JX (2013) Dopamine D1 receptors mediate dopamine-induced duodenal epithelial ion transport in rats. Transl Res 161(6):486–494. https://doi.org/10.1016/j.trsl.2012.12.002

Feng XY, Zhang DN, Wang YA, Fan RF, Hong F, Zhang Y, Li Y, Zhu JX (2017) Dopamine enhances duodenal epithelial permeability via the dopamine D5 receptor in rodent. Acta Physiol (Oxf) 220(1):113–123. https://doi.org/10.1111/apha.12806

Feng XY, Yan JT, Li GW, Liu JH, Fan RF, Li SC, Zheng LF, Zhang Y, Zhu JX (2020) Source of dopamine in gastric juice and luminal dopamine-induced duodenal bicarbonate secretion via apical dopamine D2 receptors. Br J Pharmacol 177(14):3258–3272. https://doi.org/10.1111/bph.15047

Fitzgerald P, Dinan TG (2008) Prolactin and dopamine: what is the connection? A review article. J Psychopharmacol 22(2 Suppl):12–19. https://doi.org/10.1177/0269216307087148

Galligan JJ (2004) 5-hydroxytryptamine, ulcerative colitis, and irritable bowel syndrome: molecular connections. Gastroenterology 126(7):1897–1899. https://doi.org/10.1053/j.gastro.2004.04.028

Gershon MD (2013) 5-Hydroxytryptamine (serotonin) in the gastrointestinal tract. Curr Opin Endocrinol Diabetes Obes 20(1):14–21. https://doi.org/10.1097/MED.0b013e32835bc703

Greengard P (2001) The neurobiology of slow synaptic transmission. Science 294 (5544):1024–1030. https://doi.org/10.1126/science.294.5544.1024

Grivegnee AR, Fontaine J, Reuse J (1984) Effect of dopamine on dog distal colon in-vitro. J Pharm Pharmacol 36(7):454–457. https://doi.org/10.1111/j.2042-7158.1984.tb04424.x

Hakanson R, Owman C, Sjoberg NO (1969) Three different systems of monoamine-storing cells in the gastrointestinal tract of fetal and neonatal rats. Acta Physiol Scand 75(1):213–230. https://doi.org/10.1111/j.1748-1716.1969.tb04373.x

Hieble JP, Bylund DB, Clarke DE, Eikenburg DC, Langer SZ, Lefkowitz RJ, Minneman KP, Ruffolo RR Jr (1995) International Union of Pharmacology. X. Recommendation for nomenclature of alpha 1-adrenoceptors: consensus update. Pharmacol Rev 47(2):267–270

Hong F, Liu L, Fan RF, Chen Y, Chen H, Zheng RP, Zhang Y, Gao Y, Zhu JX (2014) New perspectives of vesicular monoamine transporter 2 chemical characteristics in mammals and its constant expression in type 1 diabetes rat models. Transl Res 163(2):171–182. https://doi.org/10.1016/j.trsl.2013.10.001

Iversen SD, Iversen LL (2007) Dopamine: 50 years in perspective. Trends Neurosci 30(5):188–193. https://doi.org/10.1016/j.tins.2007.03.002

Janeway CAJ, Travers P, Walport M, Shlomchik MJ (2001) Immunobiology: the immune system in health and disease, 5th edn. Garland Science, New York

Kirschstein T, Dammann F, Klostermann J, Rehberg M, Tokay T, Schubert R, Kohling R (2009) Dopamine induces contraction in the proximal, but relaxation in the distal rat isolated small intestine. Neurosci Lett 465(1):21–26. https://doi.org/10.1016/j.neulet.2009.08.080

Klein MO, Battagello DS, Cardoso AR, Hauser DN, Bittencourt JC, Correa RG (2019) Dopamine: functions, signaling, and association with neurological diseases. Cell Mol Neurobiol 39 (1):31–59. https://doi.org/10.1007/s10571-018-0632-3

Kurosawa S, Hasler WL, Torres G, Wiley JW, Owyang C (1991) Characterization of receptors mediating the effects of dopamine on gastric smooth muscle. Gastroenterology 100(5 Pt 1):1224–1231

Lee TL, Hsu CT, Yen ST, Lai CW, Cheng JT (1998) Activation of beta3-adrenoceptors by exogenous dopamine to lower glucose uptake into rat adipocytes. J Auton Nerv Syst 74 (2–3):86–90. https://doi.org/10.1016/s0165-1838(98)00120-9

Lee SP, So CH, Rashid AJ, Varghese G, Cheng R, Lanca AJ, O'Dowd BF, George SR (2004) Dopamine D1 and D2 receptor co-activation generates a novel phospholipase C-mediated calcium signal. J Biol Chem 279(34):35671–35678. https://doi.org/10.1074/jbc.M401923200

Li ZS, Schmauss C, Cuenca A, Ratcliffe E, Gershon MD (2006) Physiological modulation of intestinal motility by enteric dopaminergic neurons and the D2 receptor: analysis of dopamine receptor expression, location, development, and function in wild-type and knock-out mice. J Neurosci 26(10):2798–2807. https://doi.org/10.1523/JNEUROSCI.4720-05.2006

Li Y, Zhang Y, Zhang XL, Feng XY, Liu CZ, Zhang XN, Quan ZS, Yan JT, Zhu JX (2019) Dopamine promotes colonic mucus secretion through dopamine D5 receptor in rats. Am J Physiol Cell Physiol 316(3):C393–C403. https://doi.org/10.1152/ajpcell.00261.2017

Liu CZ, Zhang XL, Zhou L, Wang T, Quan ZS, Zhang Y, Li J, Li GW, Zheng LF, Li LS, Zhu JX (2018) Rasagiline, an inhibitor of MAO-B, decreases colonic motility through elevating colonic dopamine content. Neurogastroenterol Motil 30(11):e13390. https://doi.org/10.1111/nmo.13390

Lucchelli A, Boselli C, Grana E (1990) Dopamine-induced relaxation of the Guinea-pig isolated jejunum is not mediated through dopamine receptors. Pharmacol Res 22(4):433–444. https://doi.org/10.1016/1043-6618(90)90750-8

Malenka RC, Nicoll RA (1986) Dopamine decreases the calcium-activated afterhyperpolarization in hippocampal CA1 pyramidal cells. Brain Res 379(2):210–215. https://doi.org/10.1016/0006-8993(86)90773-0

Martel JC, Gatti McArthur S (2020) Dopamine receptor subtypes, physiology and pharmacology: new ligands and concepts in schizophrenia. Front Pharmacol 11:1003. https://doi.org/10.3389/fphar.2020.01003

Marzullo P, Di Renzo L, Pugliese G, De Siena M, Barrea L, Muscogiuri G, Colao A, Savastano S, Obesity Programs of nutrition, Education, Research and Assessment (OPERA) Group (2020) From obesity through gut microbiota to cardiovascular diseases: a dangerous journey. Int J Obes Suppl 10(1):35–49. https://doi.org/10.1038/s41367-020-0017-1

Masato A, Plotegher N, Boassa D, Bubacco L (2019) Impaired dopamine metabolism in Parkinson's disease pathogenesis. Mol Neurodegener 14(1):35. https://doi.org/10.1186/s13024-019-0332-6

Meiser J, Weindl D, Hiller K (2013) Complexity of dopamine metabolism. Cell Commun Signal 11 (1):34. https://doi.org/10.1186/1478-811X-11-34

Missale C, Nash SR, Robinson SW, Jaber M, Caron MG (1998) Dopamine receptors: from structure to function. Physiol Rev 78(1):189–225. https://doi.org/10.1152/physrev.1998.78.1.189

Moraga-Amaro R, Gonzalez H, Ugalde V, Donoso-Ramos JP, Quintana-Donoso D, Lara M, Morales B, Rojas P, Pacheco R, Stehberg J (2016) Dopamine receptor D5 deficiency results in a selective reduction of hippocampal NMDA receptor subunit NR2B expression and impaired memory. Neuropharmacology 103:222–235. https://doi.org/10.1016/j.neuropharm.2015.12.018

Myohanen TT, Schendzielorz N, Mannisto PT (2010) Distribution of catechol-O-methyltransferase (COMT) proteins and enzymatic activities in wild-type and soluble COMT deficient mice. J Neurochem 113(6):1632–1643. https://doi.org/10.1111/j.1471-4159.2010.06723.x

Perreault ML, Hasbi A, Alijaniaram M, Fan T, Varghese G, Fletcher PJ, Seeman P, O'Dowd BF, George SR (2010) The dopamine D1-D2 receptor heteromer localizes in dynorphin/enkephalin neurons: increased high affinity state following amphetamine and in schizophrenia. J Biol Chem 285(47):36625–36634. https://doi.org/10.1074/jbc.M110.159954

Philipp M, Hein L (2004) Adrenergic receptor knockout mice: distinct functions of 9 receptor subtypes. Pharmacol Ther 101(1):65–74. https://doi.org/10.1016/j.pharmthera.2003.10.004

Piascik MT, Perez DM (2001) Alpha1-adrenergic receptors: new insights and directions. J Pharmacol Exp Ther 298(2):403–410

Putignani L, Del Chierico F, Vernocchi P, Cicala M, Cucchiara S, Dallapiccola B, Dysbiotrack Study G (2016) Gut microbiota Dysbiosis as risk and premorbid factors of IBD and IBS along the childhood-adulthood transition. Inflamm Bowel Dis 22(2):487–504. https://doi.org/10.1097/MIB.0000000000000602

Radl D, De Mei C, Chen E, Lee H, Borrelli E (2013) Each individual isoform of the dopamine D2 receptor protects from lactotroph hyperplasia. Mol Endocrinol 27(6):953–965. https://doi.org/10.1210/me.2013-1008

Ran X, Yang Y, Meng Y, Li Y, Zhou L, Wang Z, Zhu J (2019) Distribution of D1 and D2 receptor-immunoreactive neurons in the paraventricular nucleus of the hypothalamus in the rat. J Chem Neuroanat 98:97–103. https://doi.org/10.1016/j.jchemneu.2019.04.002

Rashid AJ, So CH, Kong MM, Furtak T, El-Ghundi M, Cheng R, O'Dowd BF, George SR (2007) D1-D2 dopamine receptor heterooligomers with unique pharmacology are coupled to rapid activation of Gq/11 in the striatum. Proc Natl Acad Sci U S A 104(2):654–659. https://doi.org/10.1073/pnas.0604049104

Rubi B, Maechler P (2010) Minireview: new roles for peripheral dopamine on metabolic control and tumor growth: let's seek the balance. Endocrinology 151(12):5570–5581. https://doi.org/10.1210/en.2010-0745

Ruffolo RR Jr, Goldberg MR, Morgan EL (1984) Interactions of epinephrine, norepinephrine, dopamine and their corresponding alpha-methyl-substituted derivatives with alpha and beta adrenoceptors in the pithed rat. J Pharmacol Exp Ther 230(3):595–600

Sanchez J, Costa G, Benedito S, Rivera L, Garcia-Sacristan A, Orensanz LM (1990) Alpha 2-mediated effect of dopamine on the motility of the chicken esophagus. Life Sci 46 (2):121–126. https://doi.org/10.1016/0024-3205(90)90044-r

Segawa T, Ito H, Inoue K, Wada H, Minatoguchi S, Fujiwara H (1998) Dopamine releases endothelium-derived relaxing factor via alpha 2-adrenoceptors in canine vessels: comparisons between femoral arteries and veins. Clin Exp Pharmacol Physiol 25(9):669–675. https://doi.org/10.1111/j.1440-1681.1998.tb02274.x

Sekirov I, Russell SL, Antunes LC, Finlay BB (2010) Gut microbiota in health and disease. Physiol Rev 90(3):859–904. https://doi.org/10.1152/physrev.00045.2009

Sunahara RK, Guan HC, O'Dowd BF, Seeman P, Laurier LG, Ng G, George SR, Torchia J, Van Tol HH, Niznik HB (1991) Cloning of the gene for a human dopamine D5 receptor with higher affinity for dopamine than D1. Nature 350(6319):614–619. https://doi.org/10.1038/350614a0

Svenningsson P, Nishi A, Fisone G, Girault JA, Nairn AC, Greengard P (2004) DARPP-32: an integrator of neurotransmission. Annu Rev Pharmacol Toxicol 44:269–296. https://doi.org/10.1146/annurev.pharmtox.44.101802.121415

Taylor MR, Bristow MR (2004) The emerging pharmacogenomics of the beta-adrenergic receptors. Congest Heart Fail 10(6):281–288. https://doi.org/10.1111/j.1527-5299.2004.02019.x

Thomas Broome S, Louangaphay K, Keay KA, Leggio GM, Musumeci G, Castorina A (2020) Dopamine: an immune transmitter. Neural Regen Res 15(12):2173–2185. https://doi.org/10.4103/1673-5374.284976

Tsai LH, Cheng JT (1992) The effect of exogenous dopamine on ileal smooth muscle of Guinea-pigs. Chin J Physiol 35(2):133–141

Vallone D, Picetti R, Borrelli E (2000) Structure and function of dopamine receptors. Neurosci Biobehav Rev 24(1):125–132. https://doi.org/10.1016/s0149-7634(99)00063-9
Vieira-Coelho MA, Soares-da-Silva P (1993) Dopamine formation, from its immediate precursor 3,4-dihydroxyphenylalanine, along the rat digestive tract. Fundam Clin Pharmacol 7 (5):235–243. https://doi.org/10.1111/j.1472-8206.1993.tb00237.x
Walker JK, Gainetdinov RR, Mangel AW, Caron MG, Shetzline MA (2000) Mice lacking the dopamine transporter display altered regulation of distal colonic motility. Am J Physiol Gastrointest Liver Physiol 279(2):G311–G318. https://doi.org/10.1152/ajpgi.2000.279.2.G311
Wang Q, Ji T, Zheng LF, Feng XY, Wang ZY, Lian H, Song J, Li XF, Zhang Y, Zhu JX (2012) Cellular localization of dopamine receptors in the gastric mucosa of rats. Biochem Biophys Res Commun 417(1):197–203. https://doi.org/10.1016/j.bbrc.2011.11.084
Wang ZY, Lian H, Zhou L, Zhang YM, Cai QQ, Zheng LF, Zhu JX (2016) Altered expression of D1 and D2 dopamine receptors in vagal neurons innervating the gastric muscularis externa in a Parkinson's disease rat model. J Parkinsons Dis 6(2):317–323. https://doi.org/10.3233/JPD-160817
Wang M, Ling KH, Tan JJ, Lu CB (2020) Development and differentiation of midbrain dopaminergic neuron: from bench to bedside. Cell 9(6). https://doi.org/10.3390/cells9061489
Weihe E, Eiden LE (2000) Chemical neuroanatomy of the vesicular amine transporters. FASEB J 14(15):2435–2449. https://doi.org/10.1096/fj.00-0202rev
Wood JD (2004) Enteric neuroimmunophysiology and pathophysiology. Gastroenterology 127 (2):635–657. https://doi.org/10.1053/j.gastro.2004.02.017
Wood JD (2006) Histamine, mast cells, and the enteric nervous system in the irritable bowel syndrome, enteritis, and food allergies. Gut 55(4):445–447. https://doi.org/10.1136/gut.2005.079046
Wood JD (2016) Enteric neurobiology: discoveries and directions. Adv Exp Med Biol 891:175–191. https://doi.org/10.1007/978-3-319-27592-5_17
Yang YL, Ran XR, Li Y, Zhou L, Zheng LF, Han Y, Cai QQ, Wang ZY, Zhu JX (2019) Expression of dopamine receptors in the lateral hypothalamic nucleus and their potential regulation of gastric motility in rats with lesions of bilateral substantia Nigra. Front Neurosci 13:195. https://doi.org/10.3389/fnins.2019.00195
Yap YA, Marino E (2018) An insight into the intestinal web of mucosal immunity, microbiota, and diet in inflammation. Front Immunol 9:2617. https://doi.org/10.3389/fimmu.2018.02617
Zhang W, Klimek V, Farley JT, Zhu MY, Ordway GA (1999) alpha2C adrenoceptors inhibit adenylyl cyclase in mouse striatum: potential activation by dopamine. J Pharmacol Exp Ther 289(3):1286–1292
Zhang WP, Ouyang M, Thomas SA (2004) Potency of catecholamines and other L-tyrosine derivatives at the cloned mouse adrenergic receptors. Neuropharmacology 47(3):438–449. https://doi.org/10.1016/j.neuropharm.2004.04.017
Zhang XH, Zhang XF, Zhang JQ, Tian YM, Xue H, Yang N, Zhu JX (2008) Beta-adrenoceptors, but not dopamine receptors, mediate dopamine-induced ion transport in late distal colon of rats. Cell Tissue Res 334(1):25–35. https://doi.org/10.1007/s00441-008-0661-1
Zhang XH, Ji T, Guo H, Liu SM, Li Y, Zheng LF, Zhang Y, Zhang XF, Duan DP, Zhu JX (2010) Expression and activation of beta-adrenoceptors in the colorectal mucosa of rat and human. Neurogastroenterol Motil 22(11):e325–e334. https://doi.org/10.1111/j.1365-2982.2010.01598.x
Zhang X, Guo H, Xu J, Li Y, Li L, Zhang X, Li X, Fan R, Zhang Y, Duan Z, Zhu J (2012) Dopamine receptor D1 mediates the inhibition of dopamine on the distal colonic motility. Transl Res 159(5):407–414. https://doi.org/10.1016/j.trsl.2012.01.002
Zhang X, Li Y, Liu C, Fan R, Wang P, Zheng L, Hong F, Feng X, Zhang Y, Li L, Zhu J (2015) Alteration of enteric monoamines with monoamine receptors and colonic dysmotility in 6-hydroxydopamine-induced Parkinson's disease rats. Transl Res 166(2):152–162. https://doi.org/10.1016/j.trsl.2015.02.003

Zhang X, Liu Q, Liao Q, Zhao Y (2017) Potential roles of peripheral dopamine in tumor immunity. J Cancer 8(15):2966–2973. https://doi.org/10.7150/jca.20850

Zheng LF, Song J, Fan RF, Chen CL, Ren QZ, Zhang XL, Feng XY, Zhang Y, Li LS, Zhu JX (2014) The role of the vagal pathway and gastric dopamine in the gastroparesis of rats after a 6-hydroxydopamine microinjection in the substantia nigra. Acta Physiol (Oxf) 211(2):434–446. https://doi.org/10.1111/apha.12229

Zhou L, Wang ZY, Lian H, Song HY, Zhang YM, Zhang XL, Fan RF, Zheng LF, Zhu JX (2014) Altered expression of dopamine receptors in cholinergic motoneurons of the hypoglossal nucleus in a 6-OHDA-induced Parkinson's disease rat model. Biochem Biophys Res Commun 452(3):560–566. https://doi.org/10.1016/j.bbrc.2014.08.104

Zhou L, Ran XR, Hong F, Li GW, Zhu JX (2019) Downregulated dopamine receptor 2 and upregulated corticotrophin releasing hormone in the paraventricular nucleus are correlated with decreased glucose tolerance in rats with bilateral substantia Nigra lesions. Front Neurosci 13:751. https://doi.org/10.3389/fnins.2019.00751

Zizzo MG, Mule F, Mastropaolo M, Serio R (2010) D1 receptors play a major role in the dopamine modulation of mouse ileum contractility. Pharmacol Res 61(5):371–378. https://doi.org/10.1016/j.phrs.2010.01.015

Chapter 2
Synthesis and Metabolism of Gut Dopamine

Chen-Zhe Liu, Xiao-Yan Feng, Sumei Liu, Xiao-Li Zhang, and Jin-Xia Zhu

Abstract Not only the central nervous system can produce dopamine (DA), the gastrointestinal tract is also an important source of DA, such as gastric mucosa and pancreas. Additionally, colonic lumen also contains substantial DA. The rapid degradation of DA ensures its proper function in various locations in the body. Monoamine oxidase (MAO) and catechol-O-methyltransferase (COMT) are the main metabolic enzymes of DA and are widely distributed in the gut. The altered contents or activities of these enzymes could cause the changes of DA levels in the gut, thereby might be correlated to some gastrointestinal disorders. Although the function and degradation of luminal DA are still unknown, a number of studies have reported that gut microbiota is involved in the progress of many diseases, such as Parkinson's disease, irritable bowel syndrome (IBS), and inflammatory bowel disease (IBD). Besides, the relationship between luminal DA and gut microbiome has been reported. In the present chapter, we focus on the sources and metabolism of gut DA, the distribution and function of MAO and COMT in various regions of gut, as well as the relationship among gut microbiome, luminal DA, and gastrointestinal disorders.

Keywords Dopamine · Monoamine oxidase · Catechol-O-methyltransferase · Tyrosine hydroxylase · Gastrointestinal tract

---

C.-Z. Liu
Department of Exercise Physiology, Beijing Sport University, Beijing, China
e-mail: liuchenzhe@bsu.edu.cn

X.-Y. Feng · X.-L. Zhang · J.-X. Zhu (✉)
Department of Physiology and Pathophysiology, School of Basic Medical Science, Capital Medical University, Beijing, China
e-mail: fengxy@ccmu.edu.cn; xiaoli3.23@163.com; zhu_jx@ccmu.edu.cn

S. Liu
Department of Biology, College of Science and Health, University of Wisconsin-La Crosse, La Crosse, WI, USA
e-mail: sliu@uwlax.edu

J.-X. Zhu (ed.), *Dopamine in the Gut*, https://doi.org/10.1007/978-981-33-6586-5_2

Abbreviations

3-MT	3-methoxytyramine
CD	Crohn's disease
CNS	Central nervous system
COMT	Catechol-*O*-methyltransferase
DA	Dopamine
DAT	DA transporter
DDC	DOPA decarboxylase
DOPAC	3,4-Dihydroxyphenylacetic acid
DOPAL	3,4-Dihydroxyphenylacetaldehyde
ENS	Enteric nervous system
GF	Germ-free
GFAP	Glial fibrillary acidic protein
GI	Gastrointestinal
HDC	Histidine decarboxylase
HVA	Homovanillic acid
IBD	Inflammatory bowel disease
IBS	Irritable bowel syndrome
MAO	Monoamine oxidase
α-MDL	α-Methyl-DL-tyrosine methyl ester hydrochloride
MPP+	1-Methyl-4-phenylpyridine
MPTP	1-Methyl-4-phenyl-1,2,3,6-teterahydropyridine
NF	Neurofilament
PD	Parkinson's disease
SCFA	Short-chain fatty acid
SN	Substantia nigra
SPF	Specific pathogen-free
TH	Tyrosine hydroxylase
UC	Ulcerative colitis

2.1 Introduction

Although dopamine (DA) is well known as a catecholamine neurotransmitter in the central nervous system (CNS), a substantial amount of dopamine has been found in the peripheral tissues. In addition, dopamine and its metabolites have been detected in the hepatic portal blood (Eisenhofer et al. 1997), suggesting that DA is produced and metabolized in the gut. DA in the gut inhibits motility, stimulates ion and mucus secretion, modulates mucosal blood flow, and enhances intestinal barrier. The rate-limiting enzyme for DA synthesis, tyrosine hydroxylase (TH), and the DA transporter (DAT) are located along the whole gastrointestinal tract, from the stomach to distal colon (Li et al. 2004) (Fig. 2.1). DA is synthesized within specific parenchyma

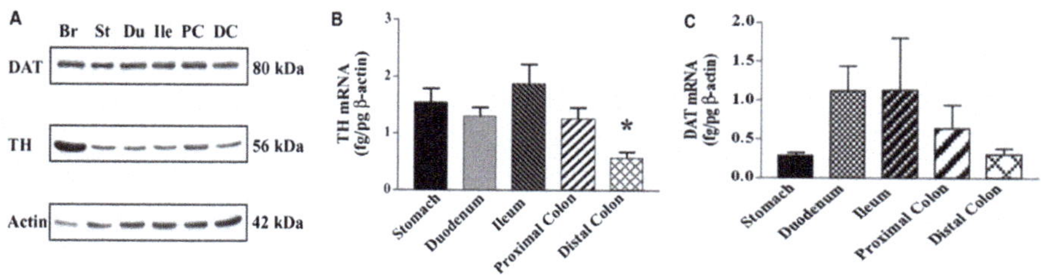

Fig. 2.1 The expression of TH and DAT in the mouse gut. (Li Z S, et al., J Neurosci 2004, 24, 1330–1339.) Protein levels of DAT and TH (**a**), the abundance of transcripts encoding TH (**b**), and DAT (**c**) in each region of the bowel. Brain (Br), stomach (St), duodenum (Du), ileum (Ile), proximal colon (PC), distal colon (DC). *: compare with other regions of the bowel. This figure is reprinted with permission from Copyright [2004] Society for Neuroscience

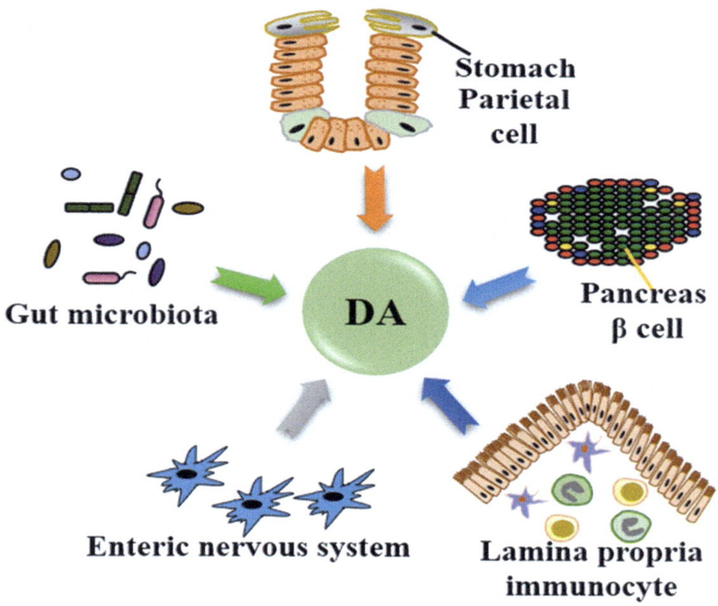

Fig. 2.2 The sources of dopamine in the digestive tract

of the gut, such as the gastric parietal cells (Christensen and Brandsborg 1974; Eisenhofer et al. 1997; Haggendal 1967; Tian et al. 2008), pancreatic β cells (Mezey et al. 1996; Ustione and Piston 2012), enteric neurons (Eisenhofer et al. 1997; Li et al. 2004), immune cells, and some microbes (Shishov et al. 2009; Tsavkelova et al. 2000; Villageliu and Lyte 2018; Wall et al. 2014) (Fig. 2.2). The enzymes for DA metabolism, monoamine oxidase (MAO), and catechol-O-methyl transferase (COMT) are expressed not only in the CNS but also in the periphery, including the gut (Karhunen et al. 1994; Billett 2004). DA concentrations in the blood and

Table 2.1 The concentration of DA in gut and circulation

Region	DA concentration (M)
Circulation	10^{-11}—10^{-8}
Pancreas	10^{-7}—10^{-5}
Small intestines	10^{-9}—10^{-5}
Colon	10^{-8}—10^{-4}
Gut microbiome	10^{-7}—10^{-4}

(Matt S M et al., J Neuroimmune Pharmacol 2020, 5, 114–164) This table is reprinted with permission from Journal of NeuroImmune Pharmacology

tissues depend on a dynamic balance between its synthesis and metabolism. Matt and Gaskill have summarized the range of DA concentrations throughout the periphery based on the literature (Table 2.1), DA concentration is about 10^{-11}–10^{-8} M in the circulation, 10^{-7}–10^{-5} M in the pancreas, 10^{-9}–10^{-5} M in the small intestines, and 10^{-8}–10^{-4} M in the colon. Moreover, gut microbiome also contains 10^{-7}–10^{-4} M DA (Matt and Gaskill 2020).

2.2 Sources of Dopamine in the Gut

2.2.1 Dopamine Produced by the Gastric Mucosa

The presence of TH and DAT in H^+-K^+-ATPase-positive gastric mucosal cells in human (Eisenhofer et al. 1997) and the presence of very high DA concentrations in the gastric juice after pyloric ligation in rats (Mezey et al. 1998) suggest that DA might be produced by gastric parietal cells. Recently Dr. Zhu JX's lab has also demonstrated TH and DOPA decarboxylase (DDC) are co-localized in the same cells (Fig. 2.3a), TH and DAT are co-localized with the parietal cell marker H^+-K^+-ATPase in rat gastric corpus mucosa (Fig. 2.3b) using immunofluorescence staining (Feng et al. 2020).

The group has reported that the DA content of the fluid incubated with the gastric corpus mucosa in vitro is approximately 60% of DA content in the substantia nigra (SN), but much higher than that of plasma in rats (Fig. 2.4a). Tyrosine induces a concentration-dependent increase of DA in the in vitro gastric mucosa-incubated fluid (Fig. 2.4b), which is largely blocked by the TH inhibitor α-methyl-DL-tyrosine methyl ester hydrochloride (α-MDL) (Fig. 2.4c), suggesting that the in vitro gastric mucosa can take up tyrosine and then synthesize and release DA into the incubation fluid. The muscarinic receptor agonist, bethanechol chloride, increases gastric acid secretion and DA content of the incubation fluid, while the proton pump inhibitor omeprazole markedly decreases gastric acid production and DA content of the fluid (Fig. 2.4d). In in vivo experiments, intravenous infusion of histamine also increases gastric acid secretion (Fig. 2.4e) and enhances the DA content of the effluent

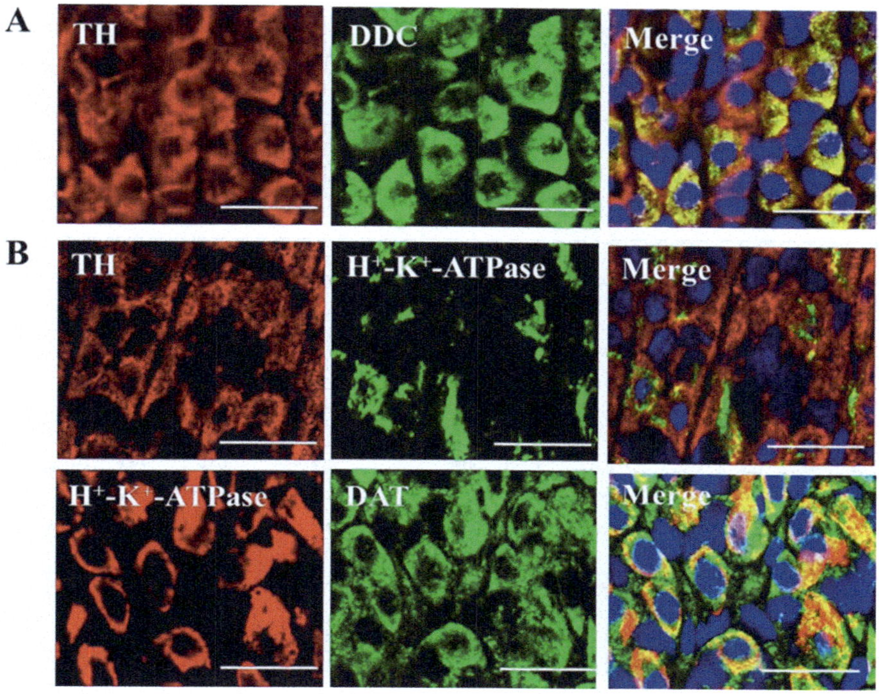

Fig. 2.3 Parietal cells are the main source of dopamine in the gastric mucosa. (**a**) The labeling immunofluorescence of tyrosine hydroxylase (TH) and DOPA decarboxylase (DDC) in the gastric corpus mucosa of rat. (**b**) The labeling immunofluorescence of H^+-K^+-ATPase and TH/dopamine transporter (DAT) in the gastric corpus mucosa of rat. A nuclear marker, 4′,6-diamidino-2-phenylindole (DAPI, blue), is used in the present study. Scale bars: 25 μm

perfusate (Fig. 2.4f). These results strongly support that gastric acid secretion is accompanied by the secretion of gastric-derived DA (Feng et al. 2020).

There is no significant difference in the amount of DA between the muscular and mucosal layers in the corpus region of the rat stomach (Tsai and Cheng 1995). However, the amount of DA in the corpus region is the lowest. The amount of DA is much higher in the antrum, followed by the rumen (Table 2.2) (Tsai and Cheng 1995).

The gastric juice of the rat also contains a significantly higher amount of DA than plasma from cardiac blood samples after pyloric ligation. The blood from hepatic portal vein contains higher level of DA than the arterial supply to the stomach. However, the DA content of the stomach wall has no significant change after chemical sympathectomies (Mezey et al. 1998). In Western blots, immunoreactive bands of the same apparent molecular weight are recognized by an antibody to the TH in gastric mucosa, submucosa, and adrenal glands. However, the activity of TH in stomach is decreased after chemical sympathectomy, which reflects the disappearance of sympathetic nerve fibers, while the remaining 36% of the TH activity

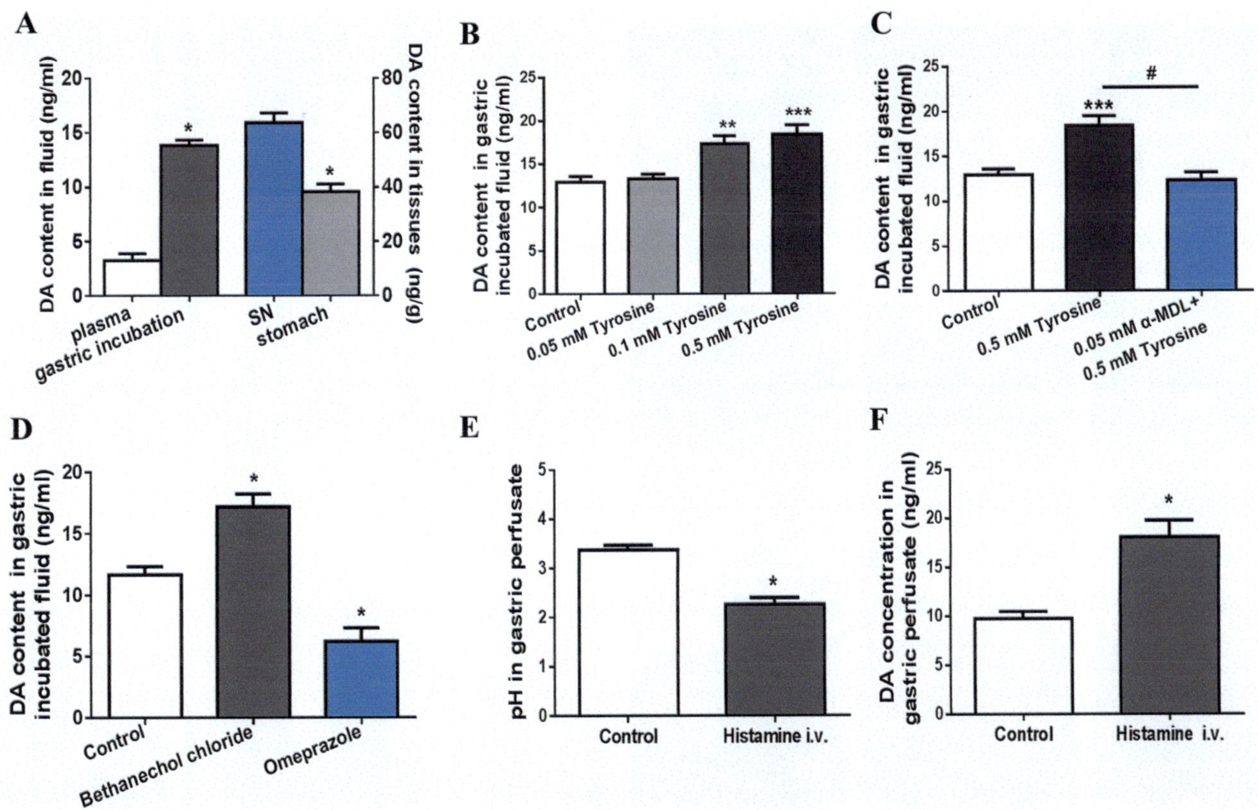

Fig. 2.4 Dopamine content of the stomach measured by ultra-performance liquid chromatography tandem mass spectrometry. (Feng XY, et al., BJP 2020, 177:3258–3272.) (**a**) DA content of gastric incubation fluid, plasma, gastric mucosa, and substantia nigra (SN) in rats. (**b**) DA content after the application of

Table 2.2 Regional distribution of DA contents in the male rat stomach

Region	DA concentration (nmol/mg protein)	Sample size
Rumen	1.062 ± 0.205	8
Mucosal layer of corpus	0.283 ± 0.110	8
Muscular layer of corpus	0.207 ± 0.052	8
Antrum	2.920 ± 0.509	8

The data are expressed as mean ± S.E.M. Mucosal layer included mucosa, submucosal plexus, and a small amount of adherent submucosal connective tissues. This table is reprinted with permission from Neurosci Res. (Tsai L H & Cheng J T. Neurosci Res, 1995, 21(3): 235–240)

most likely reflects TH in parietal cells by immunohistochemistry (Mezey et al. 1998).

2.2.2 *Dopamine Produced by the Pancreas*

DA can be produced by the pancreatic tissue. The β cells in pancreatic islets are key sources of peripheral DA. This part will be further discussed in Chap. 6.

2.2.3 *Dopamine Produced by the Enteric Nervous System and Gut Immune Cells*

TH and DAT are expressed in the murine bowel, including the enteric nervous system (ENS) and immune cells. ENS contains intrinsic dopaminergic neurons arising from a *mash*-1-independent lineage of noncatecholaminergic precursors (Li et al. 2004). TH-positive neurons are found in both submucous plexus and myenteric plexus of the ileum and colon (Li et al. 2004) (Fig. 2.5).

DA is an important molecule bridging the nervous and immune systems. DA can modulate the functions of immune effector cells by binding to its receptors expressed in many types of immune cells (Sarkar et al. 2010). Under certain conditions, immune cells can release DA to the extracellular milieu, leading to autocrine and paracrine effects (Bergquist et al. 1994; Ferrari et al. 2004; Josefsson et al. 1996; Musso et al. 1996; Taraskina et al. 2015). It is reported that peripheral blood or bone marrow-derived immune cells, such as regulatory T cells (Cosentino et al. 2007),

Fig. 2.4 (continued) tyrosine. (**c**) DA content after application of tyrosine or pretreatment with TH inhibitor α-MDL. (**d**) DA content after application of bethanechol chloride and omeprazole. (**e**) The pH of gastric perfusate after intravenous (i.v.) infusion of histamine (2 mg kg^{-1} h^{-1}). (**f**) DA content of effluent perfusate after i.v. infusion of histamine. $*p < 0.05$ $**p < 0.01$, $***p < 0.001$. This figure is reprinted with permission from Br J Pharmacol

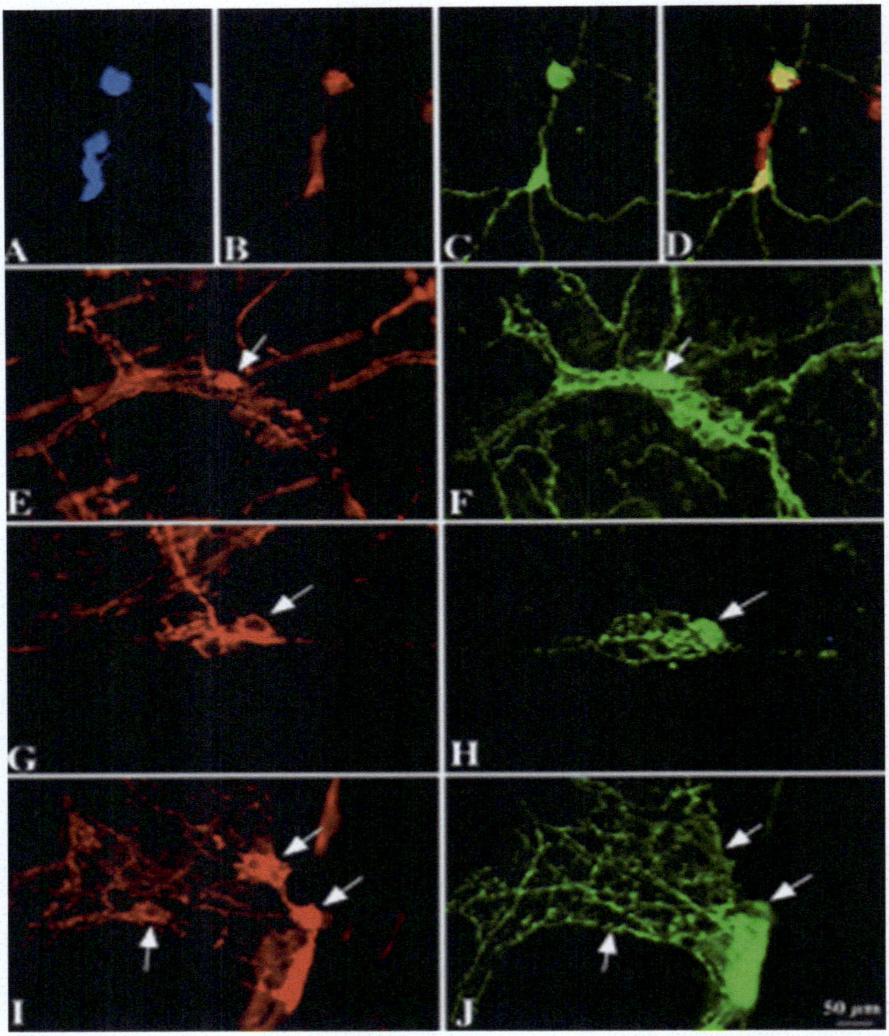

Fig. 2.5 The distribution of TH in the mouse ileum and colon. (Li Z S, et al., J Neurosci 2004, 24, 1330–1339.) (**a–d**) Submucosal plexus of the ileum. (**e** and **f**) Myenteric plexus of the ileum. (**g** and **h**) Submucosal plexus of the colon. (**i** and **j**) Myenteric plexus of the colon. Hu (a neuronal marker **a**; blue fluorescence), DAT (**b**, **e**, **g**, **i**; red fluorescence), and TH (**c**, **f**, **h**, **j**; green fluorescence). Scale bar: 50 μm. This figure is reprinted with permission from Copyright [2004] Society for Neuroscience

helper T cells (Papa et al. 2017), and dendritic cells (Prado et al. 2012), all express TH. Besides, macrophages, neutrophils (Cosentino et al. 2000), and B cells also contain DA (Beck et al. 2004).

Bergquist et al. have revealed that lymphocyte and peripheral human T cells contain DA. The level of DA is reduced after treatment with TH inhibitor but increased after incubating with the precursor L-DOPA, indicating that lymphocytes are able to synthesize DA (Bergquist et al. 1994; Musso et al. 1996). The production of DA in human lymphocytes is promoted by protein kinase C activation of TH (Ferrari et al. 2004). Moreover, mouse spleen cells and macrophages also contain DA on average 7×10^{-17} and 2×10^{-17} mole per cell examined by a highly sensitive capillary electrophoresis assay, respectively (Josefsson et al. 1996). Human and mouse follicular helper T cells can produce and release DA (Papa et al. 2017). Dendritic cells express TH, MAO-A, and MAO-B, not dopamine β-hydroxylase, indicating that dendritic cells can produce DA (Prado et al. 2012). Similarly, another study has shown that DA is also present in the monocytes and neutrophils of human peripheral blood (Cosentino et al. 2000). However, it is not clear whether the immune cells in the gut synthesize DA or not.

2.2.4 Substantial Dopamine in the Colonic Lumen

Except for intestinal tissues, a substantial level of free DA (115 ± 14 ng/g stool) is identified in the gut lumen of specific pathogen-free (SPF) mice; however, the amount in luminal contents of germ-free (GF) mice is very low (5.0 ± 0.5 ng/g stool). Critically, the majority of DA in SPF mice is structurally determined to be free and biologically active, whereas those found in GF mice are present in biologically inactive, conjugated forms (Asano et al. 2012). Intragastric administration of the microbiota from SPF mice to GF mice results in the production of free and biologically active of DA within the gut lumen (Asano et al. 2012). β-Glucuronidase activity in microbiota plays a key role in the elevation of free DA in the gut lumen (Asano et al. 2012). Moreover, gut short-chain fatty acid (SCFA) (butyrate) can increase the synthesis of DA through induction of TH via ERK-dependent phosphorylation of CREB protein (Shah et al. 2006). In addition, several studies have reported that in vivo gut microbiota is capable of generating DA. The production of luminal DA by bacteria is discussed in Sect. 2.4.

2.3 Metabolism of Dopamine in the Gut

2.3.1 Metabolic Process of Dopamine

DA is degraded into inactive metabolites by several enzymes, including MAO, COMT, and aldehyde dehydrogenase (Eisenhofer et al. 2004; Finberg 2014), and these enzymes are localized intracellularly. The rapid degradation of DA ensures the proper functioning of synaptic neurotransmission and gut function. DA is deaminated to 3,4-dihydroxyphenylacetaldehyde (DOPAL) by MAO, and then aldehyde dehydrogenase metabolizes DOPAL, leading to the production of 3,4-dihydroxyphenylacetic acid (DOPAC). COMT can further convert DOPAC to homovallic acid (HVA) if it is present in the cell. DA can also be first metabolized by COMT to 3-methoxytyramine (3-MT), which is subsequently converted to HVA via MAO and AD (Bortolato et al. 2008; Eisenhofer et al. 2004; Finberg 2014). The major end-product of DA metabolism is HVA, which has no known biological activity (Eisenhofer et al. 2004). MAO has two isoforms, MAO-A and MAO-B, both can oxidize DA in most species. COMT is present in a soluble form (S-COMT) and a membrane bound form (MB-COMT), both are coded by the same single gene (Myohanen et al. 2010). All the enzymes involved in DA metabolism are present in the gut.

2.3.2 MAO and COMT in the Gut Mucosa

In mammals, MAO and COMT are present in various tissues throughout the organs of the body. In addition to be highly expressed in the liver and brain, they are widely distributed in the mucosa and muscular layer in the gastrointestinal (GI) tract (Billett 2004; Karhunen et al. 1994).

In addition to the CNS, most peripheral tissues express both MAO-A and MAO-B (Wang et al. 2013). The two types of MAO isoforms are highly expressed in the GI tract (Table 2.3). In human duodenum, MAO-A and MAO-B are mainly expressed in the villi, crypts, muscularis externa, and mucosa, while lower-level expression is found in submucosal cells (Sivasubramaniam et al. 2003). However, MAO-A immunoreactivity is particularly high in the mucosa, both in the villi and crypts, whereas MAO-B immunoreactivity is only strong in the villi (Rodriguez et al. 2001). Indeed, the activity of MAO-A is higher than that of MAO-B in most peripheral tissues (Table 2.4). MAO-A accounts approximately 80% of total MAO activity in the GI tract (Bartl et al. 2014; Billett 2004; Liu et al. 2018; Vieira-Coelho et al. 1999), one of the potential reasons could be the high level of serotonin content in the GI tract.

In the GI system, the main role of intestinal MAO is to degrade dietary amines, some of which are potentially toxic, such as tyramine, which could have a pressor effect (Da et al. 1988; Youdim and Finberg 1987). The presence of both MAO-A and MAO-B in the GI tract might provide a better defense mechanism by covering a broader spectrum of substrates.

It is reported that MAO-B immunoreactivity (Fig. 2.6b) is localized in almost all of the histidine decarboxylase (HDC)-positive cells (Fig. 2.6a) in the rat gastric mucosa (Okauchi et al. 2004). At electron microscopic levels, MAO-B is detected in cells that show morphological characteristics of enterochromaffin-like (ECL) cells. Thus, MAO-B might be involved in the inactivation mechanism of histamine, which is released from ECL cells and can activate parietal cells to secrete gastric acid (Okauchi et al. 2004).

Table 2.3 The expression of MAO-A and MAO-B in the gut

Tissue and cell type	MAO-A protein	MAO-A mRNA	MAO-B protein	MAO-B mRNA	Species	Reference
Stomach						
Enterochromaffin-like cells			+		Rat	Okauchi et al. (2004)
Duodenum						
Enterocytes of villi	+	+	+	+	Human	Sivasubramaniam et al. (2003), Billett (2004), Rodriguez et al. (2001)
Cryptive cells	+	+	+	+		
Muscularis mucosa	+	+	+	+		
Submucosal cells	+	+	+	+		
Muscularis externa	+	+	+	+		
Brunner gland cells	–		–			
Colon						
Muscular layer	+		+		Rat	Liu et al. (2018)
Mucosal layer	+		+			

Note: – not detected, + detected

Table 2.4 The activity of MAO in the gut

Tissue	Enzyme activity		Species	Reference
	MAO-A	MAO-B		
Jejunal epithelial cells	+	+	Rat	Vieira-Coelho et al. (1999)
Caco-2 cells	+	+	Human	
Intestine	+	+	Human	Billett (2004)
Colon	+	+	Rat	Liu et al. (2018)

Note: + detected

In the rat gastrointestinal tract, COMT is distributed more abundantly on the villi and crypts at the apical membrane of the epithelial cells covering the stomach, duodenum, ileum (Karhunen et al. 1994), and colon (Li et al. 2015) (Table 2.5). In the corpus and pyloric of the stomach, COMT immunoreactivity is the strongest in epithelial cells of the gastric pits comparing to other cell types. However, endocrine cells throughout the GI tract do not display significant COMT immunoreactivity. In mice, COMT is abundantly expressed in the epithelial cells of the duodenum (Kaenmaki et al. 2009) and colon.

The expression level and activity of MB-COMT and S-COMT are different in various regions of the GI tract (Table 2.6). In rat intestine, S-COMT is the predominant form both in the mucosa and muscular layer, and the activity of S-COMT in the mucosa is approximately twice as high as muscular layer. Besides, in the small

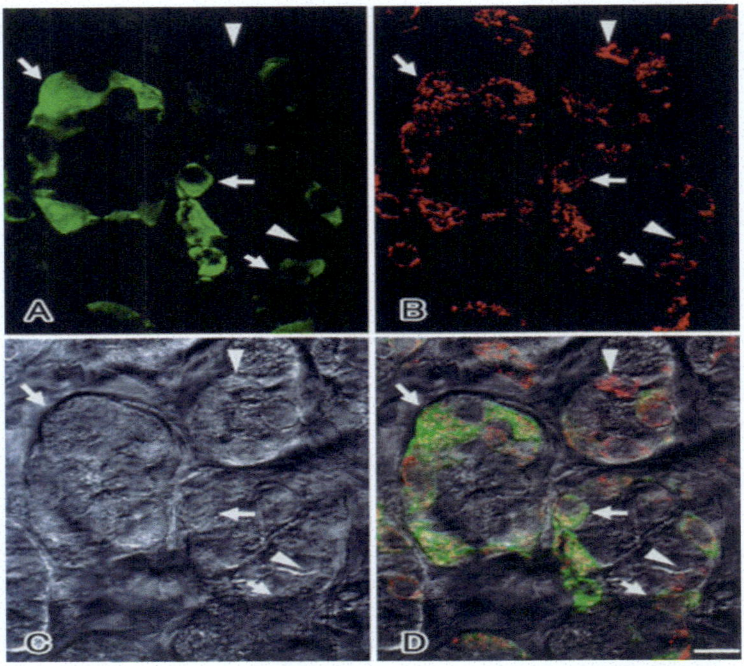

Fig. 2.6 Double-labeling histidine decarboxylase (HDC) and MAO-B in the oxyntic gland of the rat gastric mucosa. (Okauchi H et al., Histochem Cell Biol, 2004, 121, 181–188.) (**a**) HDC staining, (**b**) MAO-B staining, (**c**) differential interference contrast, (**d**) superimposition of the images in **a**, **b**, and **c**. Scale bar: 10 μm. This figure is reprinted with permission from Histochem Cell Biol

intestine, the activity of the S-COMT is highest in the duodenal mucosal layer and slightly lower in the jejunal and ileal mucosal layers, which might be associated with the high level of luminal DA from gastric secretion. Whereas, the activity of MB-COMT is almost equal in the mucosa and muscular layer as well as in the different regions of the GI (Nissinen et al. 1988a, b).

It is reported that 3-O-methyldopa level in the plasma or other tissues is not decreased after removing the liver from rats, suggesting that intestinal COMT is involved in the metabolism of exogenous catechols (Nissinen et al. 1988a, b). The results of ours and other researchers indicate that gut lumen has a high concentration of DA and noradrenaline (Feng et al. 2020; Asano et al. 2012). Hence, COMT located in the gut epithelial cells might be responsible for degrading exogenous catechols to limit their entry to the body. COMT may serve some unique functions in the gastrointestinal tract by modulating dopamine levels to keep dopaminergic tone.

Because the amount of COMT is high in the peripheral tissues, such as liver, kidney, and intestinal mucosa (Kaenmaki et al. 2009; Karhunen et al. 1994; Myohanen et al. 2010), the inhibition of peripheral COMT can increase brain penetration of L-DOPA. Thus, the COMT inhibitors are adjuncts to L-DOPA in the treatment of Parkinson's disease (PD) to enhance DA level in the brain (Nissinen

Table 2.5 The expression of COMT in the gut

Tissue and cell type	COMT protein	Species	Reference
Stomach			
Epithelial cells of corpus	+	Rat	Karhunen et al. (1994)
Pylorus	+		
Endocrine cells	–		
Duodenum			
Epithelial cells	+	Mouse	Kaenmaki et al. (2009)
Microvilli	+		
Ileum			
Epithelial cells	+	Rat	Karhunen et al. (1994)
Colon			
Mucosal layer	+	Rat	Li et al. (2015)

Note: – not detected, + detected

Table 2.6 The activity of COMT in the gut

	Enzyme activity				
Tissue	S-COMT	MB-COMT	COMT	Species	Reference
Duodenum				Human	Nissinen et al. (1988a, b)
Mucosal layer	+	+			
Muscular layer	+	+			
Jejunum					
Mucosal layer	+	+			
Muscular layer	+	+			
Ileum					
Mucosal layer	+	+			
Muscular layer	+	+			
Duodenum	+	+		Rat	
Jejunum	+	+			
Ileum	+	+			
Jejunal epithelial cells			+	Rat	Vieira-Coelho et al. (1999)
Caco-2 cells			–	Human	

Note: – not detected, + detected

et al. 1988a, b). This has been proved to be very effective for more patients with advanced PD stages (Hristova and Koller 2000).

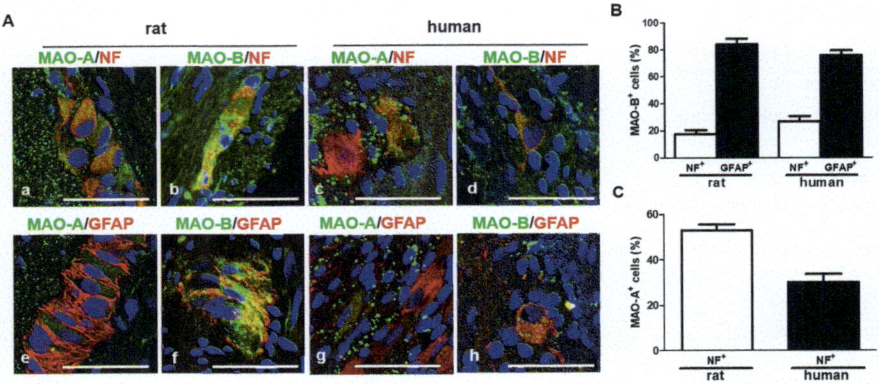

Fig. 2.7 The cellular localization of MAO-A (green) and MAO-B (green) in the colonic myenteric plexus of rat and human (**A**). (Liu C Z et al., Neurogastroenterol Motil, 2018, 30, e13390.) *NF* a neuronal marker, *GFAP* a marker of enteric glial cell. The percentage of MAO-B (**B**) and MAO-A (**C**) colocalization with NF- or GFAP-positive cells in the colon of rat and human. Scale bar: 50 μm. This figure is reprinted with permission from Neurogastroenterol Motil

2.3.3 MAO and COMT in the Enteric Nervous System

Previously, COMT expression was reported in both neurons and glial cells in the CNS (Matsumoto et al. 2003). Levels of COMT mRNA are obviously higher in neurons than in glia. In the midbrain, COMT mRNA is detected in dopaminergic neurons in both human and rat (Matsumoto et al. 2003). However, the distribution of COMT in ENS is not yet reported. MAO-A and MAO-B are also widely distributed in the brain. MAO-B is expressed in the glial cells in the CNS (Westlund et al. 1985).

Recently, Dr. Zhu JX's Lab has reported that both MAO-A and MAO-B are expressed in the colonic myenteric plexus of the rat and human (Liu et al. 2018). MAO-B immunoreactivity is present in the neurofilament (NF)-positive neurons and glial fibrillary acidic protein (GFAP, a marker of glial cell)-positive glial cells (Fig. 2.7A), whereas MAO-A immunoreactivity is only observed in the NF-positive neurons, not in the GFAP-positive glial cells (Fig. 2.7A) (Liu et al. 2018). Most of GFAP-positive cells express MAO-B, but only a small proportion of NF-positive neurons have MAO-B. In the rat colonic myenteric plexus, the percentage of NF-positive neurons expressing MAO-B is approximately 17%, while the percentage of GFAP-positive cells occupied by MAO-B is about 83% (Fig. 2.7B). In the human colonic myenteric plexus, the ratio is similar to that in rat, it is about 26% and 75%, respectively (Fig. 2.7B). The percentage of NF-positive neurons expressing MAO-A in colonic myenteric plexus is 52% in rat and 30% in human (Fig. 2.7C) (Liu et al. 2018).

In glial cells of the CNS, MAO-B plays an important role in converting 1-methyl-4-phenyl-1,2,3,6-teterahydropyridine (MPTP) to the toxic metabolite, 1-methyl-4-phenylpyridine (MPP+), which selectively destroys nigral DAergic neurons

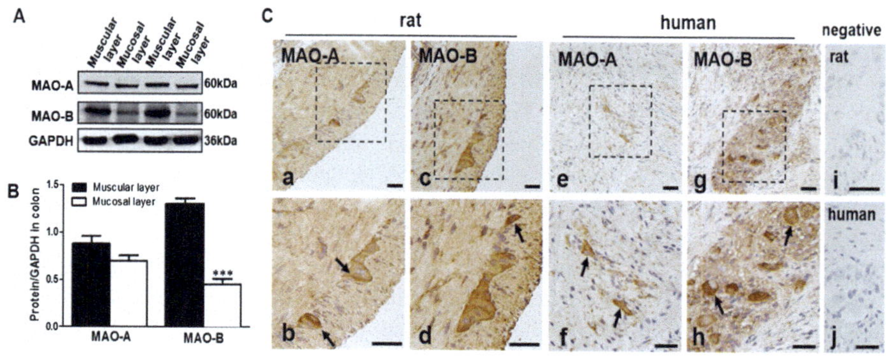

Fig. 2.8 MAO-A and MAO-B are expressed in the colonic muscular layers. (Liu C Z et al., Neurogastroenterol Motil, 2018, 30, e13390.) (**A**, **B**) The protein expressions of MAO-A and MAO-B in the muscular layer and mucosal layer. (**C**) The distribution of MAO-A and MAO-B immunoreactivity in the colon of rat and human. Scale bar: 50 μm. $^{***}p < 0.001$. This figure is reprinted with permission from Neurogastroenterol Motil

(Smeyne and Jackson-Lewis 2005). The neurodegeneration induced by MPTP is similar to the neuronal damage in PD; therefore, MAO-B appears to be associated with the pathogenesis of MPTP-induced Parkinsonism (Smeyne and Jackson-Lewis 2005; Srivastav et al. 2019). MPTP administration can also damage the peripheral dopaminergic system (Anderson et al. 2007; Cote et al. 2011) and induce a selective decrease in TH-expressing neurons and DA level in the mouse gut (Anderson et al. 2007; Natale et al. 2010) Intestinal glial cells also express MAO-B, suggesting that MPTP can also be converted into MPP^+ in the intestine, thereby damaging DA neurons in the intestine, which may be the cause of decreased intestinal DA content and rapid impairment of enteric motility in MPTP mice (Anderson et al. 2007; Cote et al. 2011; Natale et al. 2010).

2.3.4 MAO and COMT in the Gut Smooth Muscles

The activity of MB-COMT is almost equal in the different regions of the GI tract. But in the muscle layers, the S-COMT activity is the highest in the duodenum and lowest in the ileum (Karhunen et al. 1994). There are no significant differences between the expression of MB-COMT and S-COMT in the human muscle layers of the jejunum and ileum (Karhunen et al. 1994).

MAO-A and MAO-B are expressed in both muscular and mucosal layers of the rat colon (Liu et al. 2018) (Fig. 2.8). Additionally, the expression of MAO-B protein in the muscular layer is much higher than that in the mucosal layer, while the MAO-A protein level shows no significant difference between the two layers (Fig. 2.8A, B). In the colon of rat and human, the immunoreactivities of both

MAO-A and MAO-B are present in the muscular layer, including the myenteric plexus (Fig. 2.8C).

The expression of COMT and MAO in the colonic smooth muscle is related to the colonic motility (Li et al. 2015). The clinical symptoms in PD patients are significantly improved when entacapone (a COMT inhibitor) or rasagiline (MAO-B inhibitor) is used in conjunction with L-DOPA (Jatana et al. 2013; Weinreb et al. 2010), but the side effects of gastrointestinal tract are also common, such as a slightly increased risk of constipation (Guay 2006; Tolosa and Stern 2012). It is reported that entacapone inhibits colon longitudinal muscle contraction in a dose-dependent manner, and this inhibitory effect can be blocked by the β_2 adrenoceptor antagonist ICI-118,551 by 67%, indicating that entacapone may inhibit colon motility through the β_2 adrenoceptor (Li et al. 2015). Therefore, this may be a potential reason for constipation seen in the PD patients taking a COMT inhibitor along with L-DOPA.

MAO-B inhibitors have become important pharmacological targets for the treatment of PD owing to their inhibition of DA metabolism and neuroprotective effect. The most commonly used drugs are rasagiline and safinamide (Fox et al. 2018; Riederer and Muller 2018). Rasagiline and safinamide are the second-generation of selective and irreversible MAO-B inhibitor, respectively. Safinamide can also inhibit glutamate release. In addition, both rasagiline and safinamide have neuroprotective effects (Teixeira et al. 2018; Weinreb et al. 2010). Clinically, rasagiline is used as a monotherapy in the early stages of PD to alleviate the disease process or used in the advanced stages of PD to assist L-DOPA to improve motor symptoms (Dezsi and Vecsei 2017; Weintraub et al. 2016). When used as an adjunct to L-DOPA or in combination with other PD drugs, safinamide increases the daily "on" time and significantly improves motor symptoms, clinical fluctuations, and quality of life in PD patients at advanced stage (Cattaneo et al. 2019). However, there are also clinical reports that long-term combination of L-DOPA and MAO-B inhibitors (rasagiline and safinamide) increases the risk of constipation in PD patients (Borgohain et al. 2014; Guay 2006; Tolosa and Stern 2012). MAO may have influence on the colonic motility by impacting the level of monoamine neurotransmitters. In a recent report from Dr. JX Zhu's lab, administration of rasagiline for 4 weeks can increase the level of colonic DA by directly suppressing colonic MAO-B activity in rats (Fig. 2.9a, b). Therefore, the rat exhibits decreased motility and a lengthened colonic transit time (Fig. 2.9c, d). This may be a potential reason that long-term application of rasagiline is associated with a slightly increased risk of constipation in PD patients (Liu et al. 2018).

2.4 Gut Microbiome and Dopamine

The most abundant microbiota is found in the gut, where the bacterial density reaches 10^{11}–10^{12} cells/g in the human distal colon (Barko et al. 2018). The number of bacteria maybe ten times higher than the number of eukaryotic cells, and the biomass of the gut microbiota may reach up to 1.5 kg. Therefore, the gut microbiota

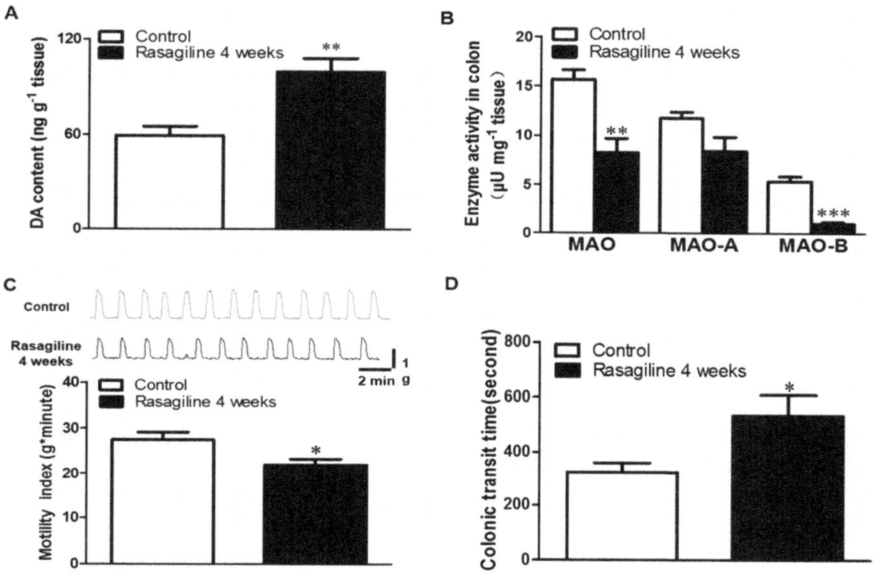

Fig. 2.9 Long-term application of rasagiline increases DA content and attenuates colonic motility. (Liu C Z et al., Neurogastroenterol Motil, 2018, 30, e13390.) (**a**, **b**) DA content and MAO-B activity in long-term rasagiline-treated rats. (**c**, **d**) The in vitro colonic motility and colonic transit time. $*p < 0.05$ $**p < 0.01$, $***p < 0.001$. This figure is reprinted with permission from Neurogastroenterol Motil

could be considered as a "forgotten organ." It is reported that gut microbiota has many benefits (Barko et al. 2018; Moffa et al. 2019; Stange and Schroeder 2019). Intestinal microbiota protects the mucosa from adherence and invasion by exogenous pathogens. Some probiotic bacteria release nutrients, such as vitamins, SCFA, and sugars. Intestinal absorption of SCFA translates into reinforcement of the intestinal epithelial cells. Gut microbes also modulate innate and adaptive immunity by activating release of macrophages and cytokines including IL-4, IL-5, IL-6, and IL-10.

2.4.1 Dopamine Generated by the Gut Microbiota

Recent data indicate that bacteria can produce DA (Strandwitz 2018). Asano et al. (2012) quantitate the levels of DA in the gastrointestinal lumen in SPF mice, GF mice, and gnotobiotic mice (reconstituted with a mixture of various bacterial species). The substantial physiological amount of DA exists in the gut lumen in the SPF mice, but only a trace amount of DA in the GF mice. Similar results are also obtained from the rats (Fig. 2.10). Besides, DA is determined to be free and biologically active in the SPF mice. Whereas, in GF mice, 90% of DA is present in a biologically

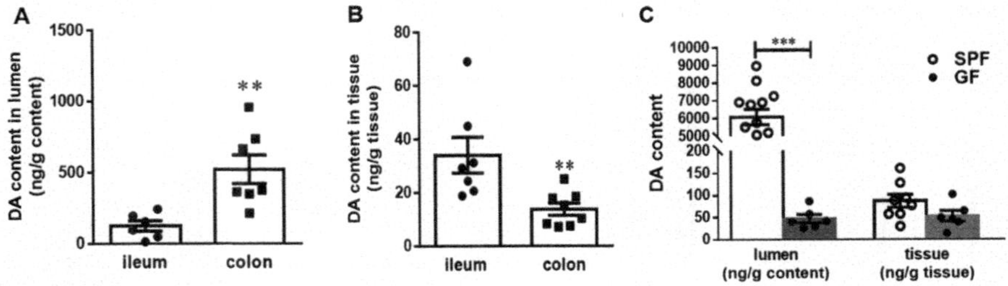

Fig. 2.10 DA levels in the lumen and tissue of the ileum and colon of specific pathogen-free (SPF) and germ-free (GF) rats. DA levels in the lumen (**a**) and tissue (**b**) of ileum and colon of specific pathogen-free (SPF) rats, luminal, and colonic tissue DA content of SPF and GF rats (**c**). $**p < 0.05$, $***p < 0.001$

inactive form that is conjugated with glucuronide. After transplanting the microbiota of SPF mice to GF mice, the free, biologically active DA within the gut lumen is obviously increased in GF mice (Asano et al. 2012). The reason is that some bacteria themselves, in particular *Clostridium* species, express β-glucuronidase, which hydrolyzes glucuronide from glucorinide-conjugated DA and significantly increases free DA in the gut lumen (Asano et al. 2012). This study suggests that in vivo microbiota is able to increase the content of luminal DA. Currently, several studies have reported that in vivo microbiota is capable of generating DA, such as the members of the genera *Escherichia, Saccharomyces, Streptococcus, Enterococcus*, and *Bacillus* (Nicholson et al. 2012; Tsavkelova et al. 2000; Wall et al. 2014) (Table 2.7). The culture fluid of *E. coli* K-12 contains micromolar concentration of DOPA and nanomolar concentration of DA (Shishov et al. 2009). A recent study has reported that the *Enterococcus* sp. can produce DA in a gastrointestinal-like environment when supplied with the DA precursor L-DOPA, suggesting that *Enterococci* has the enzymatic capacity to produce DA and maybe they can do so with resources derived from the host diet (Villageliu and Lyte 2018). Besides free DA, DA sulfate also exists in the gastrointestinal tract, which is mainly produced from both dietary and locally synthesized DA (Eisenhofer et al. 2004). Apart from producing and releasing DA, the microbiota also influences the concentration of tyrosine (the precursor of DA) in the nervous system. The level of tyrosine in the total brain of GF mice is lower than that of the mice having been re-colonized by the gut microbiota (Matsumoto et al. 2013).

2.4.2 *Metabolism of Gut Luminal Dopamine*

In the biomass of *Bacillus cereus, B. subtilis, Proteus vulgaris, B. mycoides, Serratia marcescens, S. aureus*, and *E. coli*, the concentrations of DA is from 0.45 to 2.13 mmol/L (Tsavkelova et al. 2000). Interestingly, the content of DA in bacteria is higher than that in the gut tissue, where the physiological concentration of DA is

Table 2.7 DA-producing or -releasing bacterial strains

Bacterial strains	References
Enterococcus faecium	Villageliu and Lyte (2018)
Bacillus cereus	Tsavkelova et al. (2000)
Bacillus mycoides	
Bacillus subtilis	
Escherichia coli	
Proteus vulgaris	
Serratia marcescens	
Staphylococcus aureus	
Escherichia coli (K-12)	Shishov et al. (2009)
Hafnia alvei	Özoğul (2004)
Klebsiella pneumoniae	
Morganella morganii	

about 1 μmol/L (Yan et al. 2015), suggesting that the gut lumen has a substantial amount of DA.

DAT is the main transporter for DA. It is reported that DAT is predominantly distributed in the apical side of crypts and epithelial cells in rat colonic mucosa (Tian et al. 2008). A great amount of MAO and COMT is present in the colonic mucosa. However, what is the function of DAT, and whether the luminal DA could be transported into colonic mucosa and then degraded by mucosa COMT and MAO are still unknown. Whether the important role of epithelial COMT and MAO is to metabolize exogenous DA, thus maintain the homeostasis of endogenous DA still needs further research.

2.4.3 Gut Luminal Dopamine, Microbiota, and Gastrointestinal Disorders

Luminal DA has been suggested to play important roles in regulating gastrointestinal function. The luminal DA is able to increase colonic water absorption and stimulate active ileal ion absorption (Asano et al. 2012; Barry et al. 1994, 1995).

In regard to the mucosal barrier, DA has been reported to enhance duodenal epithelial permeability via the dopamine D_5 receptor (Feng et al. 2017) and stimulate colonic mucus secretion by the D1-like receptors (D_1 and D_5) (Li et al. 2019). However, the role of luminal DA in modulating mucosal barrier, particularly tight junction and mucus secretion, remains elusive.

It is possible that the DA secreted by bacteria in the intestinal lumen may induce epithelial cells to release molecules that in turn modulate neural signaling within the ENS or act directly on primary afferent axons (Forsythe and Kunze 2013), thus modulates the functions of the CNS (Roshchina 2016). It has been also reported that probiotic bacteria can produce neurochemicals and have been considered as a novel treatment for neuropsychiatric diseases (Lyte 2011).

Over the last decade, the number of reports demonstrating the ability of bacteria to respond to host neuroendocrine hormones, especially during times of stress, has steadily increased. In 1992, Lyte M reported that the stress-related neurohormones noradrenaline and DA could increase the growth of human intestinal bacterial pathogens by over six orders of magnitude within hours (Lyte 1992; Lyte and Ernst 1992). After the administration of the adrenergic neurotoxin 6-hydroxydopamine, which can induce the systemic wide release of catecholamines, gut bacterial populations shift from predominantly Gram-positive to Gram-negative. Besides, the number of *E. coli* is increased nearly 7 log-fold within 24 h following neurotoxin administration. Furthermore, after the recovery of the adrenergic nerves, the composition of the gut microbiota is returned to the normal pre-neurotoxin distribution (Lyte and Bailey 1997). This demonstrates that the host-derived neuroendocrine hormones can alter the composition of the gut microbiota.

Numerous studies have reported that gut microbiota is related to PD and its clinical phenotypes (Aho et al. 2019; Pietrucci et al. 2019; Sherwin et al. 2017). The abundance of *Prevotellaceae* in the feces of PD patients is reduced by 77.6% (Scheperjans et al. 2015; Unger et al. 2016). A recent clinical study has also discovered a reduction in the abundance of *Prevotella* species, along with the increase in *Akkermansia muciniphila* in PD patients (Bedarf et al. 2017). Fecal SCFA level is significantly reduced in PD patients (Unger et al. 2016), since the SCFA-producing microbiota (*Prevotellaceae* and *Lachnospiraceae*) are decreased in PD patients (Barichella et al. 2019). Moreover, plasma SCFA is paradoxically increased in PD and is associated with disease severity and antiparkinsonian medications (Shin et al. 2020).

GI dysfunctions, including gastroparesis and constipation, are an important feature of PD. In the ascending colon of PD patients, the dopaminergic neurons are decreased in the myenteric plexus (Singaram et al. 1995). Furthermore, increased DA content in the colonic muscular layer is involved in the GI dysfunctions in the rats with the lesions of bilateral substantial nigras by 6-hydroxydopamine, a kind of PD model (Zhang et al. 2015).

It is reported that the tyrosine decarboxylase from fecal and DA dehydroxylase from Eggerthella lenta A2 are able to metabolize L-DOPA into 3-tyramine (Maini et al. 2019). If the gut microbiota metabolizes L-DOPA before it passes through the blood–brain barrier, the medication is ineffective. Moreover, the host-targeted drug carbidopa (a peripheral DOPA decarboxylase inhibitor, reduces the production of DA in peripheral tissues, and then increases the amount of L-DOPA that enters the CNS) did not affect gut bacterial L-DOPA decarboxylation. Thus, the effect of L-DOPA on PD is very different among individuals, depending on the composition of their microbiota (Gray et al. 2014; Spanogiannopoulos et al. 2016). A recent research reports that (S)-α-fluoromethyltyrosine can inhibit the activity in complex human gut microbiota (Maini et al. 2019), which suggests the possibilities for developing combinations of Parkinson's drugs to eliminate the effect of microbiota on metabolism L-DOPA (Lubomski et al. 2019).

Table 2.8 The alteration of DA and MAO activity in gastrointestinal disorders

Subject	Tissue affected	Alteration	Reference
CD and UC patients	Inflamed gut mucosa	In IBD patients, L-DOPA level is increased, the ratio of DA/L-DOPA and DA level are reduced	Magro et al. (2002)
2,4,6-Trinitrobenzene sulphonic acid (TNBS)-induced rats	Inflamed gut mucosa	Decreased DA and L-DOPA levels	Magro et al. (2004)
Patients with the irritable colon syndrome	Colonic tissue	MAO activity was decreased by 34% at the acute stage of the disease and increased by 44% in those with nonspecific ulcerative colitis	Akopian et al. (1994)

Gut microbiota is also related to the progress of functional bowel diseases (Chong et al. 2019; Knox et al. 2019). Meanwhile, the change of DA content in the gut tissue of some gastrointestinal disorders has also been reported (Akopian et al. 1994; Magro et al. 2002, 2004) (Table 2.8). When intestinal inflammation resulting from an autoimmune reaction occurs, it is often referred to as inflammatory bowel disease (IBD). IBD has two major types, ulcerative colitis (UC) and Crohn's disease (CD). Increased levels of L-DOPA and decreased levels of colonic DA are observed in both diseases, resulting in a reduction in DA/L-DOPA tissue ratios, which indicates low L-amino acid decarboxylase activity (Akopian et al. 1994; Magro et al. 2002, 2004). In addition, the patients with irritable bowel syndrome (IBS) have a lowered MAO activity, which is decreased by 34% during the acute stage of the disease. In those with nonspecific ulcerative colitis, MAO activity is increased by 44% (Akopian et al. 1994). Thus, MAO activity may serve as an indicator of the adequacy and efficiency of the treatment performed. However, whether the microbes producing DA are altered in IBD and IBS is worthy to give more attention.

2.5 Perspectives

In addition to the CNS, the digestive tract is an important source of DA. High levels of MAO and COMT are distributed in the gastrointestinal tract, which plays an important role in the regulation of DA content and DA function in the digestive tract. Therefore, MAO and COMT may become therapeutic targets for some diseases. The dysbiosis of gut microbiota has been found in many diseases, such as IBS, IBD, metabolic diseases, and neurodegenerative diseases. Some intestinal microorganisms can produce DA or metabolize DA. The relationship among the gut DA, microbiota, and diseases are largely unknown. It is necessary to put more effort on the investigation of the roles of gut microbiota and DA in normal gut function and functional disorders or diseases.

References

Aho V, Pereira P, Voutilainen S, Paulin L, Pekkonen E, Auvinen P, Scheperjans F (2019) Gut microbiota in Parkinson's disease: temporal stability and relations to disease progression. EBioMedicine 44:691–707. https://doi.org/10.1016/j.ebiom.2019.05.064

Akopian AA, Arutiunian MV, Agavelian AM (1994) The role of monoamine oxidase in large intestine pathology. Vopr Med Khim 40:54–57

Anderson G, Noorian AR, Taylor G, Anitha M, Bernhard D, Srinivasan S, Greene JG (2007) Loss of enteric dopaminergic neurons and associated changes in colon motility in an MPTP mouse model of Parkinson's disease. Exp Neurol 207:4–12. https://doi.org/10.1016/j.expneurol.2007.05.010

Asano Y, Hiramoto T, Nishino R, Aiba Y, Kimura T, Yoshihara K, Koga Y, Sudo N (2012) Critical role of gut microbiota in the production of biologically active, free catecholamines in the gut lumen of mice. Am J Physiol Gastrointest Liver Physiol 303:G1288–G1295. https://doi.org/10.1152/ajpgi.00341.2012

Barichella M, Severgnini M, Cilia R, Cassani E, Bolliri C, Caronni S, Ferri V, Cancello R, Ceccarani C, Faierman S, Pinelli G, De Bellis G, Zecca L, Cereda E, Consolandi C, Pezzoli G (2019) Unraveling gut microbiota in Parkinson's disease and atypical parkinsonism. Mov Disord 34:396–405. https://doi.org/10.1002/mds.27581

Barko PC, McMichael MA, Swanson KS, Williams DA (2018) The gastrointestinal microbiome: a review. J Vet Intern Med 32:9–25. https://doi.org/10.1111/jvim.14875

Barry MK, Aloisi JD, Pickering SP, Yeo CJ (1994) Luminal adrenergic agents modulate ileal transport: discrimination between alpha 1 and alpha 2 receptors. Am J Surg 167:156–162

Barry MK, Maher MM, Gontarek JD, Jimenez RE, Yeo CJ (1995) Luminal dopamine modulates canine ileal water and electrolyte transport. Dig Dis Sci 40:1738–1743

Bartl J, Muller T, Grunblatt E, Gerlach M, Riederer P (2014) Chronic monoamine oxidase-B inhibitor treatment blocks monoamine oxidase-a enzyme activity. J Neural Transm (Vienna) 121:379–383. https://doi.org/10.1007/s00702-013-1120-z

Beck G, Brinkkoetter P, Hanusch C, Schulte J, van Ackern K, van der Woude FJ, Yard BA (2004) Clinical review: immunomodulatory effects of dopamine in general inflammation. Crit Care 8:485–491. https://doi.org/10.1186/cc2879

Bedarf JR, Hildebrand F, Coelho LP, Sunagawa S, Bahram M, Goeser F, Bork P, Wullner U (2017) Functional implications of microbial and viral gut metagenome changes in early stage L-DOPA-naive Parkinson's disease patients. Genome Med 9(39). https://doi.org/10.1186/s13073-017-0428-y

Bergquist J, Tarkowski A, Ekman R, Ewing A (1994) Discovery of endogenous catecholamines in lymphocytes and evidence for catecholamine regulation of lymphocyte function via an autocrine loop. Proc Natl Acad Sci U S A 91:12912–12916

Billett EE (2004) Monoamine oxidase (MAO) in human peripheral tissues. Neurotoxicology 25:139–148. https://doi.org/10.1016/S0161-813X(03)00094-9

Borgohain R, Szasz J, Stanzione P, Meshram C, Bhatt MH, Chirilineau D, Stocchi F, Lucini V, Giuliani R, Forrest E, Rice P, Anand R (2014) Two-year, randomized, controlled study of safinamide as add-on to levodopa in mid to late Parkinson's disease. Mov Disord 29:1273–1280. https://doi.org/10.1002/mds.25961

Bortolato M, Chen K, Shih JC (2008) Monoamine oxidase inactivation: from pathophysiology to therapeutics. Adv Drug Deliv Rev 60:1527–1533. https://doi.org/10.1016/j.addr.2008.06.002

Cattaneo C, Jost WH, Bonizzoni E (2019) Long-term efficacy of safinamide on symptoms severity and quality of life in fluctuating Parkinson's disease patients. J Parkinsons Dis. https://doi.org/10.3233/JPD-191765

Chong PP, Chin VK, Looi CY, Wong WF, Madhavan P, Yong VC (2019) The microbiome and irritable bowel syndrome - a review on the pathophysiology, current research and future therapy. Front Microbiol 10:1136. https://doi.org/10.3389/fmicb.2019.01136

Christensen NJ, Brandsborg O (1974) Dopamine in human gastric juice determined by a sensitive double-isotope-derivative technique Scand. J Clin Lab Invest 34:315–320

Cosentino M, Bombelli R, Ferrari M, Marino F, Rasini E, Maestroni GJ, Conti A, Boveri M, Lecchini S, Frigo G (2000) HPLC-ED measurement of endogenous catecholamines in human immune cells and hematopoietic cell lines. Life Sci 68:283–295. https://doi.org/10.1016/s0024-3205(00)00937-1

Cosentino M, Fietta AM, Ferrari M, Rasini E, Bombelli R, Carcano E, Saporiti F, Meloni F, Marino F, Lecchini S (2007) Human CD4+CD25+ regulatory T cells selectively express tyrosine hydroxylase and contain endogenous catecholamines subserving an autocrine/paracrine inhibitory functional loop. Blood 109:632–642. https://doi.org/10.1182/blood-2006-01-028423

Cote M, Drouin-Ouellet J, Cicchetti F, Soulet D (2011) The critical role of the MyD88-dependent pathway in non-CNS MPTP-mediated toxicity. Brain Behav Immun 25:1143–1152. https://doi.org/10.1016/j.bbi.2011.02.017

Da PM, Zurcher G, Wuthrich I, Haefely WE (1988) On tyramine, food, beverages and the reversible MAO inhibitor moclobemide. J Neural Transm Suppl 26:31–56

Dezsi L, Vecsei L (2017) Monoamine oxidase B inhibitors in Parkinson's disease. CNS Neurol Disord Drug Targets 16:425–439. https://doi.org/10.2174/1871527316666170124165222

Eisenhofer G, Aneman A, Friberg P, Hooper D, Fandriks L, Lonroth H, Hunyady B, Mezey E (1997) Substantial production of dopamine in the human gastrointestinal tract. J Clin Endocrinol Metab 82:3864–3871. https://doi.org/10.1210/jcem.82.11.4339

Eisenhofer G, Kopin IJ, Goldstein DS (2004) Catecholamine metabolism: a contemporary view with implications for physiology and medicine. Pharmacol Rev 56:331–349. https://doi.org/10.1124/pr.56.3.1

Feng XY, Zhang DN, Wang YA, Fan RF, Hong F, Zhang Y, Li Y, Zhu JX (2017) Dopamine enhances duodenal epithelial permeability via the dopamine D5 receptor in rodent. Acta Physiol (Oxf) 220:113–123. https://doi.org/10.1111/apha.12806

Feng XY, Yan JT, Li GW, Liu JH, Fan RF, Li SC, Zheng LF, Zhang Y, Zhu JX (2020) Source of dopamine in gastric juice and luminal dopamine-induced duodenal bicarbonate secretion via apical dopamine D2 receptors. Br J Pharmacol 177:3258–3272. https://doi.org/10.1111/bph.15047

Ferrari M, Cosentino M, Marino F, Bombelli R, Rasini E, Lecchini S, Frigo G (2004) Dopaminergic D1-like receptor-dependent inhibition of tyrosine hydroxylase mRNA expression and catecholamine production in human lymphocytes. Biochem Pharmacol 67:865–873

Finberg JP (2014) Update on the pharmacology of selective inhibitors of MAO-A and MAO-B: focus on modulation of CNS monoamine neurotransmitter release. Pharmacol Ther 143:133–152. https://doi.org/10.1016/j.pharmthera.2014.02.010

Forsythe P, Kunze WA (2013) Voices from within: gut microbes and the CNS. Cell Mol Life Sci 70:55–69. https://doi.org/10.1007/s00018-012-1028-z

Fox SH, Katzenschlager R, Lim SY, Barton B, de Bie R, Seppi K, Coelho M, Sampaio C (2018) International Parkinson and movement disorder society evidence-based medicine review: update on treatments for the motor symptoms of Parkinson's disease. Mov Disord. https://doi.org/10.1002/mds.27372

Gray R, Ives N, Rick C, Patel S, Gray A, Jenkinson C, McIntosh E, Wheatley K, Williams A, Clarke CE (2014) Long-term effectiveness of dopamine agonists and monoamine oxidase B inhibitors compared with levodopa as initial treatment for Parkinson's disease (PD MED): a large, open-label, pragmatic randomised trial. Lancet 384:1196–1205. https://doi.org/10.1016/S0140-6736(14)60683-8

Guay DR (2006) Rasagiline (TVP-1012): a new selective monoamine oxidase inhibitor for Parkinson's disease. Am J Geriatr Pharmacother 4:330–346. https://doi.org/10.1016/j.amjopharm.2006.12.001

Haggendal J (1967) The presence of dopamine in human gastric juice. Acta Physiol Scand 71:127–128. https://doi.org/10.1111/j.1748-1716.1967.tb03718.x

Hristova AH, Koller WC (2000) Early Parkinson's disease: what is the best approach to treatment. Drugs Aging 17:165–181. https://doi.org/10.2165/00002512-200017030-00002
Jatana N, Apoorva N, Malik S, Sharma A, Latha N (2013) Inhibitors of catechol-O-methyltransferase in the treatment of neurological disorders. Cent Nerv Syst Agents Med Chem 13:166-194
Josefsson E, Bergquist J, Ekman R, Tarkowski A (1996) Catecholamines are synthesized by mouse lymphocytes and regulate function of these cells by induction of apoptosis. Immunology 88:140–146
Kaenmaki M, Tammimaki A, Garcia-Horsman JA, Myohanen T, Schendzielorz N, Karayiorgou M, Gogos JA, Mannisto PT (2009) Importance of membrane-bound catechol-O-methyltransferase in L-DOPA metabolism: a pharmacokinetic study in two types of Comt gene modified mice. Br J Pharmacol 158:1884–1894. https://doi.org/10.1111/j.1476-5381.2009.00494.x
Karhunen T, Tilgmann C, Ulmanen I, Julkunen I, Panula P (1994) Distribution of catechol-O-methyltransferase enzyme in rat tissues. J Histochem Cytochem 42:1079–1090
Knox NC, Forbes JD, Peterson CL, Van Domselaar G, Bernstein CN (2019) The gut microbiome in inflammatory bowel disease: lessons learned from other immune-mediated inflammatory diseases. Am J Gastroenterol 114:1051–1070. https://doi.org/10.14309/ajg.0000000000000305
Li ZS, Pham TD, Tamir H, Chen JJ, Gershon MD (2004) Enteric dopaminergic neurons: definition, developmental lineage and effects of extrinsic denervation. J Neurosci 24:1330–1339. https://doi.org/10.1523/JNEUROSCI.3982-03.2004
Li LS, Liu CZ, Xu JD, Zheng LF, Feng XY, Zhang Y, Zhu JX (2015) Effect of entacapone on colon motility and ion transport in a rat model of Parkinson's disease. World J Gastroenterol 21:3509–3518. https://doi.org/10.3748/wjg.v21.i12.3509
Li Y, Zhang Y, Zhang XL, Feng XY, Liu CZ, Zhang XN, Quan ZS, Yan JT, Zhu JX (2019) Dopamine promotes colonic mucus secretion through dopamine D5 receptor in rats. Am J Physiol Cell Physiol 316:C393–C403. https://doi.org/10.1152/ajpcell.00261.2017
Liu CZ, Zhang XL, Zhou L, Wang T, Quan ZS, Zhang Y, Li J, Li GW, Zheng LF, Li LS, Zhu JX (2018) Rasagiline, an inhibitor of MAO-B, decreases colonic motility through elevating colonic dopamine content. Neurogastroenterol Motil 30:e13390. https://doi.org/10.1111/nmo.13390
Lubomski M, Davis RL, Sue CM (2019) The gut microbiota: a novel therapeutic target in Parkinson's disease? Parkinsonism Relat Disord 66:265–266. https://doi.org/10.1016/j.parkreldis.2019.08.010
Lyte M (1992) The role of catecholamines in gram-negative sepsis. Med Hypotheses 37:255–258
Lyte M (2011) Probiotics function mechanistically as delivery vehicles for neuroactive compounds: microbial endocrinology in the design and use of probiotics. BioEssays 33:574–581. https://doi.org/10.1002/bies.201100024
Lyte M, Bailey MT (1997) Neuroendocrine-bacterial interactions in a neurotoxin-induced model of trauma. J Surg Res 70:195–201. https://doi.org/10.1006/jsre.1997.5130
Lyte M, Ernst S (1992) Catecholamine induced growth of gram negative bacteria. Life Sci 50:203–212. https://doi.org/10.1016/0024-3205(92)90273-r
Magro F, Vieira-Coelho MA, Fraga S, Serrao MP, Veloso FT, Ribeiro T, Soares-da-Silva P (2002) Impaired synthesis or cellular storage of norepinephrine, dopamine, and 5-hydroxytryptamine in human inflammatory bowel disease. Dig Dis Sci 47:216–224
Magro F, Fraga S, Ribeiro T, Soares-da-Silva P (2004) Decreased availability of intestinal dopamine in transmural colitis may relate to inhibitory effects of interferon-gamma upon L-DOPA uptake. Acta Physiol Scand 180:379–386. https://doi.org/10.1111/j.1365-201X.2004.01260.x
Maini RV, Bess EN, Bisanz JE, Turnbaugh PJ, Balskus EP (2019) Discovery and inhibition of an interspecies gut bacterial pathway for levodopa metabolism. Science 364:eaau6323. https://doi.org/10.1126/science.aau6323
Matsumoto M, Weickert CS, Akil M, Lipska BK, Hyde TM, Herman MM, Kleinman JE, Weinberger DR (2003) Catechol O-methyltransferase mRNA expression in human and rat brain: evidence for a role in cortical neuronal function. Neuroscience 116:127–137. https://doi.org/10.1016/s0306-4522(02)00556-0

Matsumoto M, Kibe R, Ooga T, Aiba Y, Sawaki E, Koga Y, Benno Y (2013) Cerebral low-molecular metabolites influenced by intestinal microbiota: a pilot study. Front Syst Neurosci 7(9). https://doi.org/10.3389/fnsys.2013.00009

Matt SM, Gaskill PJ (2020) Where is dopamine and how do immune cells see it?: dopamine-mediated immune cell function in health and disease. J Neuroimmune Pharmacol 15:114–164. https://doi.org/10.1007/s11481-019-09851-4

Mezey E, Eisenhofer G, Harta G, Hansson S, Gould L, Hunyady B, Hoffman BJ (1996) A novel nonneuronal catecholaminergic system: exocrine pancreas synthesizes and releases dopamine. Proc Natl Acad Sci U S A 93:10377–10382. https://doi.org/10.1073/pnas.93.19.10377

Mezey E, Eisenhofer G, Hansson S, Hunyady B, Hoffman BJ (1998) Dopamine produced by the stomach may act as a paracrine/autocrine hormone in the rat. Neuroendocrinology 67:336–348. https://doi.org/10.1159/000054332

Moffa S, Mezza T, Cefalo C, Cinti F, Impronta F, Sorice GP, Santoro A, Di Giuseppe G, Pontecorvi A, Giaccari A (2019) The interplay between immune system and microbiota in diabetes. Mediat Inflamm 2019:9367404. https://doi.org/10.1155/2019/9367404

Musso NR, Brenci S, Setti M, Indiveri F, Lotti G (1996) Catecholamine content and in vitro catecholamine synthesis in peripheral human lymphocytes. J Clin Endocrinol Metab 81:3553–3557. https://doi.org/10.1210/jcem.81.10.8855800

Myohanen TT, Schendzielorz N, Mannisto PT (2010) Distribution of catechol-O-methyltransferase (COMT) proteins and enzymatic activities in wild-type and soluble COMT deficient mice. J Neurochem 113:1632–1643. https://doi.org/10.1111/j.1471-4159.2010.06723.x

Natale G, Kastsiushenka O, Fulceri F, Ruggieri S, Paparelli A, Fornai F (2010) MPTP-induced parkinsonism extends to a subclass of TH-positive neurons in the gut. Brain Res 1355:195–206. https://doi.org/10.1016/j.brainres.2010.07.076

Nicholson JK, Holmes E, Kinross J, Burcelin R, Gibson G, Jia W, Pettersson S (2012) Host-gut microbiota metabolic interactions. Science 336:1262–1267. https://doi.org/10.1126/science.1223813

Nissinen E, Linden IB, Schultz E, Kaakkola S, Mannisto PT, Pohto P (1988a) Inhibition of catechol-O-methyltransferase activity by two novel disubstituted catechols in the rat. Eur J Pharmacol 153:263–269

Nissinen E, Tuominen R, Perhoniemi V, Kaakkola S (1988b) Catechol-O-methyltransferase activity in human and rat small intestine. Life Sci 42:2609–2614

Okauchi H, Nakajima S, Tani T, Ito A, Arai R (2004) Immunocytochemical localization of monoamine oxidase type B in enterochromaffin-like cells of rat oxyntic mucosa. Histochem Cell Biol 121:181–188. https://doi.org/10.1007/s00418-004-0622-z

Özoğul F (2004) Production of biogenic amines by Morganella morganii, Klebsiella pneumoniae and hafnia alvei using a rapid HPLC method. Eur Food Res Technol 219:465–469

Papa I, Saliba D, Ponzoni M, Bustamante S, Canete PF, Gonzalez-Figueroa P, McNamara HA, Valvo S, Grimbaldeston M, Sweet RA, Vohra H, Cockburn IA, Meyer-Hermann M, Dustin ML, Doglioni C, Vinuesa CG (2017) TFH-derived dopamine accelerates productive synapses in germinal centres. Nature 547:318–323. https://doi.org/10.1038/nature23013

Pietrucci D, Cerroni R, Unida V, Farcomeni A, Pierantozzi M, Mercuri NB, Biocca S, Stefani A, Desideri A (2019) Dysbiosis of gut microbiota in a selected population of Parkinson's patients. Parkinsonism Relat Disord. https://doi.org/10.1016/j.parkreldis.2019.06.003

Prado C, Contreras F, Gonzalez H, Diaz P, Elgueta D, Barrientos M, Herrada AA, Lladser A, Bernales S, Pacheco R (2012) Stimulation of dopamine receptor D5 expressed on dendritic cells potentiates Th17-mediated immunity. J Immunol 188:3062–3070. https://doi.org/10.4049/jimmunol.1103096

Riederer P, Muller T (2018) Monoamine oxidase-B inhibitors in the treatment of Parkinson's disease: clinical-pharmacological aspects. J Neural Transm (Vienna). https://doi.org/10.1007/s00702-018-1876-2

Rodriguez MJ, Saura J, Billett EE, Finch CC, Mahy N (2001) Cellular localization of monoamine oxidase a and B in human tissues outside of the central nervous system. Cell Tissue Res 304:215–220

Roshchina VV (2016) New trends and perspectives in the evolution of neurotransmitters in microbial, plant and animal cells. Adv Exp Med Biol 874:25–77. https://doi.org/10.1007/978-3-319-20215-0_2

Sarkar C, Basu B, Chakroborty D, Dasgupta PS, Basu S (2010) The immunoregulatory role of dopamine: an update. Brain Behav Immun 24:525–528. https://doi.org/10.1016/j.bbi.2009.10.015

Scheperjans F, Aho V, Pereira PA, Koskinen K, Paulin L, Pekkonen E, Haapaniemi E, Kaakkola S, Eerola-Rautio J, Pohja M, Kinnunen E, Murros K, Auvinen P (2015) Gut microbiota are related to Parkinson's disease and clinical phenotype. Mov Disord 30:350–358. https://doi.org/10.1002/mds.26069

Shah P, Nankova BB, Parab S, La Gamma EF (2006) Short chain fatty acids induce TH gene expression via ERK-dependent phosphorylation of CREB protein. Brain Res 1107:13–23. https://doi.org/10.1016/j.brainres.2006.05.097

Sherwin E, Dinan TG, Cryan JF (2017) Recent developments in understanding the role of the gut microbiota in brain health and disease. Ann N Y Acad Sci. https://doi.org/10.1111/nyas.13416

Shin C, Lim Y, Lim H, Ahn TB (2020) Plasma short-chain fatty acids in patients with Parkinson's disease. Mov Disord 35:1021–1027. https://doi.org/10.1002/mds.28016

Shishov VA, Kirovskaia TA, Kudrin VS, Oleskin AV (2009) Amine neuromediators, their precursors, and oxidation products in the culture of Escherichia coli K-12. Prikl Biokhim Mikrobiol 45:550–554

Singaram C, Ashraf W, Gaumnitz EA, Torbey C, Sengupta A, Pfeiffer R, Quigley EM (1995) Dopaminergic defect of enteric nervous system in Parkinson's disease patients with chronic constipation. Lancet 346:861–864. https://doi.org/10.1016/s0140-6736(95)92707-7

Sivasubramaniam SD, Finch CC, Rodriguez MJ, Mahy N, Billett EE (2003) A comparative study of the expression of monoamine oxidase-a and -B mRNA and protein in non-CNS human tissues. Cell Tissue Res 313:291–300. https://doi.org/10.1007/s00441-003-0765-6

Smeyne RJ, Jackson-Lewis V (2005) The MPTP model of Parkinson's disease. Brain Res Mol Brain Res 134:57–66. https://doi.org/10.1016/j.molbrainres.2004.09.017

Spanogiannopoulos P, Bess EN, Carmody RN, Turnbaugh PJ (2016) The microbial pharmacists within us: a metagenomic view of xenobiotic metabolism. Nat Rev Microbiol 14:273–287. https://doi.org/10.1038/nrmicro.2016.17

Srivastav S, Neupane S, Bhurtel S, Katila N, Maharjan S, Choi H, Hong JT, Choi DY (2019) Probiotics mixture increases butyrate, and subsequently rescues the nigral dopaminergic neurons from MPTP and rotenone-induced neurotoxicity. J Nutr Biochem 69:73–86. https://doi.org/10.1016/j.jnutbio.2019.03.021

Stange EF, Schroeder BO (2019) Microbiota and mucosal defense in IBD: an update. Expert Rev Gastroenterol Hepatol 13:963–976. https://doi.org/10.1080/17474124.2019.1671822

Strandwitz P (2018) Neurotransmitter modulation by the gut microbiota. Brain Res 1693:128–133. https://doi.org/10.1016/j.brainres.2018.03.015

Taraskina AE, Nasyrova RF, Grunina MN, Zabotina AM, Ivashchenko DV, Ershov EE, Sosin DN, Kirnichnaya KA, Ivanov MV, Krupitsky EM (2015) Dopamine neurotransmission of peripheral blood lymphocytes is a potential biomarker of psychiatric and neurological disorders. Zh Nevrol Psikhiatr Im S S Korsakova 115:65–69

Teixeira FG, Gago MF, Marques P, Moreira PS, Magalhaes R, Sousa N, Salgado AJ (2018) Safinamide: a new hope for Parkinson's disease? Drug Discov Today 23:736–744. https://doi.org/10.1016/j.drudis.2018.01.033

Tian YM, Chen X, Luo DZ, Zhang XH, Xue H, Zheng LF, Yang N, Wang XM, Zhu JX (2008) Alteration of dopaminergic markers in gastrointestinal tract of different rodent models of Parkinson's disease. Neuroscience 153:634–644. https://doi.org/10.1016/j.neuroscience.2008.02.033

Tolosa E, Stern MB (2012) Efficacy, safety and tolerability of rasagiline as adjunctive therapy in elderly patients with Parkinson's disease. Eur J Neurol 19:258–264. https://doi.org/10.1111/j.1468-1331.2011.03484.x

Tsai LH, Cheng JT (1995) Stimulatory effect of dopamine on acid secretion from the isolated rat stomach. Neurosci Res 21:235–240. https://doi.org/10.1016/0168-0102(94)00854-9

Tsavkelova EA, Botvinko IV, Kudrin VS, Oleskin AV (2000) Detection of neurotransmitter amines in microorganisms with the use of high-performance liquid chromatography. Dokl Biochem 372:115–117

Unger MM, Spiegel J, Dillmann KU, Grundmann D, Philippeit H, Burmann J, Fassbender K, Schwiertz A, Schafer KH (2016) Short chain fatty acids and gut microbiota differ between patients with Parkinson's disease and age-matched controls. Parkinsonism Relat Disord 32:66–72. https://doi.org/10.1016/j.parkreldis.2016.08.019

Ustione A, Piston DW (2012) Dopamine synthesis and D3 receptor activation in pancreatic beta-cells regulates insulin secretion and intracellular [ca(2+)] oscillations. Mol Endocrinol 26:1928–1940. https://doi.org/10.1210/me.2012-1226

Vieira-Coelho MA, Teixeira VL, Guimaraes JT, Serrao MP, Soares-da-Silva P (1999) Caco-2 cells in culture synthesize and degrade dopamine and 5-hydroxytryptamine: a comparison with rat jejunal epithelial cells. Life Sci 64:69–81

Villageliu D, Lyte M (2018) Dopamine production in Enterococcus faecium: a microbial endocrinology-based mechanism for the selection of probiotics based on neurochemical-producing potential. PLoS One 13:e207038. https://doi.org/10.1371/journal.pone.0207038

Wall R, Cryan JF, Ross RP, Fitzgerald GF, Dinan TG, Stanton C (2014) Bacterial neuroactive compounds produced by psychobiotics. Adv Exp Med Biol 817:221–239. https://doi.org/10.1007/978-1-4939-0897-4_10

Wang CC, Billett E, Borchert A, Kuhn H, Ufer C (2013) Monoamine oxidases in development. Cell Mol Life Sci 70:599–630. https://doi.org/10.1007/s00018-012-1065-7

Weinreb O, Amit T, Bar-Am O, Youdim MB (2010) Rasagiline: a novel anti-parkinsonian monoamine oxidase-B inhibitor with neuroprotective activity. Prog Neurobiol 92:330–344. https://doi.org/10.1016/j.pneurobio.2010.06.008

Weintraub D, Hauser RA, Elm JJ, Pagan F, Davis MD, Choudhry A (2016) Rasagiline for mild cognitive impairment in Parkinson's disease: a placebo-controlled trial. Mov Disord 31:709–714. https://doi.org/10.1002/mds.26617

Westlund KN, Denney RM, Kochersperger LM, Rose RM, Abell CW (1985) Distinct monoamine oxidase a and B populations in primate brain. Science 230:181–183

Yan Y, Jiang W, Liu L, Wang X, Ding C, Tian Z, Zhou R (2015) Dopamine controls systemic inflammation through inhibition of NLRP3 inflammasome. Cell 160:62–73. https://doi.org/10.1016/j.cell.2014.11.047

Youdim MB, Finberg JP (1987) Monoamine oxidase B inhibition and the "cheese effect". J Neural Transm Suppl 25:27–33

Zhang X, Li Y, Liu C, Fan R, Wang P, Zheng L, Hong F, Feng X, Zhang Y, Li L, Zhu J (2015) Alteration of enteric monoamines with monoamine receptors and colonic dysmotility in 6-hydroxydopamine–induced Parkinson's disease rats. Transl Res 166:152–162. https://doi.org/10.1016/j.trsl.2015.02.003

Chapter 3
Dopamine Receptors in the Gastrointestinal Tract

Xiao-Li Zhang, Sumei Liu, Qi Sun, and Jin-Xia Zhu

Abstract Dopamine (DA) regulates physiological functions through dopamine receptors (DARs), including D_1, D_2, D_3, D_4, and D_5. All of the five subtypes mediate neuronal and non-neuronal functions in the central nervous system (CNS) and peripheral organs including gastrointestinal (GI) tract. DA plays a wide range of physiological roles in the GI tract after binding to the DARs and/or the adrenergic receptors, which includes protection of GI mucosa, regulation of GI motility, secretion and absorption, etc. Here we predominately review the distribution of DARs on the GI mucosa, muscularis externa, and enteric nervous system.

Keywords Dopamine receptors (DARs) · Gastrointestinal (GI) tract · Mucosa · Muscularis externa · Enteric nervous system (ENS)

Abbreviations

6-OHDA	6-Hydroxydopamine
cAMP	Cyclic adenosine monophosphate
CD	Crohn's disease
CNS	Central nervous system
D_1R	Dopamine D_1 receptor
D_2R	Dopamine D_2 receptor
D_3R	Dopamine D_3 receptor
D_4R	Dopamine D_4 receptor
D_5R	Dopamine D_5 receptor
DA	Dopamine

X.-L. Zhang · Q. Sun · J.-X. Zhu (✉)
Department of Physiology and Pathophysiology, School of Basic Medical Science, Capital Medical University, Beijing, China
e-mail: zhu_jx@ccmu.edu.cn

S. Liu
Department of Biology, College of Science and Health, University of Wisconsin-La Crosse, La Crosse, WI, USA
e-mail: sliu@uwlax.edu

J.-X. Zhu (ed.), *Dopamine in the Gut*, https://doi.org/10.1007/978-981-33-6586-5_3

DARs	Dopamine receptors
DAT	DA transporter
ENS	Enteric nervous system
GI	Gastrointestinal
IBD	Inflammatory bowel disease
ICC	Interstitial cells of Cajal
IGF-IR	Insulin-like growth factor receptor-I
LMMP	Longitudinal muscle myenteric plexus
PD	Parkinson's disease
RT-PCR	Reverse transcription-PCR
UC	Ulcerative colitis
VIP	Vasoactive intestinal peptide

3.1 Introduction

Dopamine receptors (DARs), a member of the class of seven transmembrane, G-protein coupled receptors, comprise five distinct subtypes. The D1-like subtype comprises dopamine D_1 receptor (D_1R) and D_5R, whereas the D2-like subtype comprises D_2R, D_3R, and D_4R. The D1-like subtype is often coupled to the Gs family of G proteins that stimulates cyclic adenosine monophosphate (cAMP) production, whereas the D2-like subtype classically activates the Gi family of G proteins to inhibit cAMP production. Dopamine (DA) regulates body functions via activating these DAR subtypes. In the gastrointestinal (GI) tract, DA regulates many GI functions including gut motility, absorption and secretion, local blood flow, mucosa barrier, etc. Each subtype of DARs also has a unique distribution throughout the whole gut.

GI tract is a continuous hollow, twisting tube from the mouth to the anus, including the esophagus, stomach, duodenum, jejunum, ileum, and colon. Generally, the wall of the digestive tract includes four layers which are the mucosa, submucosa, muscularis externa, and serosa. The enteric nervous system (ENS) is located within the wall of the gut and consists of two plexuses, submucosal plexus and myenteric plexus. The submucosal plexus is located within the dense connective tissue between the muscularis externa and the mucosa, and the myenteric plexus is situated between the inner circular and outer longitudinal layers of the muscularis externa. In this chapter, we focus on the distribution of DARs in the whole GI tract.

3.2 Dopamine Receptors in the Gastrointestinal Mucosa

DARs are distributed in the mucosa of GI tract in mice, rat, and human (Hernandez et al. 1987; Li et al. 2006a, b; Wang et al. 2012), and play important roles in the transepithelial ion transport, mucosal blood flow, secretion and absorption, mucosal protection, anti-inflammation, etc. (Feng et al. 2020; Xu et al. 2020). Since the report on the distribution of DARs in the esophageal mucosa is rare, this section focuses on the distribution of DARs in the mucosa of the stomach, small intestine, and colon.

3.2.1 Dopamine Receptors in the Gastric Mucosa

In 1987, Hermandez et al. demonstrated the presence of DARs in human gastric mucosa via ^{3}H-dopamine binding assay (Hernandez et al. 1987). The mRNA expressions of all five DAR subtypes have been detected in the gastric mucosa of rat by in situ hybridization, and the order of signal intensity for DARs is $D_5R > D_4R > D_3R \approx D_1R > D_2R$ (Deng et al. 1997). Mezey et al. have also investigated the expression of D_1R–D_5R mRNA in the gastric mucosa of rat using the in situ hybridization histochemistry technique and found that the D_5R mRNA is present in the entire gastric epithelium at very high level; D_3R and D_4R mRNAs are present in some cells of the lamina propria; whereas D_1R and D_2R mRNAs are not expressed in the gastric mucosa (Mezey et al. 1998, 1999). However, Dr. Zhu JX's lab has reported that the relative mRNA expression of the all five DARs in the rat gastric mucosa is $D_2R > D_3R > D_1R > D_4R > D_5R$ by real-time reverse transcription (RT)-PCR, and the mRNA level of the D_2R is the highest among all five subtypes of DARs (Wang et al. 2012) (Fig. 3.1A). Moreover, the immunoreactivities of the all DARs, except D_4R, are observed in the gastric mucosa, especially in the basal gastric glands by immunofluorescence staining (Wang et al. 2012) (Fig. 3.1B). The differential results could be due to different detective methods, tissue preparation conditions, or tissue preparation processes. In Mezey's study, the results were based on in situ hybridization histochemistry technique, and they did not distinguish the different regions of the rat gastric mucosa, while Wang's study employed tissue of gastric corpus mucosa only with real-time PCR and immunofluorescence staining. Moreover, Basu and Dasgupta have demonstrated the presence of D_2R in the mucosa of human normal and malignant stomach tissues, and the expression of D_2R is decreased in the malignant stomach tissue (Basu and Dasgupta 1997). Besides, all five DARs in the whole stomach tissue of mice have been reported by using RT-PCR technique (Li et al. 2006b). Distribution of D_1R (also called as D1A receptor) throughout the GI tract of rat has been investigated via RT-PCR, Western blot, and in situ hybridization histochemistry. An identical band for D_1R has been observed in the stomach by Western blot and RT-PCR. Immunohistochemical staining for D_1R is present in the gastric glands throughout the cardiac, fundal, and pyloric regions, and blood vessels within the gastric mucosa and muscularis

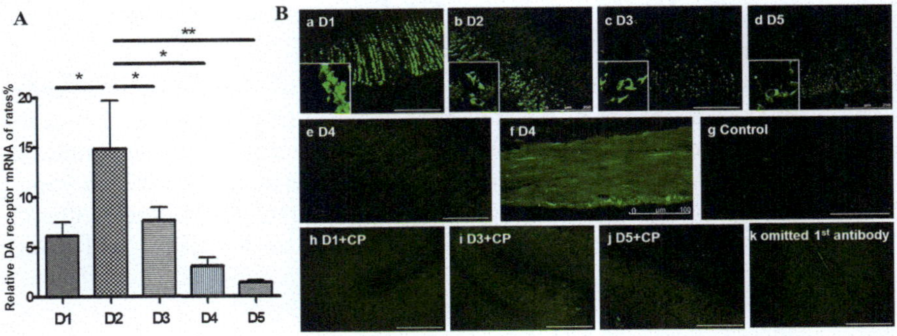

Fig. 3.1 Expression of five subtypes of dopamine receptors in the gastric mucosa of rats. (**A**) The mRNA expression of DARs in gastric corpus mucosa was analyzed by real-time PCR. The β-actin was used as an internal control. Data were presented as means (±SD). $P < 0.05$; $P < 0.01$. (**B**) Rat gastric mucosa stained with antibody against D_1R–D_5R (a–e), and the D_4R immunoreactivity was observed in gastric muscular layer (f). The primary antibodies were omitted in g, k, and pre-absorbed with corresponding control peptides (CP) in h–j. Bars 250 μm (a–e, h–k); Bars 100 μm (f and g). (Wang Q et al., Biochem and Biophys Res Commun, 2012, 417, 197–203)

mucosa (Vaughan et al. 2000). The distribution of DARs in the gastric mucosa is summarized in Table 3.1.

Some pharmacological studies have also suggested that DARs are located in the gastric mucosa and play important role in gastric mucosal secretion in vivo and in vitro. Under acute stress condition, D_1R agonist A68930 is reported to inhibit gastric H^+-K^+-ATPase activity (Rasheed et al. 2010). Administration of D_1R agonist fenoldopam, D_2R antagonist sulpiride (Desai et al. 1999; Glavin and Hall 1995), or D_3R agonist 7-OHDPAT (Glavin 1995) elicits a decrease in gastric acid secretion, pepsin secretion, and histamine content in the pylorus-ligated rats. Besides, D_3R agonist 7-OHDPAT (intraperitoneal injection, i.p.) also significantly reduces basal gastric acid secretion in conscious rats (Glavin 1995). Moreover, D_2R agonist quinpirole (0.1 μg/kg to 0.5 mg/kg, i.p.) can dose-dependently suppress gastric acid secretion stimulated by histamine, pentagastrin, or carbachol, and the effect of quinpirole is completely blocked by D_2R antagonist domperidone. These results suggest that peripheral D_2R has an inhibitory effect on histaminergic-, pentagastrin-, or cholinergic-stimulated gastric acid secretion (Eliassi et al. 2008). Interestingly, using an isolated and everted whole rat stomach model, which has ruled out the systematic role of DA, DA has been found to induce acid secretion in a dose-dependent manner from 10 nM to 10 μM, and this effect is antagonized by D_1R antagonist Scheme 23390 (Tsai and Cheng 1995). This suggests that DA induces gastric acid secretion via activation of the D_1R located on the rat stomach. Recently, D1-like family receptor (DAR-1) antagonist LE300 is reported to prevent the ability of peroxisome proliferator-activated receptor-α agonist fenofibrate-increased gastrin secretion from G cells isolated from human stomach. These results suggest that dopamine, via the D_1R and PPAR-alpha, is involved in gastrin secretion from G cells (Xu et al. 2020).

Table 3.1 Distribution of dopamine receptors in the gastric mucosa

Distribution of DARs	Species	Methods and reference
DARs in gastric mucosa	Human	^{3}H-dopamine binding assay (Hernandez et al. 1987)
mRNA expression in gastric mucosa: $D_5R > D_4R > D_3R \approx D_1R > D_2R$	Rat	In situ hybridization (Deng et al. 1997)
D_2R in normal human gastric mucosa, which decreases in malignant stomach tissues	Human	Immunohistochemistry (Basu and Dasgupta 1997)
High amounts of D_5R mRNA in epithelium, D_3R and D_4R mRNA in lamina propria, no D_1R and D_2R mRNA in the gastric mucosa	Rat	In situ hybridization histochemistry (Mezey et al. 1998, 1999)
D_1R in epithelium, muscularis mucosa, vessels of the submucosa	Rat	RT-PCR, Western blot, in situ hybridization histochemistry (Vaughan et al. 2000)
All five DARs in whole stomach tissue	Mouse	RT-PCR, Western blot (Li et al. 2006b)
D_2R in gastric adenocarcinoma	Human gastric cancer AGS cell line	Flow cytometric analysis and Western blot (Ganguly et al. 2010)
mRNA expression in gastric mucosa: $D_2R > D_3R > D_1R > D_4R > D_5R$, D_1R, D_2R, and D_5R in chief cell	Rat	RT-PCR and immunofluorescence (Wang et al. 2012)
Decreased D_2R expression	Human gastric cancer cell line: MKN28, SGC-7901, BGC-823, and MGC-803	Western blot (Huang et al. 2016)
Higher D_2R expression in the gastric cancerous tissues and respective adjacent non-cancerous tissues	Human	Immunohistochemical (Mu et al. 2017)
D_5R protein in gastric cancer cell line	Human gastric cancer cell line SCG7901	Immunofluorescence, Western blot (Leng et al. 2017)

Importantly, some studies performed in animal models have highlighted the importance of DARs in gastric ulcer. ^{3}H-dopamine binding is increased in the gastric mucosa induced by cold-restraint stress, indicating the upregulation of mucosal DARs in gastric ulcers (Hernandez et al. 1987). It appears that both D_1R and D_2R are involved in the gastric ulcerations, but have opposite effects. In this regard, it has been shown that D_1R agonists, such as SKF38393 (Desai et al. 1995), fenoldopam (Desai et al. 1999), and A 68930 (Rasheed et al. 2010), are effective in relieving experimental gastric ulcers, but D_1R antagonist SCH 39166 can aggravate the gastric ulcer in rats (Desai et al. 1999), which suggests that D_1R might be able to inhibit gastric ulcer. Whereas D_2R agonist quinpirole can increase the gastric ulcer-index in rats (Desai et al. 1999). Notably, sulpiride, the D_2R antagonist, has gastroprotective effects at higher doses, but has ulcerogenic effects at lower doses (Desai et al. 1999; Puri et al. 1994). Activating D_3R with 7-OHDPAT can also reduce restraint stress-induced gastric mucosal injury (Glavin 1995), indicating the involvement of D_3R in protecting the mucosa from injury.

The more accurate cellular localization of DARs in the epithelial cells of gastric mucosa has been reported (Wang et al. 2012). The double-labeling immunofluorescence of each DAR subtype, except D_4R, combined with the following three cell lineage markers, pepsin C (chief cells), H^+/K^+-ATPase (parietal cells), and mucin 6 (mucous neck cells), are performed to identify the cellular location of the DARs in the gastric mucosa. A majority of the D_1R-immunoreactive (IR) cells (84%) express pepsin C-IR. And, about 77% of D_2R-IR cells are also pepsin C-IR. Conversely, approximately 66% and 40% of pepsin C-IR cells are D_1R-IR and D_2R-IR, respectively. Approximately 20% of pepsin C-positive cells express D_5R-IR, which is 50% of the total number of D_5R-IR cells. These results suggest that some pepsin C-positive cells express more than one subtype of DAR. No colocalization of D_3R-IR with pepsin C-IR can be observed in the gastric epithelial cells (Fig. 3.2). H^+/K^+-ATPase-IR cells are distributed along most of the length of the gastric gland. Mucin 6, a well-established marker for mucous neck cells, is heavily stained in the neck region of the gastric glands. None of H^+/K^+-ATPase-positive cells or mucin 6-IR cells express D_1R-, D_2R-, D_3R-, or D_5R-IR (Figs. 3.3 and 3.4) (Wang et al. 2012), which provides important morphologic evidence for the regulation of pepsinogen output by DA and DARs. The local DA (the stomach-derived DA, discussed further in Chap. 2) may directly act on the D_1R, D_2R, and/or D_5R to regulate the function of chief cells, such as the pepsinogen secretion in a paracrine manner. However, the differential role of these DAR subtypes on the gastric chief cells is still unclear, and further study is needed to delineate it.

3.2.2 Dopamine Receptors in the Small Intestinal Mucosa

In the small intestine, the mRNA expressions of D_1R, D_3R, D_4R, and D_5R are identified in the mucosa by RT-PCR technique, and the D_4R mRNA is only expressed in the mucosa, while the D_2R mRNA is only in the ENS (Li et al. 2006b). The D_1R-IR in small intestine has been reported in the duodenum, jejunum, and ileum by immunohistochemistry. It is present across the wall of the small intestine, including the epithelium of intestinal villi and crypts, lamina propria, muscularis mucosa, muscularis externa, and myenteric plexus (Vaughan et al. 2000).

3.2.2.1 Dopamine Receptors in the Duodenal Mucosa

^3H-dopamine binding assay demonstrates the presence of DARs in human duodenal mucosa (Hernandez et al. 1987, 1989). The relative mRNA expression levels of all the five subtypes of DARs in the duodenal mucosa of rat are $D_5R > D_4R > D_3R > D_1R > D_2R$, which are detected by in situ hybridization, and the five DAR mRNA subtypes are spread over the entire layer of the duodenal

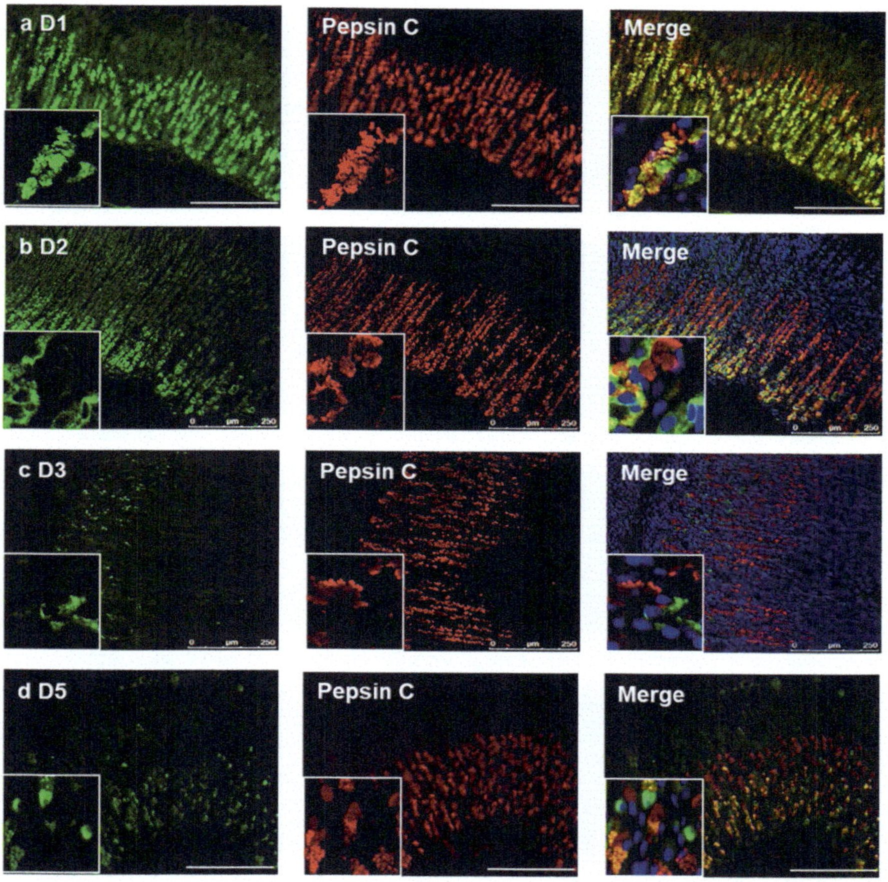

Fig. 3.2 Double-label immunofluorescence of dopamine receptors with pepsin C in the rat gastric mucosa. D_1R-, D_2R-, and D_5R-IR were expressed in some pepsin C-positive cells in the gastric glands (**a**, **b**, and **d**). No colocalization of D_3-IR with pepsin C-IR was observed in the gastric epithelial cells (**c**). Each figure is magnified underneath. Nuclei were stained with 4′6-diamidino-2-phenylindole (DAPI, blue). Bars, 250 μm (**a**–**d**). (Wang Q et al., Biochem and Biophys Res Commun, 2012, 417, 197–203)

mucosa (Deng et al. 1997). Similarly, based on the in situ hybridization histochemistry technique, Mezey et al. have also revealed that D_5R is the most abundant form of DAR, whereas D_3R and D_4R mRNAs are just at the detectable level in the duodenal mucosa and pancreas of rat (Mezey et al. 1999). Recently, Dr. Zhu JX's lab has revealed that the relative mRNA expression levels of five DAR subtypes in the rat duodenum mucosa are $D_2R > D_3R > D_4R > D_5R > D_1R$ by real-time RT-PCR. The mRNA level of the D_2R is the highest among all five subtypes of DARs (Feng et al. 2020). The different results may be due to the varying detective

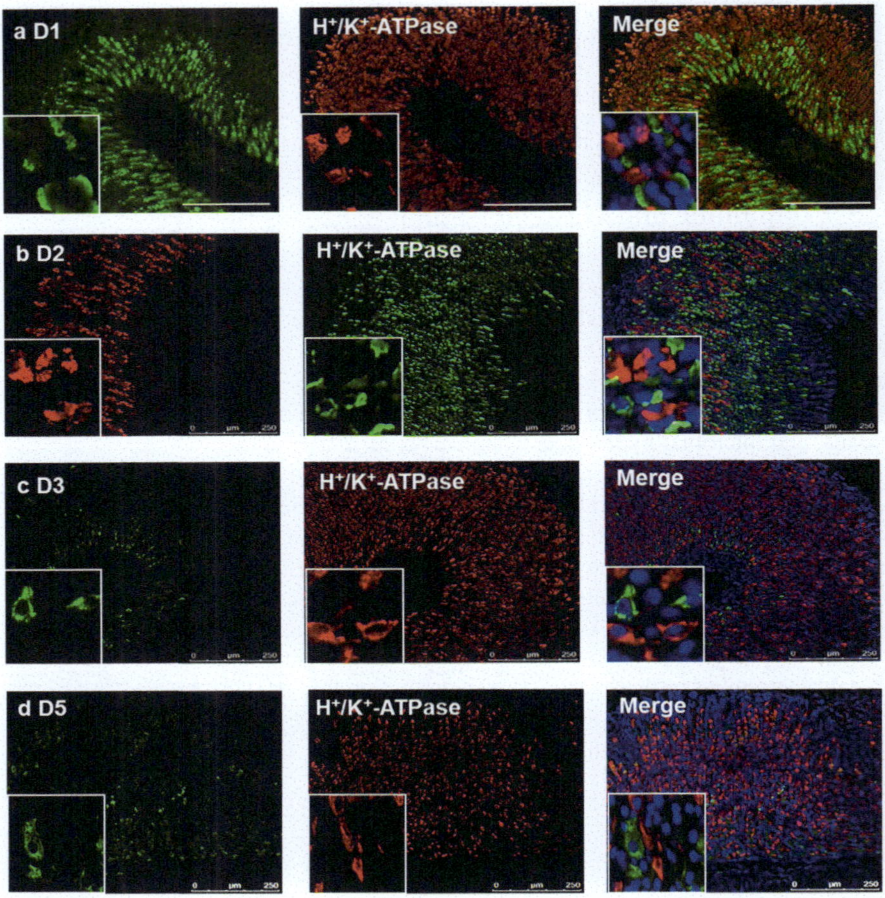

Fig. 3.3 Double-label immunofluorescence of dopamine receptors with H^+/K^+-ATPase in the rat gastric mucosa. None of the DAR subtypes was colocalized with H^+/K^+-ATPase in the gastric mucosa (**a**–**d**). Each figure is magnified underneath. Nuclei were stained with DAPI (blue). Bars 250 μm (**a**–**d**). (Wang Q et al., Biochem and Biophys Res Commun, 2012, 417, 197–203)

methods or tissue preparation. Using immunofluorescence staining, Feng et al. have demonstrated that D_1R-IR is predominant in the apical membrane of duodenal villi but weaker in the Brunner's glands; D_5R-IR is at both the apical and the basolateral sides of cells in the Brunner's glands and intestinal crypts in the rat duodenum, whereas D_2R-IR is predominantly distributed in the apical membrane of cells in both Brunner's glands and intestinal crypts (Fig. 3.5) (Feng et al. 2013). D_1R in the duodenal mucosa mediates the DA-stimulated duodenal cAMP-dependent K^+ secretion (DARs-mediated ion transport will be discussed further in Chap. 4) (Feng et al. 2013). A strong D_5R-IR signal is observed in both Brunner's glands and duodenal

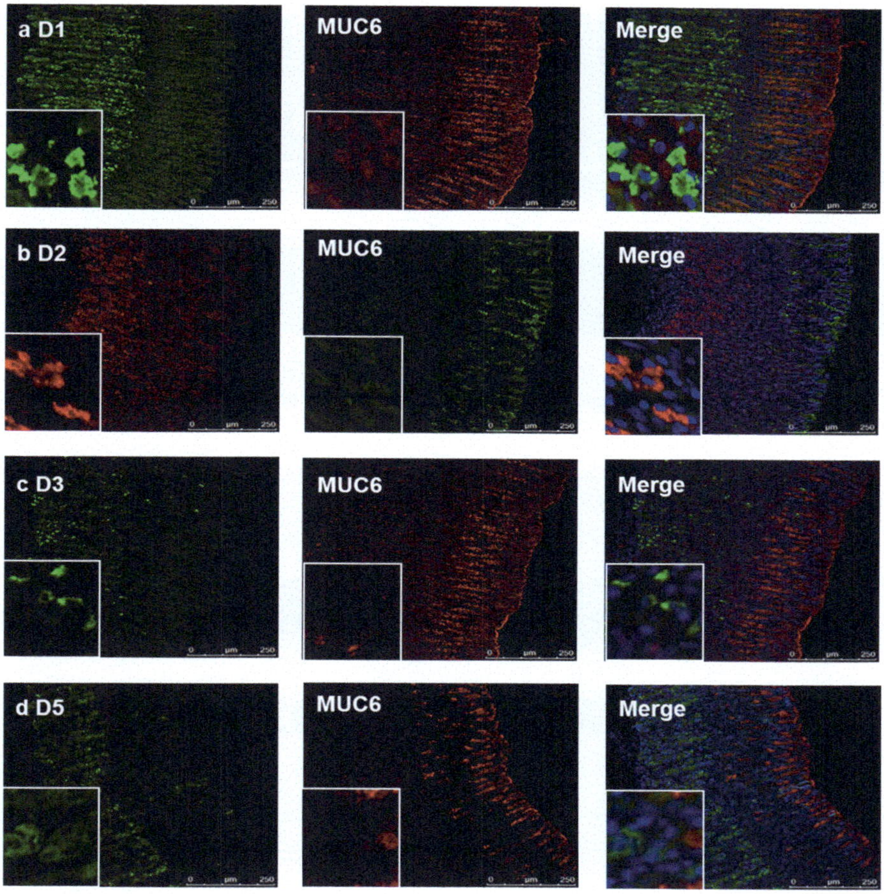

Fig. 3.4 Double-label immunofluorescence of dopamine receptors with mucin 6 in the rat gastric mucosa. None of the DAR subtypes was colocalized with mucin 6 in the gastric mucosa (**a–d**). Each figure is magnified underneath. Nuclei were stained with DAPI (blue). (**a–d**) Bars 250 μm. (Wang Q et al., Biochem and Biophys Res Commun, 2012, 417, 197–203)

villi in a hyperenteric DA rats (HEnD rats), a disease model for Parkinson's disease (PD), which are microinjected with 6-hydroxydopamine (6-OHDA) into the bilateral substantia nigra. D_5R is significantly increased in the duodenal mucosal preparations from this model by Western blot analysis. Moreover, the D_5R is also detected in the human duodenal epithelium from patients who underwent painless gastroendoscopic examination (Fig. 3.6). The D_5R in the duodenal mucosa plays an important role in barrier function (discussed in Chap. 4) (Feng et al. 2017).

^{3}H-dopamine binding is increased in the duodenal mucosa of duodenal ulcer patients, suggesting the role of DARs in the duodenal ulcer (Hernandez et al. 1989). Some animal models have also indicated the involvement of DARs in the duodenal

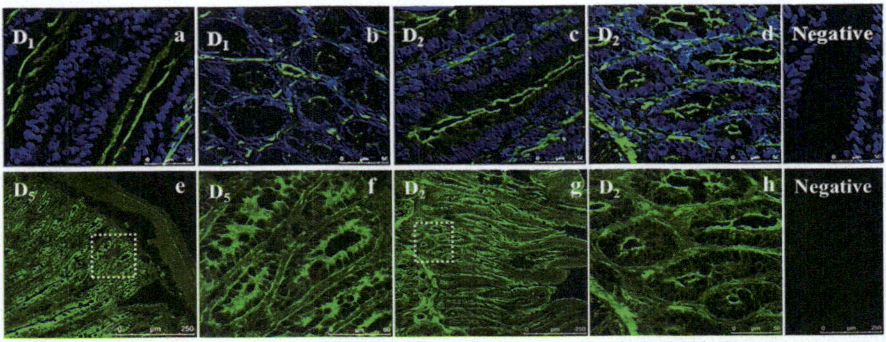

Fig. 3.5 The distribution of D_1R, D_2R, and D_5R in the rat duodenum. D_1 was predominantly distributed in the apical membrane of villi (**a**, **b**). D_2 was predominantly distributed in the apical membrane of cells in both Brunner's glands and intestinal crypts (**c**, **d**, **g**, **h**). D_5 was at the both apical and basolateral sides of Brunner's glands and intestinal crypts (**e**, **f**). The negative control was performed without primary antibody. Nuclei were stained with DAPI (blue). Bars 250 μm (**a–h**). (Feng XY et al., Transl Res, 2013, 161, 486–94)

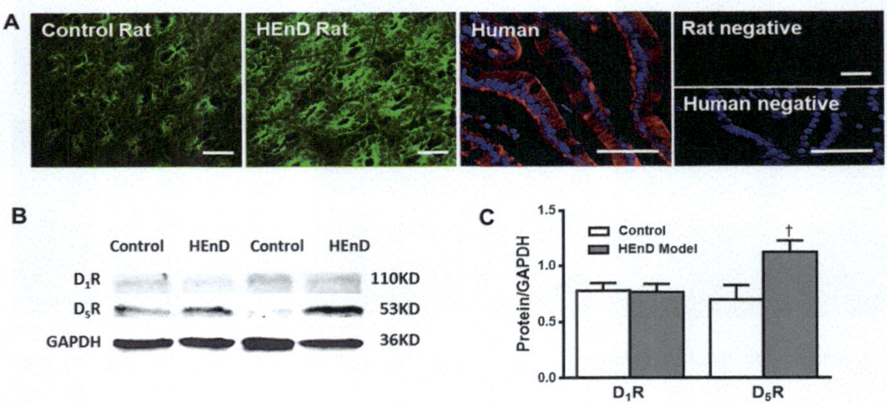

Fig. 3.6 The D_5R in the duodenum of rat and human. (**a**) The distribution of D_5R in the duodenum isolated from rat and human. The negative control was performed without primary antibody. Nuclei were stained with DAPI (blue). Scale bar: 50 μm. (**b**, **c**) Protein expression of D_1R and D_5R in the duodenum of control and HEnD rats (GAPDH, D-glyceraldehyde-3-phosphate dehydrogenase). Values are mean ± SEM. HEnD rats, hyperenteric DA rats. (Feng XY et al., Acta Physiol, 2017, 220, 113–123)

ulcer. Treatment with D_1R agonist SKF 38393 (Desai et al. 1995) or fenoldopam (Desai et al. 1999) can reduce the duodenal lesions induced by cysteamine in rat, while D_2R antagonist such as sulpiride produces an anti-ulcer effect (Desai and Parmar 1994; Desai et al. 1999). Furthermore, D_2R agonist quinpirole can increase the ulcer area and the score of duodenal ulcer induced by cysteamine in rat (Desai et al. 1999). Thus, it appears that D_1R and D_2R have opposite effects on duodenal ulcers. Duodenal mucosal bicarbonate secretion is one main mechanism in the protection of the epithelium against luminal acid. DA and D_1R agonist SKF-38393 increase bicarbonate secretion from duodenal mucosa distal to the Brunner's glands area cannulated in situ in anesthetized rats (Flemstrom et al. 1993) and in isolated duodenal villus and crypt cells of rat (Flemstrom and Safsten 1994). Interestingly, apical application of DA increases bicarbonate secretion in a concentration-dependent manner, which is mimicked by D_2R agonist quinpirole. The effect of apical application of DA is inhibited by D_2R antagonist L741626 and sulpiride in isolated duodenal mucosa from rat. And, the effect of apical application of quinpirole disappeared in isolated duodenal mucosa from mice treated by in vivo D_2R siRNA and in D_2R knockout mice (Feng et al. 2020). DA and quinpirole also enhance the intracellular pH in duodenal enterocyte (Feng et al. 2020). All the above-mentioned morphological and functional studies provide a new perspective of targeting on DARs for potential therapeutic treatment of duodenal mucosal disorders.

3.2.2.2 Dopamine Receptors in the Jejunal and Ileal Mucosa

The mRNA expressions of D_1R, D_3R, D_4R, and D_5R are also present in the ileal mucosa of mice by RT-PCR technique. The D_4R mRNA is only detected in the mucosa, while the D_2R mRNA is present only in the ENS (Fig. 3.7) (Li et al. 2006b). The gene of D_1R is detected in the small intestinal tissue of rats using a ribonuclease protection assay. Furthermore, D_1R is detected at the base cells of the jejunal and ileal crypts of adult rats via emulsion autoradiography performed with 125I-Scheme 23982 on antimesenteric sections (Marmon et al. 1993). Besides, the expression of DARs and the transmembrane dopamine transporter (DAT) are also detected in the human neuroendocrine pancreatic cell line BON and in the neuroendocrine gut cell

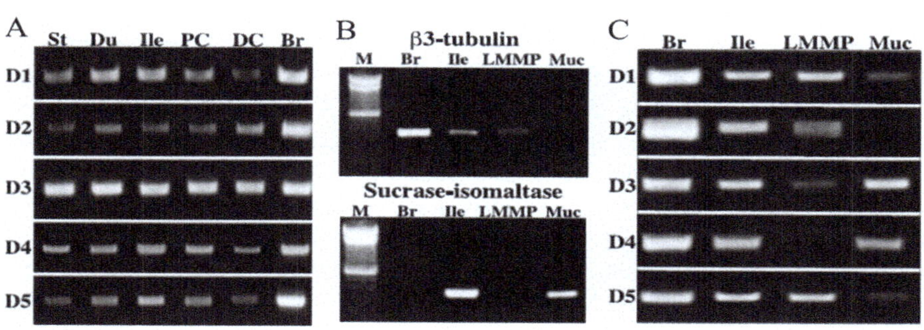

Fig. 3.7 Dopamine receptors are expressed in the gut. Transcripts encoding the D_1R–D_5R were analyzed regionally in the whole gut wall (**a**) and in dissected layers of the wall of the ileum (**b**, **c**). The presence of transcripts encoding the neural marker β3-tubulin and mucosal epithelial marker sucrase-isomaltase was studied to assess the potential contamination of mucosal preparations with RNA from neurons and LMMP preparations with RNA from the mucosal epithelium (**b**). M, Molecular marker; St, stomach; Du, duodenum; Ile, ileum; PC, proximal colon; DC, distal colon; LMMP, longitudinal muscle myenteric plexus. (Li ZS et al., J Neurosci, 2006; 26, 2798–807)

line STC-1. The mRNA and protein expression of D_2R-D_5R and DAT are present in both BON and STC-1 cells by RT-PCR and immunocytochemistry. D_1R mRNA is only detected in BON cells. Using the flow cytometry technique with fluorescent probes, the presence of D_1R and D_2R is assessed in the isolated villus, crypt enterocytes, and intraepithelial lymphocytes of the guinea pig jejunum with the fluorescent D_1R (BODIP FL Scheme 23390) and D_2R (BODIPY FL NAPS) ligands. Approximately 70–80% of enterocytes and approximately 25–50% of intraepithelial lymphocytes express D_1R. And, approximately 70–80% of enterocytes and 50% of intraepithelial lymphocytes express D_2R (Baglole et al. 2005) (Table 3.2). The presence of DARs on the enterocytes or intraepithelial lymphocytes suggests that they may modulate intestinal secretion, absorption, and immune function. The distribution of DARs in the mucosa of small intestine is summarized in Table 3.2.

3.2.3 Dopamine Receptors in the Colonic Mucosa

The mRNA expressions of D_1R–D_5R are also present in the proximal and distal colon of mice by RT-PCR technique (Baglole et al. 2005). Moreover, Dr. Zhu JX's lab has detected the mucosal expressions of D_1R-D_5R in the proximal and distal colon of rats by means of semi-quantitative real-time RT-PCR, Western blot, and immunofluorescence (Li et al. 2019). All the proteins of D_1R–D_5R are expressed in the colonic mucosa including proximal and distal colon, but only D_5R manifests segmental difference with a much higher level expression in the distal colonic mucosa. The mucosal expression of D_4R mRNA is the lowest among all the DARs in the distal colon of rat (Fig. 3.8). D_1R-IR is observed only at the epithelial surface of colonic mucosa, whereas D_2R-IR and D_4R-IR are predominantly located in the colonic lamina propria. D_3R-IR is barely observed in the colonic mucosa, but D_5R-IR is abundantly distributed in the colonic epithelium, including the cells along the crypts. Similarly, the D_5R-IR is abundantly distributed in the colonic epithelium of human and mice. Most of the Mucin 2-marked goblet cells along each crypt are D_5R-IR positive, indicating the colocalization of D_5R with goblet cells (Fig. 3.9) (Li et al. 2019). This finding provides the morphological evidence for the role of DA and DARs in the regulation of colonic mucus secretion. D_1R-IR has a transmural distribution in the rat colon. Strong D_1R immunoreactivity is detected in colonic epithelial crypts and myocytes in the muscularis mucosa. Signals for D_1R are also observed in the colonic epithelium by using in situ hybridization. The muscularis mucosa shows a very intense hybridization signal for D_1R. A positive hybridization signal for D_1R is also detected in the muscularis externa, although the signal is weaker than that in the muscularis mucosa (Vaughan et al. 2000).

Importantly, DA and DARs in the colonic mucosa are involved in the pathogenesis of some GI diseases in human and animal models. Basu and Dasgupta have demonstrated the significant decreases of DA content, DARs, and intracellular cAMP in both mucosal and muscle layers in human malignant colon tissues by ^{3}H-dopamine binding assay (Basu and Dasgupta 1999). Moreover, polymorphisms

Table 3.2 Distribution of dopamine receptors in the small intestinal mucosa

Tissue	Distribution of DAR	Species	Methods and reference
Duodenum	DARs in duodenal mucosa	Human	^{3}H-dopamine binding assay (Hernandez et al. 1987, 1989)
	mRNA expression in duodenal mucosa: $D_5R > D_4R > D_3R \approx D_1R > D_2R$	Rat	In situ hybridization (Deng et al. 1997)
	High levels of D_5R mRNA in duodenal mucosa, followed by D_3R and D_4R mRNA; no D_1R and D_2R mRNA detected	Rat	In situ hybridization histochemistry (Mezey et al. 1999)
	D_1R in duodenal mucosa	Rat	RT-PCR, Western blot, in situ hybridization histochemistry (Vaughan et al. 2000)
	mRNA and protein of D_1R, D_3R-D_5R in duodenal mucosa	Mouse	RT-PCR, Western blot (Li et al. 2006b)
	D_1R in the apical membrane of villi and Brunner's glands; D_5R on both the apical and the basolateral sides of Brunner's glands and intestinal crypts; D_2R in the apical membrane of cells in both Brunner's glands and intestinal crypts	Rat	Immunofluorescence, Western blot (Feng et al. 2013, 2017, 2020)
	D_5R in the epithelium of duodenal mucosa	Human	Immunofluorescence (Feng et al. 2017)
	mRNA expression in duodenal mucosa: $D_2R > D_3R > D_4R > D_5R > D_1R$	Rat	Real-time PCR (Feng et al. 2020)
Jejunum	D_1R on the base of cells in crypts of jejunal mucosa	Rat	Emulsion autoradiography (Marmon et al. 1993)
	D_1R and D_2R in enterocytes and intraepithelial lymphocytes in jejunal mucosa	Guinea pig	Flow cytometry with specific fluorescent probes (Baglole et al. 2005)
	mRNA and protein of D_2R-D_5R	Gut cell line STC-1	Real-time PCR and Western blot (Lemmer et al. 2002)
Ileum	D_1R in ileal mucosa	Rat	RT-PCR, Western blot, in situ hybridization histochemistry (Vaughan et al. 2000)
	D_1R, D_3R–D_5R in ileal mucosa	Mouse	RT-PCR, Western blot (Li et al. 2006b)

in D_2R have been reported to be associated with an increased risk of colorectal cancer (Gemignani et al. 2005). DARs, especially D_2R, have been reported to be involved in the development of inflammatory bowel disease (IBD), which has been discussed in Chap. 2 (Sect. 2.5). Future studies on DAR-mediated signal transduction may provide new insight into the mechanisms and new therapeutic strategies of colonic related diseases, such as colorectal cancer and IBD.

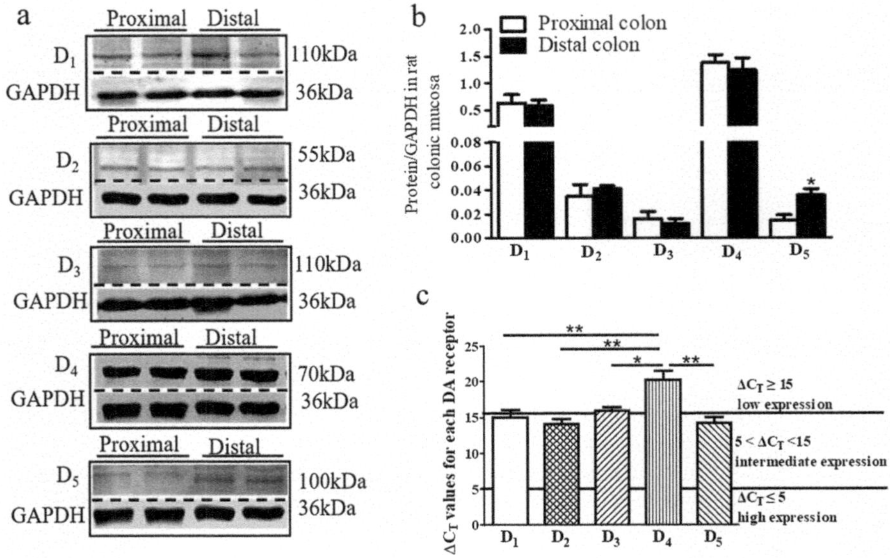

Fig. 3.8 Expression of five dopamine receptor subtypes in the colonic mucosa of rats. (**a**, **b**) Protein expression of D_1R–D_5R in the proximal and distal colon detected by Western blot; C, $\triangle C_T$ values for each dopamine receptor (DAR). The basal mRNA expression of DARs is estimated by determining the difference between the C_T of the interest *DAR* gene and the C_T of the endogenous control gene β-actin. In this graph, the highest $\triangle C_T$ value of DAR means the lowest mRNA expression of the DAR. (Li Y et al., Am J Physiol Cell Physiol, 2019, 316: C393–C403)

3.3 Dopamine Receptors in the Muscularis Externa of Gastrointestinal Tract

3.3.1 Dopamine Receptors in the Muscularis Externa of Esophagus

In the gastroesophageal junction, D_1R has been detected in the muscle fibers of the rat lower esophageal sphincter (LES) by in situ hybridization, Western blot, and RT-PCR (Fig. 3.10) (Vaughan et al. 2000). The DARs including D_1R, D_2R, and D_5R at mRNA and protein levels are also detected in the sling and clasp fibers of human LES by real-time PCR and Western blot, and the mRNA level of D_1R is the highest. The D_3R and D_4R are not identified (Liu and Liu 2012). The distribution of DARs in the muscularis externa of esophagus is summarized in Table 3.3.

Pharmacological evidence suggests that DARs are present in the muscular layer of the esophagus and play important roles in esophageal contraction. In LES of opossum, D_2R antagonist domperidone or metoclopramide does not antagonize DA-induced relaxation, while D_1R antagonist bulbocapnine and SKF 83566 significantly inhibit the relaxation induced by DA (Lombardi et al. 1986). In human LES,

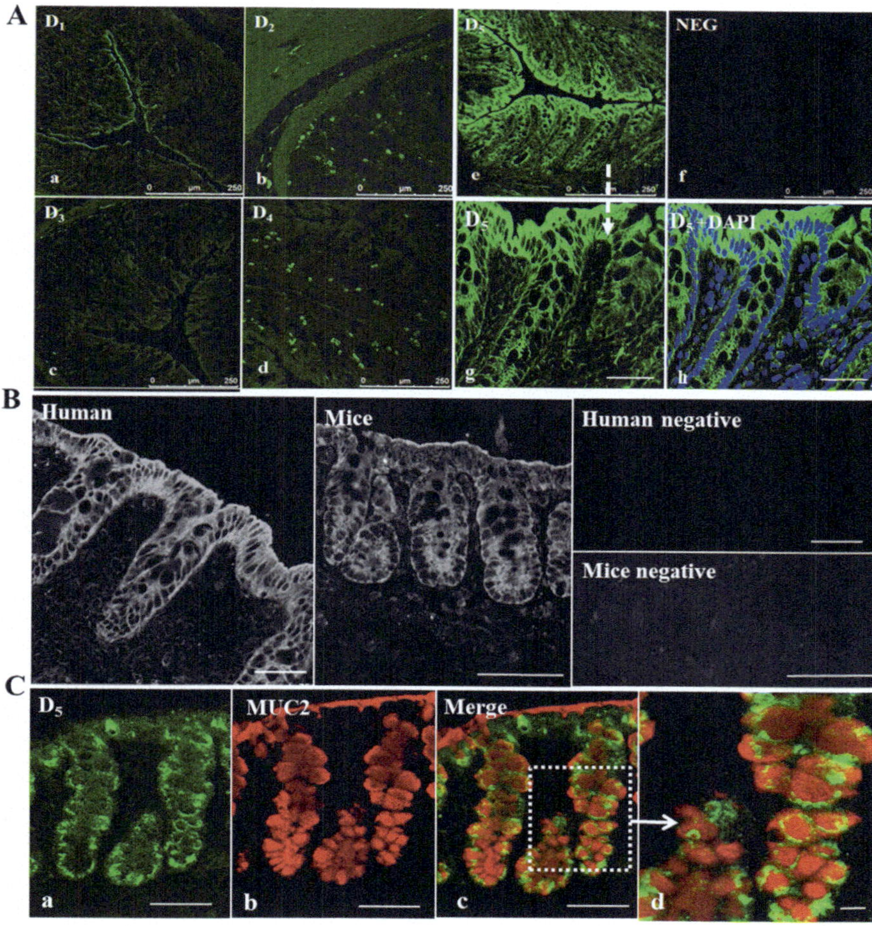

Fig. 3.9 Location of five dopamine receptor subtypes in the colonic mucosa of rats. (**A**) Distribution of D_1R–D_5R in the rat distal colonic mucosa (a–e, scale bars, 250 μm). Higher-power magnification images of e are shown in g and h (scale bars, 50 μm). (**B**) Distribution of D_5R in the colonic mucosa of human and mice. (**C**) Double-label immunofluorescence staining with the D_5R and Mucin 2 antibodies in distal colonic mucosa (a–c, scale bars, 50 μm). Higher-power magnifications in c are shown in d; scale bars, 10 μm. Negative control was not incubated in any of the D_1R–D_5R primary antibodies. (Li Y et al., Am J Physiol Cell Physiol, 2019, 316: C393–C403)

the selective D_1R agonist fenoldopam increases cAMP level in the LES, which is prevented by the selective D_1R antagonists Scheme 23390 and d-butaclamol. Bromocriptine, a selective D_2R agonist, inhibits adenylate cyclase activity in the LES, an effect blocked by the D_2R antagonist sulpiride (Missale et al. 1990). These data indirectly suggest the presence of D_1R and D_2R in the smooth muscle of LES. D_1R is reported to play stimulatory effects on the sphincter motility in rat LES by means of

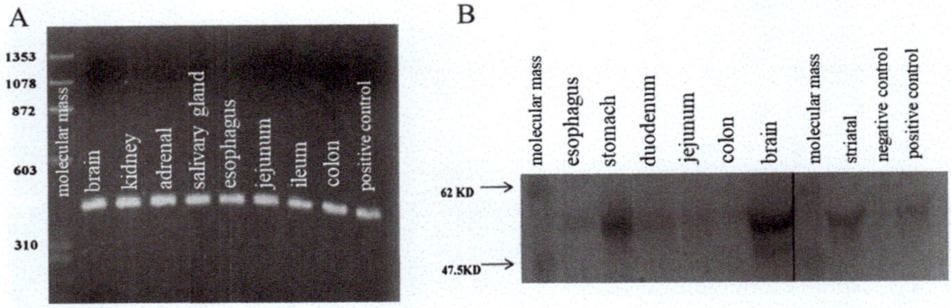

Fig. 3.10 RT-PCR and Western blot analysis of D_1R expression in the rat gastrointestinal tract. A, An amplification product of predicted 425-bp size for D_1R was detectable for RNA extracted from the brain, kidney, adrenal glands, salivary gland, and different regions of the GI tract in the rat. Positive control, a murine fibroblast cell line (LTK2) that had been stably transfected with a full-length rat D_1R cDNA. B, Western blot analysis demonstrated a single 50 kDa band for the D_1R transfected LTK2 fibroblasts (positive control), but not in wild-type (nontransfected) LTK2 cells (negative control). The same 50 kDa band for the D_1R was also detected in different regions of the rat digestive tract and the brain. (Vaughan CJ et al., Am J Physiol Regul Integr Comp Physiol, 2000, 279, R599–609)

Table 3.3 Distribution of dopamine receptors in the muscularis externa of esophagus

Tissue	Distribution of DA receptor	Species	Methods and references
Esophagus	D_1R in lower esophageal sphincter (LES)	Opossum	D_1R antagonist (bulbocapnine and SKF 83566) inhibits the relaxation induced by DA in LES (Lombardi et al. 1986)
	D_1R and D_2R in LES	Human	D_1R agonist fenoldopam stimulate cAMP level in the LES; D_2R stimulator bromocriptine inhibits adenylate cyclase activity in the LES (Missale et al. 1990)
	D_1R and D_2R in LES	Rat	D_1R is involved in LES contraction, while D_2R mediates the relaxation of the LES (Sigala et al. 1994)
	D_1R in muscle layer in LES	Rat	Immunohistochemistry, in situ hybridization, Western blot and RT-PCR (Vaughan et al. 2000)
	D_1R, D_2R, and D_5R in the sling and clasp fibers of LES	Human	Western blot and RT-PCR (Liu and Liu 2012)

in vitro motility recording and in vivo measurement of intraluminal LES pressure. However, D_2R plays an inhibitory role on the sphincter motility (Sigala et al. 1994). The different results reported by different research groups could be due to the different species, drug specificity, and the varying detective methods or tissue preparations etc.

3.3.2 Dopamine Receptors in the Gastric Muscularis Externa

A few studies directly investigate the expression of DARs in the stomach. D_1R-IR has been detected in the gastric muscularis externa of rat by immunohistochemical staining (Vaughan et al. 2000). Dr. Zhu JX's lab has demonstrated the mRNA and protein expression of D_1R and D_2R in the gastric muscularis externa of rat (Fig. 3.11), and the increased DA content and enhanced D_2R expression are involved in the delayed gastric empty of the rats with bilaterally microinjection of 6-OHDA in the substantia nigra (6-OHDA rats, a rat model of PD) (Zheng et al. 2014). The distribution of DARs in the gastric muscularis externa is summarized in Table 3.4.

Some pharmacological studies have also indicated that DARs play important roles in gastric motility in vivo and in vitro. In the gastric body of guinea pig, DA induces relaxation in longitudinal muscle, and the selective D_1R antagonist Scheme 23390 shows competitive antagonism, indicating that DA induces relaxation through D_1R on the smooth muscle cells (Kurosawa et al. 1991). However, domperidone, a specific antagonist of D_2R, suppresses DA-induced inhibition of gastric movement in the corpus and antrum of rat. LY171555, a specific agonist of D_2R, can mimic the effect of DA (Nagahata et al. 1992, 1995). Dr. Zhu JX's lab has also reported that the administration of D_2R antagonist domperidone relieves impaired gastric motility in 6-OHDA rats, but the D_1R antagonist SCH23390 fails to do so (Zheng et al. 2014).

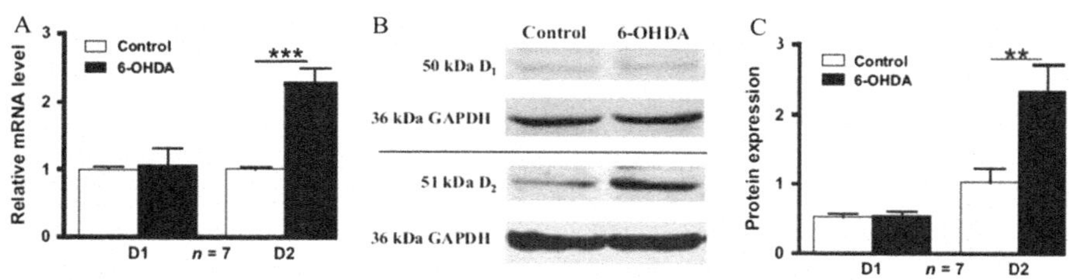

Fig. 3.11 Expression of D_1R and D_2R in the muscularis externa of rat gastric corpus. The mRNA (**a**) and protein levels (**b**, **c**) of D_2 receptors in the gastric muscularis externa were increased in 6-OHDA rats compared with control rats. (Zheng LF et al., Acta Physiol, 2014, 211, 431–46)

Table 3.4 Distribution of dopamine receptors in the gastric muscularis externa

Tissue	Distribution of DA receptor	Species	Methods and references
Stomach	D_1R in muscularis externa in the stomach	Rat	Immunohistochemistry, in situ hybridization, Western blot, and RT-PCR (Vaughan et al. 2000)
	D_1R in the gastric body	Guinea pig	D_1R antagonist (Scheme 23390) blocks DA-induced relaxation in longitudinal muscle (Kurosawa et al. 1991)
	D_1R and D_2R in the muscularis externa of gastric corpus	Rat	Real-time PCR and Western blot (Zheng et al. 2014)

These results indicate that DA inhibits gastric motility through D_2R. Furthermore, the total GI transit time and the colonic transit time are decreased in D_2R knockout mice and in D_2R and D_3R double-knockout mice, but not in D_3R knockout mice, which also suggests that endogenous DA exerts a net inhibitory effect on intestinal motility primarily through the activation of D_2R (Li et al. 2006b). Moreover, there are also some studies showing the involvement of D_2R and D_3R in the regulation of gastric emptying in rats (Kashyap et al. 2009; Yoshikawa and Yoshida 2010). Subcutaneous injection (s.c.) of D_3R agonists S(+)-PD 128,907 (0.01–1 mg/kg) and R(+)-7-OH-DPAT (0.03–1 mg/kg), and D_2R agonist quinpirole (0.01–1 mg/kg) dose-dependently delay gastric emptying in rats. Both the selective D_1R and D_5R agonist SKF38393 and the selective D_4R agonist PD 168,077 fail to delay the gastric emptying in rats. The selective D_3R antagonist (+)-S 14297 (10 mg/kg, s.c.) partially inhibits the S(+)-PD 128,907-induced delay in gastric emptying. Although administration of S(+)-PD 128,907 (1–100 μg/kg) into the fourth cerebral ventricle can partially and dose-dependently delay rat gastric emptying, its administration into the lateral cerebral ventricle does not affect gastric emptying. The results suggest that peripheral D_2R and, at least in part, D_3R, and central D_2/D_3R play an important role in the regulation of gastric motility in rats (Yoshikawa and Yoshida 2010).

3.3.3 Dopamine Receptors in the Muscularis Externa of Small Intestine

Transcripts encoding of the five DARs are detected in the whole wall of duodenum and ileum in mice (Yoshikawa and Yoshida 2010). In the small intestine of rat, D_1R protein and mRNA signals are detectable in the duodenum, jejunum, and ileum, with the strongest signals in the muscularis externa than the other layers of the small intestinal wall (Vaughan et al. 2000). The distribution of DARs in the muscularis externa of small intestine is summarized in Table 3.5.

DA exerts differential effects on smooth muscle motility via DARs in different regions of the small intestine by pharmacological studies. A biphasic effect with a rapid DA-induced contraction followed by a slower relaxation has been reported in the small intestine including duodenum, ileum, and jejunum of rats. Both the D_1R antagonist SCH23390 and D_2R antagonist raclopride can block the DA-induced contraction in the duodenum and jejunum. The relaxations in the jejunum and ileum are markedly reduced by β-adrenoceptor antagonist propranolol and α-adrenoceptor antagonist prazosin, but not influenced by DAR antagonists SCH23390 or raclopride (Kirschstein et al. 2009). In rabbit jejunum, octopamine produces relaxation via D_1R through an increase of cAMP (Cheng and Hsieh-Chen 1988). Similar result is observed in the isolated ileum of guinea pig (Tsai and Cheng 1992) and mouse (Zizzo et al. 2010), where DA induces relaxation in ileal smooth muscle via activating the D_1R and then increasing intracellular cAMP. D_1R antagonist SCH-23390 increases basal tone and amplitude of spontaneous contractions but

Table 3.5 Distribution of dopamine receptors in the muscularis externa of small intestine

Tissue	Distribution of DA receptor	Species	Methods and references
Duodenum	D_1R in muscularis externa of duodenum	Rat	RT-PCR, Western blot, in situ hybridization histochemistry (Vaughan et al. 2000)
	D_1R and D_2R in the duodenum	Rat	Both the D_1R antagonist SCH23390 and the D_2R antagonist raclopride blocked the DA-induced contraction in the duodenum (Kirschstein et al. 2009)
Jejunum	D_1R in the jejunum body	Rabbit	Octopamine acts on intestinal D_1R to produce relaxation of rabbit jejunum (Cheng and Hsieh-Chen 1988)
Ileum	D_1R in muscularis externa of ileum	Rat	RT-PCR, Western blot, in situ hybridization histochemistry (Vaughan et al. 2000)
	D_1R in the smooth muscle of ileum	Mouse	D_1R antagonist increases basal tone and amplitude of spontaneous contractions, antagonize the responses to DA, and D_1R agonist mimics DA-induced effects (Zizzo et al. 2010)

antagonizes the effect of DA, whereas the D_2R antagonists, sulpiride or domperidone, have no effects. D_1R agonist SKF-38393 can mimic DA-induced effects (Zizzo et al. 2010). The data suggests that D_1R exists in the smooth muscle of mouse ileum and mediate DA-induced the relaxation of ileal smooth muscle.

3.3.4 Dopamine Receptors in the Colonic Muscularis Externa

D_1R immunoreactivity has a transmural distribution in the rat colon. A positive signal for D_1R is present in the cells of the colonic muscularis externa by in situ hybridization, although the signal is weaker than that in the muscularis mucosa (Vaughan et al. 2000). By means of immunofluorescence and Western blot, Dr. Zhu JX's lab has reported that the D_2R is only detected in the myenteric plexus, and the D_1R, and D_5R are expressed in both the muscular layer and the myenteric plexus of rat colon (Zhang et al. 2012, 2015). In addition, D_5R-IR is also abundantly expressed in human colon (Fig. 3.12). And, D_1R, D_2R and D_5R proteins are expressed in the muscular layer of distal colon in mouse (Auteri et al. 2016). Dr. Zhu JX's lab has reported that DA content is enhanced in the colon of 6-OHDA rats, a rat model of PD, which may contribute to the formation of constipation in the PD rats (Zhang et al. 2015). DA content and D_1R expression are increased in the colon of stressed rats (cold-restraint stress rats model), which may partly mitigate the enhanced motility caused by stress (Zhang et al. 2012). These results suggest that DA and DARs may be involved in the modulation of distal colonic motility. The distribution of DARs in the colonic muscularis externa is summarized in Table 3.6.

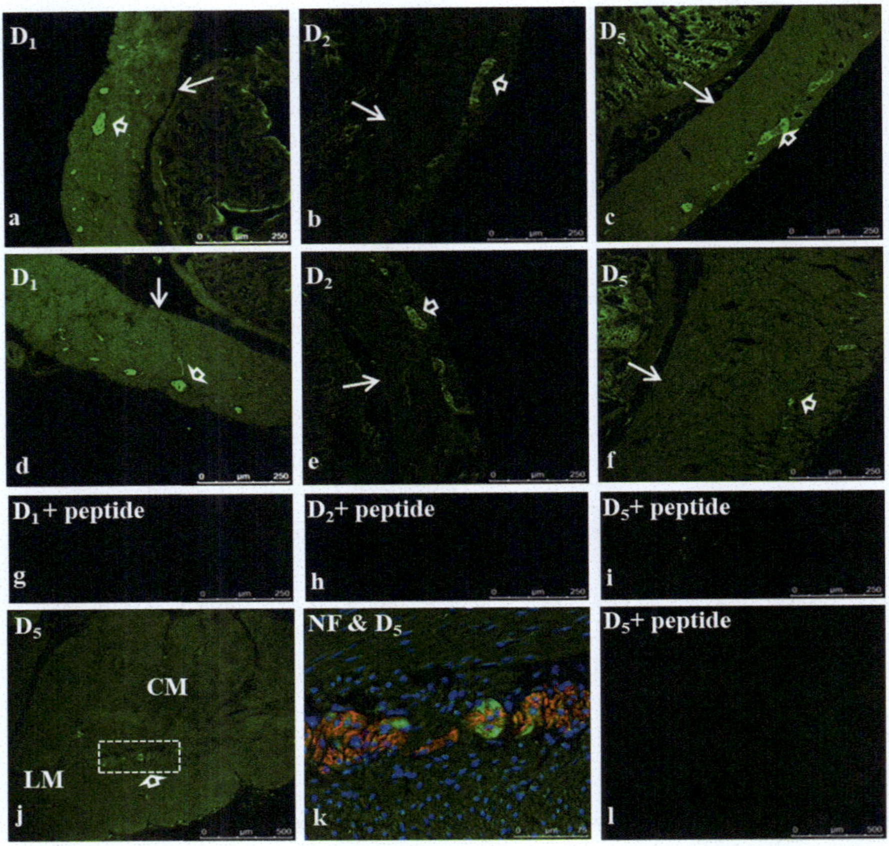

Fig. 3.12 Distribution of D_1R, D_2R, and D_5R in the colonic muscularis externa of control rats, 6-OHDA rats, and human. (**a–c**) D_1R, D_2R, and D_5R immunoreactivity in the muscularis externa of control rats. (**d–f**) D_1R, D_2R, and D_5R immunoreactivity in the muscularis externa of 6-OHDA rats. (**g–i**) Anti-D_1R, D_2R, and D_5R were blocked with their immunizing peptides. (**j**) D_5R in the human colon. (**k**) Colocalization of D_5R (green) with neurofilament (NF, red, a neuronal marker) in a higher magnification views of the frame of j. (**i**) Anti-D_5R was blocked with its immunizing peptide. Scale bars: **a–i**, 250 μm; **j–l**, 75 μm. (Zhang XL et al., Transl Res, 2015, 166, 152–62)

Pharmacological studies have shown the differential effects of the different DARs on the regulation of colonic motility. In human, D_2R antagonist domperidone can antagonize the effect of DA-stimulated rectosigmoid motility (Wiley and Owyang 1987). In murine colon, a combination of D_1R and D_2R antagonists increases the muscle tone, the amplitude of spontaneous contractions, and the neutrally evoked contractions, supporting the role of endogenous DA as a tonic negative modulator of colonic mechanical activity (Walker et al. 2000). Mice lacking the D_2R display an increased colonic propulsive activity and decreased total GI transit (Li et al. 2006b). Infusion of the D_2R antagonist itopride accelerates in vitro colonic transit in guinea pig (Lim et al. 2008) and the D_2R agonist bromocriptine reduces intestinal transit in

Table 3.6 Distribution of dopamine receptors in the colonic muscularis externa

Tissue	Distribution of DA receptor	Species	Methods and references
Colon	D_2R in the rectosigmoid	Human	Domperidone, D_2R antagonist, antagonize the effect of DA on rectosigmoid motility in humans (Wiley and Owyang 1987)
	D_1R and D_2R in muscularis externa of colon	Mouse	A combination of D_1R and D_2R antagonists increases the muscle tone, the amplitude of spontaneous contractions, and the neutrally evoked contractions (Walker et al. 2000)
	D_2R in colon	Guinea pig	D_2R antagonist itopride accelerates peristalsis in guinea pig colon (Lim et al. 2008)
	D_1R and D_5R in muscularis externa of colon	Rat	Immunofluorescence, Western blot (Zhang et al. 2012, 2015)
	D_5R in muscularis externa of colon	Human	Immunofluorescence, Western blot (Zhang et al. 2015)
	D_1R in colonic muscularis externa	Rat	RT-PCR, Western blot, in situ hybridization histochemistry (Vaughan et al. 2000)
	D_1R and D_2R in muscularis externa of distal colon	Mouse	DA inhibits colonic circular muscle contractility of mouse distal colon by activating both D_1R and D_2R (Auteri et al. 2016)

mice (Kaneko et al. 2010), strongly suggesting an important contribution of D_2R in the DA-induced inhibitory effect on colonic motility. Indeed, other studies highlight a major role for D_1R in mediating DA-induced reduction of contraction in colon (Kaneko et al. 2010; Zhang et al. 2012). D_1R antagonist SCH-23390 blocks the inhibitory role of DA on the contraction of longitudinal muscle in rat distal colon by organ-bath technique, whereas the D_2R antagonist sulpiride does not affect the response of muscle strips to DA (Zhang et al. 2012). However, Auteri et al. have demonstrated that DA inhibits colonic circular muscle contractility of mouse distal colon by activating both D_1R and D_2R by organ-bath technique (Auteri et al. 2016). Recently, Zizzo et al. have reported that DA evokes opposite effects on the mechanical activity of circular and longitudinal muscle of human colon. In the circular muscle layer, DA induces contraction by activating non-neural D_1-like receptors. Whereas, in the longitudinal muscle layer, DA induces relaxation by acting on non-neural D_2-like receptors (Zizzo et al. 2020). The reasons for the conflicting results might be related to the differential methods, muscular tissues, and species employed.

3.3.5 Dopamine Receptors and Interstitial Cells of Cajal

Interstitial cells of Cajal (ICC) serve as a pacemaker for GI contraction. Oscillations of $[Ca^{2+}]_i$ in ICC have been considered as the primary mechanism of GI pacemaking, in which Ca^{2+}-dependent ion channels are periodically activated to generate pacemaker potentials. When the D_2R antagonist domperidone is applied to nifedipine- and TTX-treated cell clusters or cultured tissue from the mouse ileum, the amplitude of $[Ca^{2+}]_i$ oscillations manifests a decrease after a transient increase. On the other hand, D_2R agonist piribedil dihydrochloride does not apparently affect the amplitude of $[Ca^{2+}]_i$ oscillations, but reduces the frequency gradually, eventually eliminating the $[Ca^{2+}]_i$ spike formation. This result indicates that the basal level activity of D_2R is crucial for maintaining $[Ca^{2+}]_i$ oscillations in ICC (Liu et al. 2018). Thus, pharmaceutical activation of the D_2R in ICC potentially provides a novel therapeutic target for the treatment of gut motility disorders. However, there is no report on the distribution of DARs on ICC. Further studies are needed to reveal the detailed molecular mechanisms of D_2R regulation of $[Ca^{2+}]_i$ oscillations.

3.4 Dopamine Receptors in the Enteric Nervous System

ENS is located in the wall of the GI tract from the esophagus to the rectum and includes the myenteric plexus (Auerbach's plexus) and submucosal plexus (Meisser's plexus). It is large, complex, and uniquely able to control the most of the GI functions in a central nervous system (CNS)-independent manner, such as GI motility, secretion and absorption of mucosa, local blood flow, and epithelial barrier function. DARs are widely distributed in the ENS.

Except D_4R, D_1R–D_3R and D_5R have been detected in the longitudinal muscle myenteric plexus (LMMP) preparations of mice ileum (Li et al. 2006b). Furthermore, Li et al. found the expression of D_2R in the LMMP but not in the muscular layer, suggesting that D_2R is only present in the myenteric plexus. Total GI transit time and colonic transit time are decreased in D_2R knockout mice (Li et al. 2006b), which suggests that D_2R in the ENS may exert an inhibitory effect on intestinal motility, although the role of brain D_2R cannot be excluded. Mezey et al. reveal that D_2R mRNA is only detected in the enteric ganglionic cells of the rat stomach and duodenum by in situ hybridization histochemistry technique (Mezey et al. 1999), whereas the others show that the mRNA and protein expressions of D_2R are also in the gastric and duodenal mucosa in rat by real-time PCR and immunofluorescence (Feng et al. 2013; Wang et al. 2012). D_1R, D_2R, and D_3R are distributed in the submucosal plexus of mice by in situ hybridization and immunocytochemistry (Li et al. 2006b) (Fig. 3.13). D_4R immunoreactivity is distributed in the myenteric plexus of rat stomach by immunofluorescence (Wang et al. 2012). In rat pylorus, D_3R immunoreactivity is colocalized with the neuronal marker βIII-tubulin in the myenteric plexus (Fig. 3.14) (Kashyap et al. 2009). And, activation of D_3R reduces electrical field stimulation-induced relaxation of pyloric strips in a dose-dependent manner and delays gastric emptying (Kashyap et al. 2009; Yoshikawa and Yoshida 2010). In the small intestine, a strong immunoreactive signal for the D_1R is present in the myenteric plexus of rat (Vaughan et al. 2000). In our study, D_1R-, D_2R-, and D_5R-IR are abundantly expressed in the myenteric plexus of rat colon, and D_5R-IR is colocalized with the neuronal marker neurofilament (NF) in the myenteric plexus of

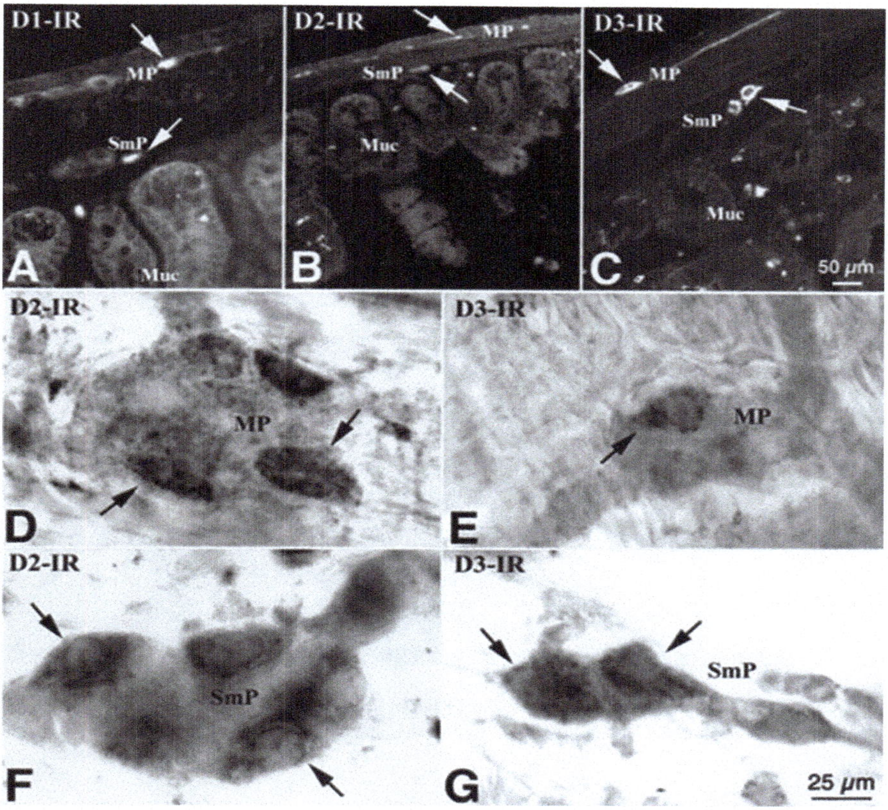

Fig. 3.13 Dopamine receptor immunoreactivity in enteric neurons of the mouse ileum. The immunoreactivity was visualized with antibodies to D_1R, D_2R, and D_3R in both frozen sections (**a–c**) and whole-mount preparations (**d–g**). MP, myenteric plexus; SmP, submucosal plexus; Muc, mucosa. Scale bars: (in **c**) A–C, 50 μm; (in **g**) **d–g**, 25 μm. (Li et al., J Neurosci, 2006; 26, 2798–807)

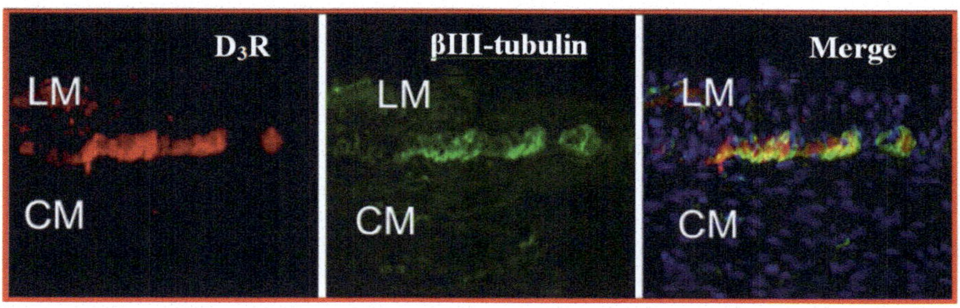

Fig. 3.14 D_3R immunoreactivity is colocalized with the neuronal marker βIII-tubulin in the myenteric plexus of rat pylorus. *CM* circular muscle, *LM* longitudinal muscle. (Kashyap et al., Dig Dis Sci, 2009, 54, 57–62)

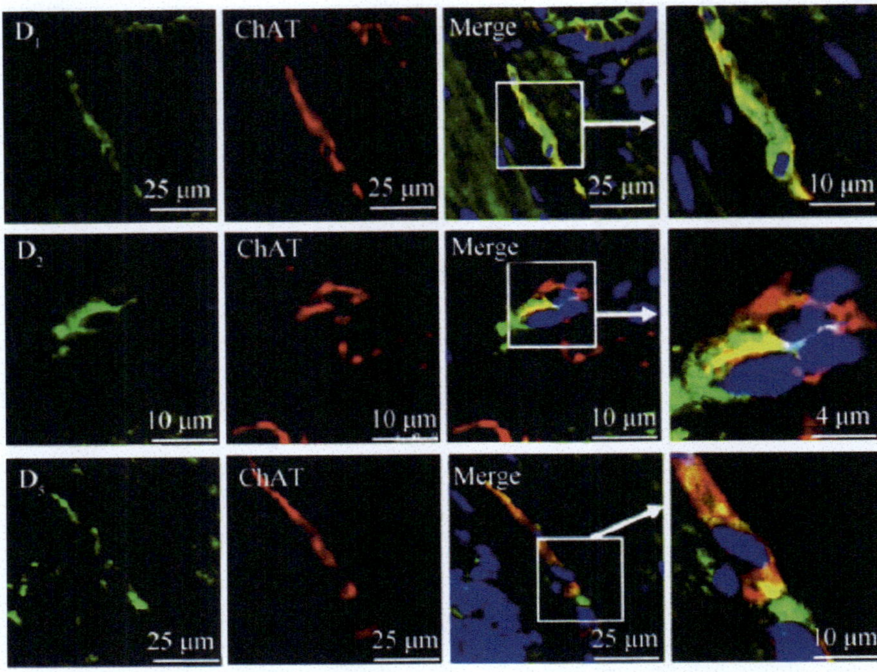

Fig. 3.15 The distribution of D_1R, D_2R, and D_5R on ChAT-IR in the submucosal plexus of rat colon. ChAT-IR: choline acetyltransferase immunoreactivity. (Zhang et al., Journal of Capital Medical University, 2017, 38, 411–416)

human colon (Zhang et al. 2015). Both the mRNA and protein of D_1R, D_2R, and D_5R are expressed in the colonic submucosal plexus of rat by real-time PCR and Western blot. Moreover, D_1R, D_2R, and D_5R are located in the vasoactive intestinal peptide (VIP)-positive neurons and cholinergic neurons in the submucosal plexus by double-labeling immunofluorescence (Figs. 3.15 and 3.16) (Zhang et al. 2017). The percentage of VIP-positive neurons expressed D_1R, D_2R, and D_5R is 58%, 52%, and 86%, respectively. The percentage of cholinergic neurons expressed D_1R, D_2R, and D_5R is 87%, 88%, and 89%, respectively (Zhang et al. 2017). There is additional evidence for the presence of dopaminergic receptors in the neuromuscular layer in small intestinal preparations from 2-day-old (P2) and adult mice (P90) by immunofluorescence. The labeling for D_1R, D_2R, D_3R, and D_5R are mainly detected in the myenteric ganglia. D_1R immunofluorescence is detected as a bright signal in P2 preparations. And, the levels of D_1R immunofluorescence in the neuromuscular layer, both smooth muscle layers, and the myenteric plexus show an age-dependent increase. A small increase of D_2R and D_3R immunoreactivity with the progress of the age is also visible (Zizzo et al. 2016) (Fig. 3.17).

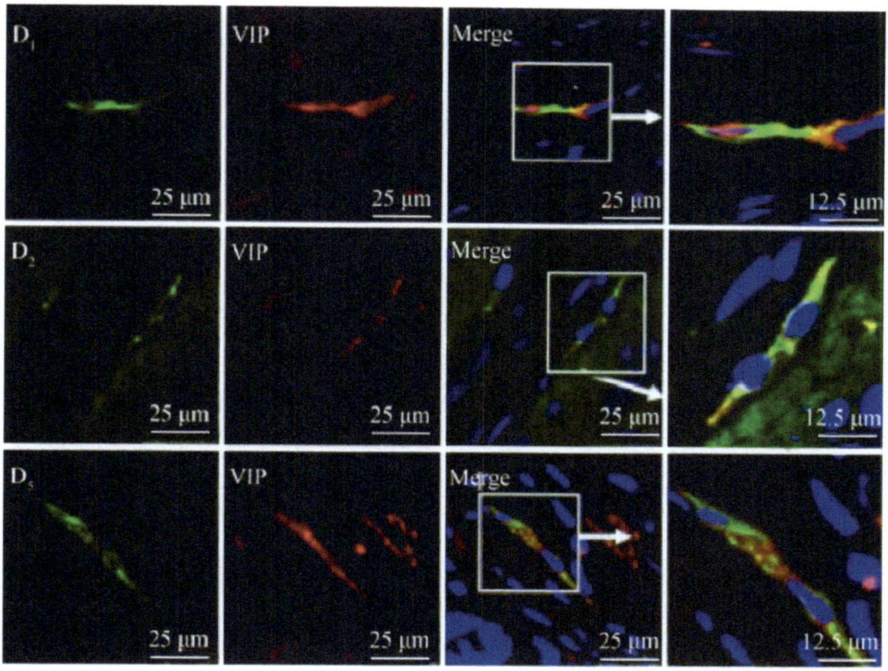

Fig. 3.16 The distribution of D_1R, D_2R, and D_5R on VIP-IR in the submucosal plexus of rat colon. VIP-IR: vasoactive intestinal peptide immunoreactivity. (Zhang et al., Journal of Capital Medical University, 2017, 38, 411–416)

3.5 Dopamine Receptors in Gastrointestinal Inflammation and Cancer

3.5.1 Dopamine Receptors and Inflammation

Increased evidence suggests that DA acts as an important regulator of immune function. Many immune cells express DARs and DA-related proteins, such as the synthetic and metabolic enzymes of DA and DAT, which may enable them to respond to DA. In the CNS, microglia are the predominant immune effector cells and they express functional D_1R and D_2R (Farber et al. 2005; Huck et al. 2015) as well as catechol-*O*-methyltransferase (Fan et al. 2018; Myohanen et al. 2010). In peripheral immune cells, T-lymphocytes are shown to express all DARs (Le Fur et al. 1980). Human myeloid cells, such as monocytes and macrophages, also express all five subtypes of DARs (Coley et al. 2015; Gaskill et al. 2012). And, monocyte-derived dendritic cells primarily express DAR-1 (Nakano et al. 2008). Moreover, all the five subtypes of DARs have been found on neutrophils (Boneberg et al. 2006; McKenna et al. 2002; Sookhai et al. 1999) and eosinophils (McKenna et al. 2002). Activation of D_2R in astrocytes can suppress neuroinflammation in the

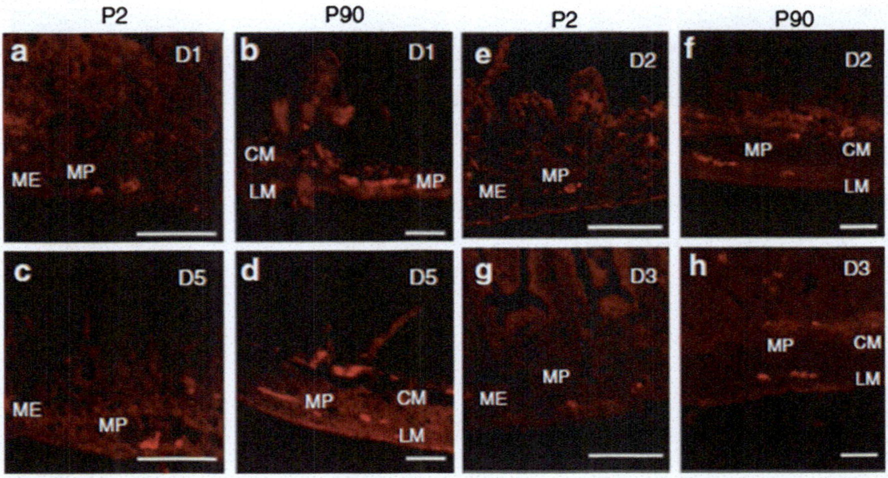

Fig. 3.17 Immunofluorescence staining of dopamine receptor in the intestinal cross-sections of P2 and P90 mice. P2 and P90 mouse intestinal cross-sections were labeled with antibodies (red) for D_1R (**a**, **b**), D_5R (**c**, **d**), D_2R (**e**, **f**), and D_3R (**g**, **h**) receptors. *CM* circular muscle, *LM* longitudinal muscle, *ME* muscularis externa, *MP* myenteric plexus; P2, 2-day-old; P90, 90-day-old. Scale bars: 25 μm. (Zizzo et al., Pediatr Res, 2016, 80, 440–7)

CNS via αB-crystallin (Shao et al. 2013). Moreover, DA, via D_1R, specifically inhibits NLRP3 inflammasome activation and related inflammation (Yan et al. 2015). And, electroacupuncture at the sciatic nerve can lead to the production of DA in the adrenal medulla by inducing a vagal activation of DOPA decarboxylase. DA has also been reported to inhibit the production of various cytokines, such as tumor necrosis factor, MCP-1, interleukin-6, and interferon-γ, and control experimental sepsis in mice via D_1R (Torres-Rosas et al. 2014). A recent study reports that DA, via D_5R, can suppress *S. aureus*-induced systemic inflammation (Wu et al. 2020).

In the GI tract, DARs are distributed in the immune cells in the lamina propria and involved in the regulation of intestinal inflammation. The mRNA of D_3R and D_4R has been found in the cells of the lamina propria in gastric mucosa (Mezey et al. 1998, 1999) and the D_2R-IR and D_4R-IR are located in the colonic lamina propria (Li et al. 2019), although the specific cellular locations of these DARs are not known. Moreover, approximately 25–50% of intraepithelial lymphocytes isolated from the guinea pig jejunum express D_1R, and 50% of them express D_2R by mean of the flow cytometry technique with fluorescent probes (Baglole et al. 2005). In biopsy specimens from inflamed human colonic mucosa, DA levels are reduced in Crohn's disease (CD) and ulcerative colitis (UC) patients compared to health volunteers (Magro et al. 2002). A similar situation has been observed in 2,4,6-trinitrobenzene sulphonic acid-induced CD (Magro et al. 2004) and iodoacetamide-induced UC in rats (Tolstanova et al. 2015). The decrease in DA levels has been attributed to the interferon-γ-mediated inhibition on L-DOPA uptake by epithelial cells (Magro et al.

2004), decreased expression levels of TH, the time-limiting enzyme that synthesizes DA, and reduced expression of the DAT in the colon (Tolstanova et al. 2015). Magro's group also reported that polymorphisms in D_2R result in decreased D_2R expression and are associated with the susceptibility to develop of refractory CD in patients (Magro et al. 2006). However, D_2R protein is increased in two animal models of IBD, interleukin-10 knockout mice (spontaneously developed a chronic IBD) and iodoacetamide-induced UC in rats (Tolstanova et al. 2015). Intragastrical application of D_2R agonist quinpirole can decrease the size of colonic lesions in the two models by attenuating the enhanced vascular permeability and preventing the excessive vascular leakage (Tolstanova et al. 2015). A similar study in experimental IBD model induced with 2,4-dinitrouorbenzene in BALB/c mice shows that the disease is enhanced by peripheral treatment with D_2R antagonist domperidone and ameliorated by the D_2R agonist bromocriptine (Herak-Perkovic et al. 2001). Whereas, Kim et al. show that gavage with D_2R antagonists sulpiride can ameliorate the severity of 2,4-dinitrobenzenesulfonicacid -induced rat colitis. Thus, D_2 may be a potential target for the treatment of IBD (Kim et al. 2019).

3.5.2 Dopamine Receptors and Cancer

Recently, the relationship of DARs and GI cancers has attracted strong attention. D_2R expression has been reported at both the mRNA and protein levels in a variety of GI cancers, which correlates with tumor progression (Mu et al. 2017). Increased D_2R expression has been reported in esophageal carcinoma (Li et al. 2006a) and pancreatic ductal adenocarcinoma (Jandaghi et al. 2016), while decreased D_2R expression has been observed in malignant stomach (Basu and Dasgupta 1997), malignant colon (Basu and Dasgupta 1999), and gastric cancer BGC-823 and MGC-803 cells (Huang et al. 2016). Moreover, polymorphisms within D_2R are associated with an increased risk of colorectal cancer (Gemignani et al. 2005). Some studies have also identified D_2R antagonists or agonists as potential anticancer therapeutics through in vitro studies utilizing cell lines and patient samples. Stimulation of D_2R with its specific D_2R agonist, quinpirole, inhibits insulin-like growth factor receptor-I (IGF-IR)-induced proliferation of gastric cancer AGS cells by upregulating KLF4, a negative regulator of the cell cycle through downregulation of IGF-IR and AKT phosphorylation (Ganguly et al. 2010). DA treatment, which activates D_2R, can also suppress gastric cancer cell invasion and migration of gastric cancer BGC-823 and MGC-803 cells via inhibition of the EGFR/AKT/MMP-13 pathway (Huang et al. 2016). In contrast, other studies conclude that a reduced proliferative effect of the D_2R antagonists and D_2 knockdown. Mu et al. have shown that patients with higher protein expression levels of D_2R in gastric cancer tissues have shorter survival durations, whereas D_2R antagonist, thioridazine, decreases the growth of gastric cancer AGS cells (Mu et al. 2017). D_2R agonist bromocriptine promotes rat gastric carcinogenesis induced by *N*-methyl-*N*′-nitro-*N*-nitrosoguanidine (Iishi et al. 1992). RNAi knockdown of D_2R or inhibition with pharmacologic antagonists (pimozide and haloperidol) reduces proliferation of pancreatic cancer cells and cell migration in mice (Jandaghi et al. 2016). These observations lead to the hypothesis that D_2R might be a potential therapeutic target for GI cancer.

Some studies have also investigated the possible role of DAR-1 (D_1R and D_5R) in GI cancers. Increased D_1R expression has been reported in esophageal carcinoma (Li et al. 2006a). In addition, Leng et al. have reported the expressions of D_5R in many human cancer cells such as glioblastomas, colon cancer, and gastric cancer. Activation of D_5R with SKF 83959 inhibits gastric cancer cell growth in nude mice injected with human gastric cancer cell line SCG7901. This study may reveal a potential use of D_5R agonists as a novel therapeutic approach for the treatment of human tumors and cancers (Leng et al. 2017).

3.6 Conclusions/Future Perspectives

Based on the wide distribution of DARs in the GI mucosa, smooth muscle, ENS, and immune cells, etc., there is no doubt that DARs play important roles in the regulation of various GI functions and gut homeostasis, such as motility, mucosal secretion, absorption, and mucosal barrier function, etc. Growing studies have shown that the dopaminergic system mediates neuroimmune communications. DA regulates a variety of immune functions including cytokine secretion, cell adhesion, cytotoxicity, and chemotaxis. Considering the altered DA levels in inflammation condition, the expression of DARs in immune cells of the GI tract, and the mechanisms by which dopaminergic signaling regulates the function of immune cells, the dopaminergic system becomes a very attractive therapeutic target for the management of IBD; however, this does require further research validation. Also, there is study showing that DARs are correlated with GI cancers. However, to verify their potential as a therapeutic target for GI cancers, a thorough understanding of DARs function and underlying mechanisms in GI cancer cells is required. The main obstacle is lack of DAR subtype selective compounds, which has hindered our knowledge of how the specific DAR subtype mediates various physiological responses and the development of novel therapeutics.

References

Auteri M, Zizzo MG, Amato A, Serio R (2016) Dopamine induces inhibitory effects on the circular muscle contractility of mouse distal colon via D1- and D2-like receptors. J Physiol Biochem 73:395–404. https://doi.org/10.1007/s13105-017-0566-0

Baglole CJ, Davison JS, Meddings JB (2005) Epithelial distribution of neural receptors in the guinea pig small intestine. Can J Physiol Pharmacol 83:389–395. https://doi.org/10.1139/y05-024

Basu S, Dasgupta PS (1997) Alteration of dopamine D2 receptors in human malignant stomach tissue. Dig Dis Sci 42:1260–1264. https://doi.org/10.1023/a:1018862309440
Basu S, Dasgupta PS (1999) Decreased dopamine receptor expression and its second-messenger cAMP in malignant human colon tissue. Dig Dis Sci 44:916–921. https://doi.org/10.1023/a:1026644110737
Boneberg EM, von Seydlitz E, Propster K, Watzl H, Rockstroh B, Illges H (2006) D3 dopamine receptor mRNA is elevated in T cells of schizophrenic patients whereas D4 dopamine receptor mRNA is reduced in CD4+ –T cells. J Neuroimmunol 173:180–187. https://doi.org/10.1016/j.jneuroim.2005.11.018
Cheng JT, Hsieh-Chen SC (1988) Octopamine relaxes rabbit jejunal smooth muscle by selective activation of dopamine D1 receptors. Naunyn Schmiedeberg's Arch Pharmacol 338:373–378. https://doi.org/10.1007/BF00172112
Coley JS, Calderon TM, Gaskill PJ, Eugenin EA, Berman JW (2015) Dopamine increases CD14 +CD16+ monocyte migration and adhesion in the context of substance abuse and HIV neuropathogenesis. PLoS One 10:e117450. https://doi.org/10.1371/journal.pone.0117450
Deng X, Zheng Z, Ye S (1997) Localization and expression of dopamine receptors in stomach and duodenum in rats. Zhonghua Yi Xue Za Zhi 77:103–105
Desai JK, Parmar NS (1994) Gastric and duodenal anti-ulcer activity of sulpiride, a dopamine D2 receptor antagonist, in rats. Agents Actions 42:149–153. https://doi.org/10.1007/BF01983482
Desai JK, Goyal RK, Parmar NS (1995) Gastric and duodenal anti-ulcer activity of SKF 38393, a dopamine D1-receptor agonist in rats. J Pharm Pharmacol 47:734–738. https://doi.org/10.1111/j.2042-7158.1995.tb06733.x
Desai JK, Goyal RK, Parmar NS (1999) Characterization of dopamine receptor subtypes involved in experimentally induced gastric and duodenal ulcers in rats. J Pharm Pharmacol 51:187–192. https://doi.org/10.1211/0022357991772123
Eliassi A, Aleali F, Ghasemi T (2008) Peripheral dopamine D2-like receptors have a regulatory effect on carbachol-, histamine- and pentagastrin-stimulated gastric acid secretion. Clin Exp Pharmacol Physiol 35:1065–1070. https://doi.org/10.1111/j.1440-1681.2008.04961.x
Fan Y, Chen Z, Pathak JL, Carneiro A, Chung CY (2018) Differential regulation of adhesion and phagocytosis of resting and activated microglia by dopamine. Front Cell Neurosci 12:309. https://doi.org/10.3389/fncel.2018.00309
Farber K, Pannasch U, Kettenmann H (2005) Dopamine and noradrenaline control distinct functions in rodent microglial cells. Mol Cell Neurosci 29:128–138. https://doi.org/10.1016/j.mcn.2005.01.003
Feng XY, Li Y, Li LS, Li XF, Zheng LF, Zhang XL, Fan RF, Song J, Hong F, Zhang Y, Zhu JX (2013) Dopamine D1 receptors mediate dopamine-induced duodenal epithelial ion transport in rats. Transl Res 161:486–494. https://doi.org/10.1016/j.trsl.2012.12.002
Feng XY, Zhang DN, Wang YA, Fan RF, Hong F, Zhang Y, Li Y, Zhu JX (2017) Dopamine enhances duodenal epithelial permeability via the dopamine D5 receptor in rodent. Acta Physiol (Oxf) 220:113–123. https://doi.org/10.1111/apha.12806
Feng XY, Yan JT, Li GW, Liu JH, Fan RF, Li SC, Zheng LF, Zhang Y, Zhu JX (2020) Source of dopamine in gastric juice and luminal dopamine-induced duodenal bicarbonate secretion via apical dopamine D2 receptors. Br J Pharmacol. https://doi.org/10.1111/bph.15047
Flemstrom G, Safsten B (1994) Role of dopamine and other stimuli of mucosal bicarbonate secretion in duodenal protection. Dig Dis Sci 39:1839–1842. https://doi.org/10.1007/BF02088112
Flemstrom G, Safsten B, Jedstedt G (1993) Stimulation of mucosal alkaline secretion in rat duodenum by dopamine and dopaminergic compounds. Gastroenterology 104:825–833. https://doi.org/10.1016/0016-5085(93)91019-e
Ganguly S, Basu B, Shome S, Jadhav T, Roy S, Majumdar J, Dasgupta PS, Basu S (2010) Dopamine, by acting through its D2 receptor, inhibits insulin-like growth factor-I (IGF-I)-induced gastric cancer cell proliferation via up-regulation of Kruppel-like factor 4 through

down-regulation of IGF-IR and AKT phosphorylation. Am J Pathol 177:2701–2707. https://doi.org/10.2353/ajpath.2010.100617

Gaskill PJ, Carvallo L, Eugenin EA, Berman JW (2012) Characterization and function of the human macrophage dopaminergic system: implications for CNS disease and drug abuse. J Neuroinflammation 9:203. https://doi.org/10.1186/1742-2094-9-203

Gemignani F, Landi S, Moreno V, Gioia-Patricola L, Chabrier A, Guino E, Navarro M, Cambray M, Capella G, Canzian F (2005) Polymorphisms of the dopamine receptor gene DRD2 and colorectal cancer risk. Cancer Epidemiol Biomark Prev 14:1633–1638. https://doi.org/10.1158/1055-9965.EPI-05-0057

Glavin GB (1995) A dopamine D3 receptor agonist, 7-hydroxy-N,N-di-n-propyl-2-aminotetralin, reduces gastric acid and pepsin secretion and experimental gastric mucosal injury in rats. Life Sci 56:287–293. https://doi.org/10.1016/0024-3205(94)00923-6

Glavin GB, Hall AM (1995) Central and peripheral dopamine D1/DA1 receptor modulation of gastric secretion and experimental gastric mucosal injury. Gen Pharmacol 26:1277–1279. https://doi.org/10.1016/0306-3623(95)00009-p

Herak-Perkovic V, Grabarevic Z, Banic M, Anic B, Novosel V, Pogacnik M (2001) Effects of dopaminergic drugs on inflammatory bowel disease induced with 2,4-dinitrofluorbenzene in BALB/c mice. J Vet Pharmacol Ther 24:267–273. https://doi.org/10.1046/j.1365-2885.2001.00343.x

Hernandez DE, Mason GA, Walker CH, Valenzuela JE (1987) Dopamine receptors in human gastrointestinal mucosa. Life Sci 41:2717–2723. https://doi.org/10.1016/0024-3205(87)90464-4

Hernandez DE, Walker CH, Valenzuela JE, Mason GA (1989) Increased dopamine receptor binding in duodenal mucosa of duodenal ulcer patients. Dig Dis Sci 34:543–547. https://doi.org/10.1007/BF01536330

Huang H, Wu K, Ma J, Du Y, Cao C, Nie Y (2016) Dopamine D2 receptor suppresses gastric cancer cell invasion and migration via inhibition of EGFR/AKT/MMP-13 pathway. Int Immunopharmacol 39:113–120. https://doi.org/10.1016/j.intimp.2016.07.002

Huck JH, Freyer D, Bottcher C, Mladinov M, Muselmann-Genschow C, Thielke M, Gladow N, Bloomquist D, Mergenthaler P, Priller J (2015) De novo expression of dopamine D2 receptors on microglia after stroke. J Cereb Blood Flow Metab 35:1804–1811. https://doi.org/10.1038/jcbfm.2015.128

Iishi H, Baba M, Tatsuta M, Okuda S, Taniguchi H (1992) Enhancement of dopaminergic agonist bromocriptine of gastric carcinogenesis induced by N-methyl-N'-nitro-N-nitrosoguanidine in Wistar rats. Br J Cancer 65:351–354. https://doi.org/10.1038/bjc.1992.71

Jandaghi P, Najafabadi HS, Bauer AS, Papadakis AI, Fassan M, Hall A, Monast A, von Knebel Doeberitz M, Neoptolemos JP, Costello E, Greenhalf W, Scarpa A, Sipos B, Auld D, Lathrop M, Park M, Buchler MW, Strobel O, Hackert T, Giese NA, Zogopoulos G, Sangwan V, Huang S, Riazalhosseini Y, Hoheisel JD (2016) Expression of DRD2 is increased in human pancreatic ductal adenocarcinoma and inhibitors slow tumor growth in mice. Gastroenterology 151:1218–1231. https://doi.org/10.1053/j.gastro.2016.08.040

Kaneko K, Iwasaki M, Yoshikawa M, Ohinata K (2010) Orally administered soymorphins, soy-derived opioid peptides, suppress feeding and intestinal transit via gut mu(1)-receptor coupled to 5-HT(1A), D(2), and GABA(B) systems. Am J Physiol Gastrointest Liver Physiol 299:G799–G805. https://doi.org/10.1152/ajpgi.00081.2010

Kashyap P, Micci MA, Pasricha S, Pasricha PJ (2009) The D2/D3 agonist PD128907 (R-(+)-trans-3,4a,10b-tetrahydro-4-propyl-2H,5H-[1]benzopyrano[4,3-b]-1,4-oxazin-9-ol) inhibits stimulated pyloric relaxation and spontaneous gastric emptying. Dig Dis Sci 54:57–62. https://doi.org/10.1007/s10620-008-0335-6

Kim D, Kim W, Jeong S, Kim D, Yoo JW, Jung Y (2019) Therapeutic switching of sulpiride, an anti-psychotic and prokinetic drug, to an anti-colitic drug using colon-specific drug delivery. Drug Deliv Transl Res 9:334–343. https://doi.org/10.1007/s13346-018-00599-7

Kirschstein T, Dammann F, Klostermann J, Rehberg M, Tokay T, Schubert R, Kohling R (2009) Dopamine induces contraction in the proximal, but relaxation in the distal rat isolated small intestine. Neurosci Lett 465:21–26. https://doi.org/10.1016/j.neulet.2009.08.080

Kurosawa S, Hasler WL, Torres G, Wiley JW, Owyang C (1991) Characterization of receptors mediating the effects of dopamine on gastric smooth muscle. Gastroenterology 100:1224–1231

Le Fur G, Phan T, Uzan A (1980) Identification of stereospecific [3H]spiroperidol binding sites in mammalian lymphocytes. Life Sci 26:1139–1148. https://doi.org/10.1016/0024-3205(80)90653-0

Lemmer K, Ahnert-Hilger G, Hopfner M, Hoegerle S, Faiss S, Grabowski P, Jockers-Scherubl M, Riecken EO, Zeitz M, Scherubl H (2002) Expression of dopamine receptors and transporter in neuroendocrine gastrointestinal tumor cells. Life Sci 71:667–678. https://doi.org/10.1016/s0024-3205(02)01703-4

Leng ZG, Lin SJ, Wu ZR, Guo YH, Cai L, Shang HB, Tang H, Xue YJ, Lou MQ, Zhao W, Le WD, Zhao WG, Zhang X, Wu ZB (2017) Activation of DRD5 (dopamine receptor D5) inhibits tumor growth by autophagic cell death. Autophagy 13:1404–1419. https://doi.org/10.1080/15548627.2017.1328347

Li L, Miyamoto M, Ebihara Y, Mega S, Takahashi R, Hase R, Kaneko H, Kadoya M, Itoh T, Shichinohe T, Hirano S, Kondo S (2006a) DRD2/DARPP-32 expression correlates with lymph node metastasis and tumor progression in patients with esophageal squamous cell carcinoma. World J Surg 30(1672–1679):1680–1681. https://doi.org/10.1007/s00268-006-0035-3

Li ZS, Schmauss C, Cuenca A, Ratcliffe E, Gershon MD (2006b) Physiological modulation of intestinal motility by enteric dopaminergic neurons and the D2 receptor: analysis of dopamine receptor expression, location, development, and function in wild-type and knock-out mice. J Neurosci 26:2798–2807. https://doi.org/10.1523/JNEUROSCI.4720-05.2006

Li Y, Zhang Y, Zhang XL, Feng XY, Liu CZ, Zhang XN, Quan ZS, Yan JT, Zhu JX (2019) Dopamine promotes colonic mucus secretion through dopamine D5 receptor in rats. Am J Physiol Cell Physiol 316:C393–C403. https://doi.org/10.1152/ajpcell.00261.2017

Lim HC, Kim YG, Lim JH, Kim HS, Park H (2008) Effect of itopride hydrochloride on the ileal and colonic motility in guinea pig in vitro. Yonsei Med J 49:472–478. https://doi.org/10.3349/ymj.2008.49.3.472

Liu XB, Liu JF (2012) Expression of dopamine receptors in human lower esophageal sphincter. J Gastroenterol Hepatol 27:945–950. https://doi.org/10.1111/j.1440-1746.2012.07100.x

Liu HN, Hirata H, Okuno Y, Okabe M, Furukawa K (2018) Dopamine and serotonin receptors cooperatively modulate pacemaker activity of intestinal cells of Cajal. Chin J Physiol 61:302–312. https://doi.org/10.4077/CJP.2018.BAH607

Lombardi DM, Grous M, Fine CF, Barone FC, Fowler PJ, Phyall WB, Rush JA, Ormsbee HS (1986) DA1 receptor mediates dopamine-induced relaxation of opossum lower esophageal sphincter in vitro. Gastroenterology 91:533–539. https://doi.org/10.1016/0016-5085(86)90619-0

Magro F, Vieira-Coelho MA, Fraga S, Serrao MP, Veloso FT, Ribeiro T, Soares-da-Silva P (2002) Impaired synthesis or cellular storage of norepinephrine, dopamine, and 5-hydroxytryptamine in human inflammatory bowel disease. Dig Dis Sci 47:216–224. https://doi.org/10.1023/a:1013256629600

Magro F, Fraga S, Ribeiro T, Soares-da-Silva P (2004) Decreased availability of intestinal dopamine in transmural colitis may relate to inhibitory effects of interferon-gamma upon L-DOPA uptake. Acta Physiol Scand 180:379–386. https://doi.org/10.1111/j.1365-201X.2004.01260.x

Magro F, Cunha E, Araujo F, Meireles E, Pereira P, Dinis-Ribeiro M, Veloso FT, Medeiros R, Soares-da-Silva P (2006) Dopamine D2 receptor polymorphisms in inflammatory bowel disease and the refractory response to treatment. Dig Dis Sci 51:2039–2044. https://doi.org/10.1007/s10620-006-9168-3

Marmon LM, Albrecht F, Canessa LM, Hoy GR, Jose PA (1993) Identification of dopamine1A receptors in the rat small intestine. J Surg Res 54:616–620. https://doi.org/10.1006/jsre.1993.1094

McKenna F, McLaughlin PJ, Lewis BJ, Sibbring GC, Cummerson JA, Bowen-Jones D, Moots RJ (2002) Dopamine receptor expression on human T- and B-lymphocytes, monocytes, neutrophils, eosinophils and NK cells: a flow cytometric study. J Neuroimmunol 132:34–40. https://doi.org/10.1016/s0165-5728(02)00280-1

Mezey E, Eisenhofer G, Hansson S, Hunyady B, Hoffman BJ (1998) Dopamine produced by the stomach may act as a paracrine/autocrine hormone in the rat. Neuroendocrinology 67:336–348. https://doi.org/10.1159/000054332

Mezey E, Eisenhofer G, Hansson S, Harta G, Hoffman BJ, Gallatz K, Palkovits M, Hunyady B (1999) Non-neuronal dopamine in the gastrointestinal system. Clin Exp Pharmacol Physiol Suppl 26:S14–S22

Missale G, Missale C, Sigala S, Cestari R, Memo M, Lojacono L, Spano P (1990) Evidence for the presence of both D-1 and D-2 dopamine receptors in human esophagus. Life Sci 47:447–455. https://doi.org/10.1016/0024-3205(90)90304-a

Mu J, Huang W, Tan Z, Li M, Zhang L, Ding Q, Wu X, Lu J, Liu Y, Dong Q, Xu H (2017) Dopamine receptor D2 is correlated with gastric cancer prognosis. Oncol Lett 13:1223–1227. https://doi.org/10.3892/ol.2017.5573

Myohanen TT, Schendzielorz N, Mannisto PT (2010) Distribution of catechol-O-methyltransferase (COMT) proteins and enzymatic activities in wild-type and soluble COMT deficient mice. J Neurochem 113:1632–1643. https://doi.org/10.1111/j.1471-4159.2010.06723.x

Nagahata Y, Urakawa T, Kuroda H, Tomonaga K, Idei H, Kawakita N, Yoshizumi K, Saitoh Y (1992) The effect of dopamine on rat gastric motility. Gastroenterol Jpn 27:482–487. https://doi.org/10.1007/BF02777783

Nagahata Y, Azumi Y, Kawakita N, Wada T, Saitoh Y (1995) Inhibitory effect of dopamine on gastric motility in rats. Scand J Gastroenterol 30:880–885. https://doi.org/10.3109/00365529509101595

Nakano K, Higashi T, Hashimoto K, Takagi R, Tanaka Y, Matsushita S (2008) Antagonizing dopamine D1-like receptor inhibits Th17 cell differentiation: preventive and therapeutic effects on experimental autoimmune encephalomyelitis. Biochem Biophys Res Commun 373:286–291. https://doi.org/10.1016/j.bbrc.2008.06.012

Puri S, Ray A, Chakravarti AK, Sen PA (1994) A differential dopamine receptor involvement during stress ulcer formation in rats. Pharmacol Biochem Behav 47:749–752. https://doi.org/10.1016/0091-3057(94)90184-8

Rasheed N, Ahmad A, Singh N, Singh P, Mishra V, Banu N, Lohani M, Sharma S, Palit G (2010) Differential response of A 68930 and sulpiride in stress-induced gastric ulcers in rats. Eur J Pharmacol 643:121–128. https://doi.org/10.1016/j.ejphar.2010.06.032

Shao W, Zhang SZ, Tang M, Zhang XH, Zhou Z, Yin YQ, Zhou QB, Huang YY, Liu YJ, Wawrousek E, Chen T, Li SB, Xu M, Zhou JN, Hu G, Zhou JW (2013) Suppression of neuroinflammation by astrocytic dopamine D2 receptors via alphaB-crystallin. Nature 494:90–94. https://doi.org/10.1038/nature11748

Sigala S, Missale G, Raddino R, Cestari R, Lojacono L, Missale C, Spano PF (1994) Opposing roles for D-1 and D-2 dopamine receptors in the regulation of lower esophageal sphincter motility in the rat. Life Sci 54:1035–1045. https://doi.org/10.1016/0024-3205(94)00414-5

Sookhai S, Wang JH, McCourt M, O'Connell D, Redmond HP (1999) Dopamine induces neutrophil apoptosis through a dopamine D-1 receptor-independent mechanism. Surgery 126:314–322

Tolstanova G, Deng X, Ahluwalia A, Paunovic B, Prysiazhniuk A, Ostapchenko L, Tarnawski A, Sandor Z, Szabo S (2015) Role of dopamine and D2 dopamine receptor in the pathogenesis of inflammatory bowel disease. Dig Dis Sci 60:2963–2975. https://doi.org/10.1007/s10620-015-3698-5

Torres-Rosas R, Yehia G, Pena G, Mishra P, Del RocioThompson-Bonilla M, Moreno-Eutimio MA, Arriaga-Pizano LA, Isibasi A, Ulloa L (2014) Dopamine mediates vagal modulation of the immune system by electroacupuncture. Nat Med 20:291–295. https://doi.org/10.1038/nm.3479

Tsai LH, Cheng JT (1992) The effect of exogenous dopamine on ileal smooth muscle of guinea-pigs. Chin J Physiol 35:133–141

Tsai LH, Cheng JT (1995) Stimulatory effect of dopamine on acid secretion from the isolated rat stomach. Neurosci Res 21:235–240. https://doi.org/10.1016/0168-0102(94)00854-9

Vaughan CJ, Aherne AM, Lane E, Power O, Carey RM, O'Connell DP (2000) Identification and regional distribution of the dopamine D(1A) receptor in the gastrointestinal tract. Am J Physiol Regul Integr Comp Physiol 279:R599–R609. https://doi.org/10.1152/ajpregu.2000.279.2.R599

Walker JK, Gainetdinov RR, Mangel AW, Caron MG, Shetzline MA (2000) Mice lacking the dopamine transporter display altered regulation of distal colonic motility. Am J Physiol Gastrointest Liver Physiol 279:G311–G318. https://doi.org/10.1152/ajpgi.2000.279.2.G311

Wang Q, Ji T, Zheng LF, Feng XY, Wang ZY, Lian H, Song J, Li XF, Zhang Y, Zhu JX (2012) Cellular localization of dopamine receptors in the gastric mucosa of rats. Biochem Biophys Res Commun 417:197–203. https://doi.org/10.1016/j.bbrc.2011.11.084

Wiley J, Owyang C (1987) Dopaminergic modulation of rectosigmoid motility: action of domperidone. J Pharmacol Exp Ther 242:548–551

Wu Y, Hu Y, Wang B, Li S, Ma C, Liu X, Moynagh PN, Zhou J, Yang S (2020) Dopamine uses the DRD5-ARRB2-PP2A signaling axis to block the TRAF6-mediated NF-kappaB pathway and suppress systemic inflammation. Mol Cell 78:42–56. https://doi.org/10.1016/j.molcel.2020.01.022

Xu P, Gildea JJ, Zhang C, Konkalmatt P, Cuevas S, Bigler Wang D, Tran HT, Jose PA, Felder RA (2020) Stomach gastrin is regulated by sodium via PPAR-alpha and dopamine D1 receptor. J Mol Endocrinol 64:53–65. https://doi.org/10.1530/JME-19-0053

Yan Y, Jiang W, Liu L, Wang X, Ding C, Tian Z, Zhou R (2015) Dopamine controls systemic inflammation through inhibition of NLRP3 inflammasome. Cell 160:62–73. https://doi.org/10.1016/j.cell.2014.11.047

Yoshikawa T, Yoshida N (2010) The possible involvement of dopamine D3 receptors in the regulation of gastric emptying in rats. Life Sci 87:638–642. https://doi.org/10.1016/j.lfs.2010.09.027

Zhang XH, Guo H, Xu JD, Li Y, Li LS, Zhang XL, Li XF, Fan RF, Zhang Y, Duan ZP, Zhu JX (2012) Dopamine receptor D1 mediates the inhibition of dopamine on the distal colonic motility. Transl Res 159:407–414. https://doi.org/10.1016/j.trsl.2012.01.002

Zhang XL, Li Y, Liu CZ, Fan RF, Wang P, Zheng LF, Hong F, Feng XY, Zhang Y, Li LS, Zhu JX (2015) Alteration of enteric monoamines with monoamine receptors and colonic dysmotility in 6-hydroxydopamine-induced Parkinson's disease rats. Transl Res 166:152–162. https://doi.org/10.1016/j.trsl.2015.02.003

Zhang Y, Li Y, Zhu JX, Zhao WM (2017) Expression and cellular distribution of dopamine receptors in the rat colonic submucosal plexus. J Capit Med Univ 38(3):411–416

Zheng LF, Song J, Fan RF, Chen CL, Ren QZ, Zhang XL, Feng XY, Zhang Y, Li LS, Zhu JX (2014) The role of the vagal pathway and gastric dopamine in the gastroparesis of rats after a 6-hydroxydopamine microinjection in the substantia nigra. Acta Physiol (Oxf) 211:434–446. https://doi.org/10.1111/apha.12229

Zizzo MG, Mule F, Mastropaolo M, Serio R (2010) D1 receptors play a major role in the dopamine modulation of mouse ileum contractility. Pharmacol Res 61:371–378. https://doi.org/10.1016/j.phrs.2010.01.015

Zizzo MG, Cavallaro G, Auteri M, Caldara G, Amodeo I, Mastropaolo M, Nuzzo D, Di Carlo M, Fumagalli M, Mosca F, Mule F, Serio R (2016) Postnatal development of the dopaminergic signaling involved in the modulation of intestinal motility in mice. Pediatr Res 80:440–447. https://doi.org/10.1038/pr.2016.91

Zizzo MG, Bellanca A, Amato A, Serio R (2020) Opposite effects of dopamine on the mechanical activity of circular and longitudinal muscle of human colon. Neurogastroenterol Motil 32: e13811. https://doi.org/10.1111/nmo.13811

Chapter 4
Dopamine and Gastrointestinal Mucosa Function

Xiao-Yan Feng, Hong Xue, Zi-Hao Guo, Jing-Ting Yan, Sumei Liu, and Jin-Xia Zhu

Abstract The mucosa layer of the gastrointestinal (GI) tract maintains structural integrity, which is essential for keeping internal balance of fluids, electrolytes, and nutrients and defending against invasion of food antigens, microorganisms, and toxins. Dopamine (DA) plays an important role in the modulation of GI mucosa functions, including GI exocrine secretion, fluid and nutrient absorption, ion transport, and barrier maintenance. The GI mucosal layer is densely populated by a myriad of distinct nerve fibers, endocrine cells, and immune cells that sense, integrate, and respond to multiple environmental cues. In this chapter, we focus on the effects of DA on the GI secretion and absorption, ion transport, mucosal barrier, blood flow, and mucosal regulation.

Keywords Dopamine · Gastrointestinal secretion and absorption · Ion transport · Mucosal barrier · Mucosal regulation

---

X.-Y. Feng · J.-T. Yan · J.-X. Zhu
Department of Physiology and Pathophysiology, School of Basic Medical Science, Capital Medical University, Beijing, China
e-mail: fengxy@ccmu.edu.cn; zhu_jx@ccmu.edu.cn

H. Xue
Digestive Laboratory of Traditional Chinese Medicine Research Institute of Spleen and Stomach Diseases, Xiyuan Hospital, China Academy of Chinese Medical Sciences, Beijing, China

Z.-H. Guo
Department of Gastroenterology, Beijing Tong Ren Hospital, Capital Medical University, Beijing, China

S. Liu (✉)
Department of Biology, College of Science and Health, University of Wisconsin-La Crosse, La Crosse, WI, USA
e-mail: sliu@uwlax.edu

J.-X. Zhu (ed.), *Dopamine in the Gut*, https://doi.org/10.1007/978-981-33-6586-5_4

Abbreviations

293B	Trans-6-cyano-4-(*N*-ethylsulfonyl-*N*-methylamino)-3-hydroxy-2,2-dimethyl-chromane
6-OHDA	6-Hydroxydopamine
AB-PAS	Alcian blue-Periodic Acid-Schiff staining
ACh	Acetylcholine
ASF	Autocrine survival factor
CAMK	Calmodulin-dependent kinase
CD	Crohn's disease
CFTR	Cystic fibrosis transmembrane conductance regulators
ChAT	Choline acetyltransferase
CSCs	Cancer stem cells
DA	Dopamine
DARs	Dopamine receptors
DAT	DA transporter
DBS	Duodenal bicarbonate secretion
ENS	Enteric nervous system
FITC-D	Fluorescein isothiocyanate conjugated-dextran
GI	Gastrointestinal
HRP	Horseradish peroxidase
i.c.v.	Intracerebroventricular
i.p.	Intraperitoneal
IBD	Inflammatory bowel disease
ICH	Intracerebral hemorrhage
ILC3s	Intestinal type-3 innate lymphoid cells
I_{SC}	Short-circuit current
JAMs	Junctional adhesion molecules
MDA	Malondialdehyde
MPTP	1-Methyl-4-phenyl-1,2,3,6-tetrahydropyridine
NA	Noradrenaline
NBC	Na^+-HCO_3^- cotransporter
NLRP3	Nucleotide-binding oligomerization domain-like receptor 3
nNOS	Neuronal nitric oxide synthase
NO	Nitric oxide
NOR	Norepinephrine
NPY	Neuropeptide Y
PGE2	Prostaglandin E2
PKA	Protein kinase A
PKC	Protein kinase C
SCO	Scopolamine
SIET	Ion-selective electrode technique
TEA	Tetraethylammonium
TER	Transepithelial resistance
TH	Tyrosine hydroxylase

TLR2	Toll-like receptor 2
TRH	Thyrotropin releasing hormone
TTX	Tetrodotoxin
UC	Ulcerative colitis
VIP	Vasoactive intestinal peptide
VIPR2	VIP receptor 2
ZOs	Zonula occludens
ΔI_{SC}	Change in the short circuit current

4.1 Introduction

The gastrointestinal (GI) mucosa from the mouth to the anus, structurally formed by the epithelium, lamina propria, and muscularis mucosa, is fundamental for the digestive juice secretion and nutritive absorption, accommodates the bodily hemodynamics, and provides host defense (Camilleri 2019). Throughout the GI tract, the mucosa provides a dynamic barrier to allow the passage of certain ions and molecules into the body and restricting the invasion of food antigens, microorganisms, and toxins (Flemstrom and Isenberg 2001).

As an important neurotransmitter and modulator in the GI tract, dopamine (DA) is not only released from enteric neurons but also synthesized within specific parenchyma of the gut (Tian et al. 2008; Feng et al. 2020). DA receptors (DARs) are widely present in the mucosa of GI tract in rodent and human (Hernandez et al. 1987; Wang et al. 2012; Feng et al. 2013). DA stimulates secretion and absorption and enhances intestinal barrier via activating the DARs (Li et al. 2019; Feng et al. 2020).

The oral cavity and esophagus are lined by an impermeable stratified squamous epithelium and receive glandular secretions. Report on the role of DA on the mucosal function in the oral cavity and esophagus is scarce. Many studies have reported that DA influences the secretion of gastric acid and pepsin (Flemstrom and Safsten 1994; Tsai and Cheng 1995). However, whether DA protects gastric mucosa or promotes ulcer is still inconclusive. DA also stimulates ion transport and enhances intestinal barrier in the small and large intestines via different DAR subtypes (Zhang et al. 2008; Li et al. 2019; Feng et al. 2020).

4.2 Dopamine and Gastrointestinal Secretion and Absorption

The GI mucosa is responsible for the secretion of digestive juice and absorption of ingested food and liquids (Martens et al. 2018). Secretion and absorption do not occur uniformly along the GI tract. The gastric mucosa is mainly responsible for the secretion of gastric juice with minimum of absorption. The small intestine is responsible for the absorption of nutrients and water and secretion of intestinal fluids. In the large intestine, there is no nutrient absorption except for some vitamins, but water is absorbed. In addition, there is some level of secretion in the large intestine. We discuss the effects of DA on gastrointestinal secretion and absorption in different regions of the GI tract, including the stomach, small intestine (duodenum, jejunum, ileum), and colon.

4.2.1 Dopamine and Gastric Acid and Pepsin Secretion

The gastric epithelium consists of several distinct cell populations including acid-secreting parietal cells, pepsinogen-producing chief cells, and mucus-producing mucous neck cells (Shao et al. 2011). All the DARs are observed in human and rat gastric mucosa (Wang et al. 2012). DA influences the volume of the gastric juice and the total acid and pepsin output (Glavin and Hall 1995). However, many discrepancies remain in the effects of DA on gastric secretion.

Both in vivo and in vitro techniques have been used in studying gastric secretion. In in vivo studies, animals are subjected to pylorus ligation under anesthesia, and then both gastric acid and pepsin secretions are determined (Dupuy and Szabo 1986). In in vitro studies, the isolated whole stomach is removed immediately after decapitation for measuring gastric secretion (Tsai and Cheng 1990). Recently, Dr. Zhu JX's lab has revealed that H^+ secretion from the mouse gastric corpus mucosa can be precisely measured using the scanning ion-selective electrode technique (SIET) (BIO-IM, Younger USA Sci. & Tech. Co., USA) and real-time pH-stat titration (Radiometer, Copenhagen, Denmark) (Zheng et al. 2020).

4.2.1.1 Dopamine and Gastric Acid Secretion

Many DARs are present in the gastric mucosa, and DA regulates gastric acid secretion in the physiological situation (Caldara et al. 1980; Schrumpf and Linnestad 1982). Both basal and stimulated gastric acid secretion is significantly influenced by DA in in vivo or in vitro settings (Caldara et al. 1978; Tsai and Cheng 1995). Because DA-induced gastric acid secretion is prevented by pretreatment with the antidopaminergic drugs, it confirms the existence of dopaminergic mechanisms in the regulation of gastric acid secretion (Ozdemir et al. 2007).

In an in vitro study, spontaneous acid secretion reaches a steady state after the isolated rat stomach has been incubated for 30–60 min. DA stimulates gastric acid secretion dose-dependently at concentrations from 10 nM to 10 μM. However, the secretion declines when the concentration of DA is higher than 10 μM. Furthermore, DA-induced gastric acid secretion is blocked by the D_1R antagonist, SCH 23390 (Fig. 4.1) (Tsai and Cheng 1995), suggesting that this effect is mediated by D_1R. However, this study has not shown the cellular localization of D_1R in gastric mucosa. In 2012, Dr. Zhu JX's lab has reported that all the subtypes of DARs are expressed in the gastric mucosa; but none of the DAR subtypes is detected in the H^+/K^+-ATPase-positive parietal cells

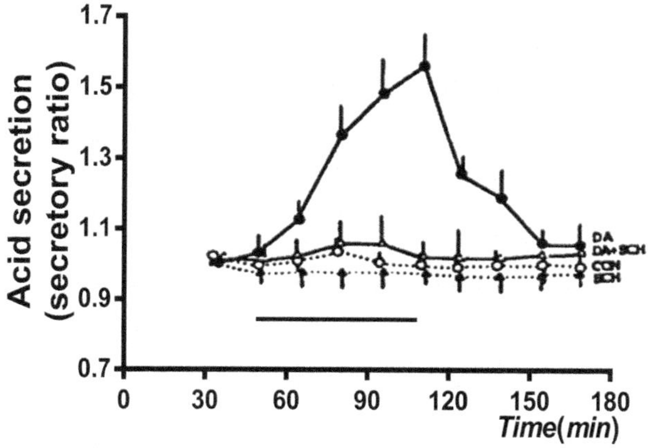

Fig. 4.1 DA-induced gastric acid secretion in the isolated rat stomach. (Tsai and Cheng, Neuroscience Research 1995, 21:235–240.) DA (1 μM)-induced acid secretion was antagonized by the D_1R antagonist, SCH 23390 (SCH, 1 μM). CON, control

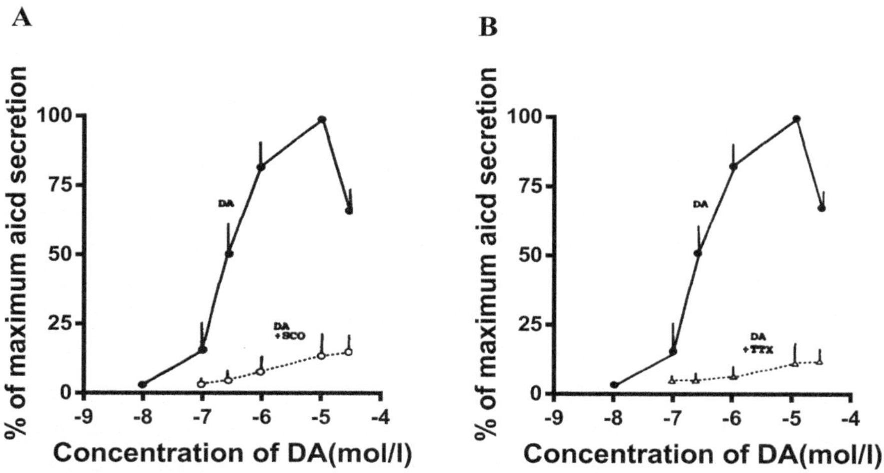

Fig. 4.2 DA-induced gastric acid secretion in the presence of scopolamine or tetrodotoxin. (Tsai and Cheng, Neuroscience Research 1995, 21:235–240.) (**a**) DA (1 μM)-induced acid secretion in the absence and presence of scopolamine (SCO, 1 μM). (**b**) DA (1 μM)-induced acid secretion in the absence and presence of tetrodotoxin (TTX, 1 μM)

(Wang et al. 2012). Thus, the effects of DA on gastric acid secretion may not be a direct consequence of DA action on the parietal cells.

It is well known that acetylcholine (ACh), histamine, and gastrin are key secretagogues for gastric secretion. However, the histamine H_2-receptor antagonist, cimetidine, and the gastrin receptor antagonist, proglumide, do not alter the gastric secretory response to DA (1 μM). The muscarinic receptor antagonist, scopolamine (Fig. 4.2a), and neuronal voltage-gated Na^+ channel blocker, tetrodotoxin

Table 4.1 Effects of D_1R agonist and antagonist on gastric acid secretion in the rat

	Central (i.c.v.) administration		
Peripheral (i.p.) administration	Vehicle (1.0 ml)	SKF38393 (1.0 mg)	SCH23390 (1.0 mg)
Vehicle (1.0 mL/kg)	22.4 ± 1.7	11.8 ± 0.8	29.3 ± 2.1[b]
SKF38393 (4.0 mg/kg)	17.4 ± 2.2[c]	10.1 ± 1.1[a]	19.8 ± 1.9
SCH23390 (4.0 mg/kg)	28.1 ± 2.3[b,d]	20.3 ± 1.8	28.8 ± 2.3[b]

Unit: mmol/h, gastric acid secretion per 2 h of pylorus ligation; $p < 0.05$. (Glavin and Hall, Gen Phormac, 1995, 26(6): 1277–1279)
[a]Less than vehicle
[b]Greater than vehicle
[c]Less than SCH23390 (i.c.v.)
[d]Greater than SKF38393 (i.c.v.)

(Fig. 4.2b), completely inhibit the acid secretion induced by DA (1 μM). Therefore, DA-induced gastric acid secretion may involve predominantly cholinergic neurons in the enteric nervous system.

However, in an in vivo study, DA inhibits, not stimulates, gastric acid secretion (Schubert 2009). Intravenous injection of DA at 40 μg/kg/min significantly inhibits gastric acid secretion in rat (Hovendal et al. 1982). Both basal and stimulated gastric acid secretions in human are significantly inhibited during DA infusion with a significant rebound to pre-infusion values after discontinuing DA (Caldara et al. 1978). Intracerebroventricular (i.c.v.) or intraperitoneal (i.p.) injection of D_1R agonist is associated with significant gastroprotective and antisecretory effects (Glavin and Hall 1995). The D_1R antagonist, SCH 23390 (i.c.v. or i.p.), augments gastric acid secretion and worsens lesion formation (Glavin and Hall 1995). When given both i.c.v. and i.p., the effects of the D_1R agonist and antagonist in gastric acid secretion were much greater than given i.c.v. or i.p. alone (Table 4.1), suggesting that both central and peripheral D_1R is involved in regulating gastric acid secretion.

In various types of experimentally induced ulcers in rats, such as pylorus ligation and water immersion/restraint stress, nonsteroidal anti-inflammatory drugs and reserpine-evoked ulcers, the anti-gastric ulcer activity of the D_2R antagonist, sulpiride, is confirmed (Desai and Parmar 1994; Desai et al. 1999). Intraperitoneal injection of quinpirole (0.0001–0.5 mg/kg), a selective agonist of D2-like family receptor, suppresses the stimulated gastric acid secretion by histamine, pentagastrin, or carbachol in a dose-dependent manner (Eliassi et al. 2008). The D_4R antagonist and putative antipsychotic, clozapine (10.0 mg/kg i.p.), is known to significantly inhibit basal gastric acid secretion (Glavin and Hall 1994). In addition, the inhibitory effect of DA (10 μg/kg/min) on the vagally stimulated gastric acid secretion is mediated by α_2 adrenoceptor and that the inhibitory effect of DA on the blood flow is partly mediated by α_1 adrenoceptor mechanisms (Wallace et al. 1989). These reports suggest that both DARs and adrenergic receptors are involved in DA-induced inhibitory effect on gastric acid secretion (Table 4.2).

High doses of bromocriptine and lergotrile, which stimulate DARs in the periphery as well as in the central nervous system, reduce gastric acid secretion in the rat (Szabo 1979), whereas lower doses have the opposite effect both in man (Caldara

Table 4.2 Effect of dopamine receptors on in vivo gastric secretion

Compounds	Receptor	Gastric secretion		Species	Reference
		Basal	Stimulated		
DA		–	–	Rat	Hovendal et al. (1982)
DA		–	–	Human	Caldara et al. (1978)
SKF38393	D_1R agonist	–		Rat	Glavin and Hall (1995)
SCH 23390	D_1R antagonist		+	Rat	
Quinpirole	D2-like receptor agonist		–	Rat	Eliassi et al. (2008)
Clozapine	D_4R antagonist	–		Rat	Glavin and Hall (1994)
Bromocriptine	DAR activator	–		Rat	Szabo (1979)
Lergotrile					
Bromocriptine	DAR activator	+		Human/Cat	Caldara et al. (1978)
Lergotrile					Hirst et al. (1976)

+: increase gastric secretion; −: decrease gastric secretion

et al. 1978) and in cat (Hirst et al. 1976) (Table 4.2). The central DAR stimulation inhibits intraventricular thyrotropin releasing hormone (TRH)-induced gastric acid secretion in rats (Maeda-Hagiwara and Watanabe 1983).

These different results of DA-induced gastric acid secretion between in vitro and in vivo studies might be due to wide expression of DAR in a variety of organs throughout the body. DA not only directly acts on gastric mucosa, but also indirectly regulates gastric acid secretion through blood flow and neurotransmitter secretion. The central and peripheral DARs are believed to be both involved in the regulation of gastric function and pathology.

4.2.1.2 Dopamine and Gastric Pepsin Secretion

Pepsin is another endogenous component in the gastric juice. However, regulation of gastric pepsin secretion has received relatively little attention. DA is one of the several endogenous mechanisms that can protect the stomach from a stress-related injury. Intracerebroventricular or intraperitoneal injection of D_1R agonist significantly reduces pepsin secretion and protects gastric damage by ethanol (Glavin and Hall 1995) (Table 4.3). Although relatively greater effects are seen upon stimulation of the central D_1R in pepsin secretion, peripheral D_1R cannot be excluded from the "brain–gut axis." The gastroprotective effects of DA are mediated through both peripheral and central D_1R.

In a study by Desai et al. (1999), administration of D_1R agonist fenoldopam or D_2R antagonist sulpiride reduces pepsin output in the gastric juice and decreases ulcer-index values in pylorus ligated rats. The D_1R antagonist SCH 39166 augments the intensity of gastric ulcers, the volume of the gastric contents, the total output of acid and pepsin, and the ulcer index. The D_2R agonist quinpirole significantly

Table 4.3 Effect of D_1R agonist and antagonist on gastric pepsin secretion in rat

Peripheral (i.p.) administration	Central (i.c.v.) administration		
	Vehicle (1.0 ml)	SKF38393 (1.0 mg)	SCH23390 (1.0 mg)
Vehicle (1.0 ml/kg)	23.8 ± 1.8	8.0 ± 0.9[a]	20.4 ± 1.6
SKF38393 (4.0 mg/kg)	10.3 ± 1.8[a,c]	9.2 ± 1.6[a]	20.4 ± 2.0
SCH23390 (4.0 mg/kg)	18.6 ± 2.1[b,d]	23.1 ± 1.9	26.3 ± 2.1

Unit: mg, gastric pepsin secretion per 2 h of pylorus ligation; $p < 0.05$. (Glavin and Hall, Gen Phormac, 1995, 26(6): 1277–1279)
[a]Less than vehicle
[b]Greater than vehicle
[c]Less than SCH23390 (i.c.v.)
[d]Greater than SKF38393 (i/c/v/)

Table 4.4 Effect of DAR agonist and antagonist on gastric pepsin secretion and intensity of gastric lesions in pylorus-ligated rats

Treatment (mg/kg × day)	Number of rats	Pepsin output (μmol/h)	Ulcer index
Fenoldopam (5 × 6)	10	−	−
SCH39166 (10 × 6)	10	+	+
Quinpirole (1 × 6)	5	+	+
Sulpiride (5 × 6)	8	−	−

+: increase gastric pepsin secretion or ulcer index; −: decrease pepsin secretion or ulcer index (Desai et al., J. Pharm. Pharmacol. 1999, 51: 187–192)

increases the ulcer index, with a concurrent increase in acid and pepsin output in pylorus-ligated rats (Table 4.4) (Desai et al. 1999). This study suggests that activation of the D_1R inhibits gastric acid and pepsin secretion, while activation of the D_2R enhances gastric acid and pepsin secretion.

Involvement of dopaminergic and adrenergic mechanisms in gastric acid and pepsin secretion has been investigated in a dog model of vagally innervated and denervated stomach (Guldvog et al. 1984). Acid and pepsin secretions are inhibited in dose–response manner by DA in vagal innervated gastric mucosa. However, phentolamine, a non-selective α adrenergic receptor antagonist, propranolol, a non-selective β adrenergic receptor antagonist, and sulpiride, a D_2R antagonist, either given alone or combined with DA, all manifest different effects on vagal innervated and denervated gastric mucosa (Guldvog et al. 1984). These results indicate that the vagal nervous system is also involved in regulating gastric acid and pepsin secretion, which suggests the existence of a functionally significant dopaminergic brain–gut axis.

Even previous studies suggest a role of DA in the regulation of pepsin secretion, and the specific mechanisms of action by DA need to be further investigated. The study from Dr. Zhu JX's lab has reported the localization of D_1R, D_2R, and D_5R on chief cells, but not on parietal cells, which provides morphological evidence for the regulation of DA on pepsinogen secretion by activating the DARs on chief cells (Wang et al. 2012). The latest research from Dr. Zhu JX's lab also confirms that DA can be derived from gastric parietal cells and released into the gastric juice. Thus, DA in the stomach may promote pepsin secretion by paracrine action on the DARs on chief cells.

4.2.2 Dopamine and Intestinal Ion Transport

The whole intestine is responsible for the transport of electrolytes. The mucosa of the intestinal inner surface is a typical electrolyte-transporting epithelium, which moves large quantity of ions between luminal side and basolateral side (Kunzelmann and Mall 2002). DA-evoked intestinal ion transports are mediated by different subtypes of DARs or adrenoceptors. DA stimulates in vitro ion transport via α_2-adrenergic receptor in the rabbit ileum (Donowitz et al. 1982), regulates ion transport in the canine ileum through D_1R (Marmon et al. 1993), and mediates colonic K^+ secretion across rat distal colon via D_2R and D_4R (Al-Jahmany et al. 2004).

Short-circuit current (I_{SC}) is an in vitro method measured using Ussing chambers. The epithelial tissue is mounted between the two halves of the Ussing chambers and continuously voltage-clamped to zero potential difference by the application of external current, with compensation for fluid resistance. The change in the short circuit current (ΔI_{SC}) is calculated on the basis of the value before and after the stimulation and is normalized as the current per unit area of epithelium (μA/cm^2). The upward deflection in I_{SC} reflects a net electrogenic anion secretion, cation absorption, or a combination of both. The downward deflection in I_{SC} reflects a net electrogenic anion absorption, cation secretion, and/or inhibition of anion secretion.

Currently, Dr. Zhu JX's lab has reported that application of DA to the basolateral side, not apical side, of rat duodenal mucosa in vitro produces a prolonged decrease in I_{SC} (Fig. 4.3a). The sustained downward I_{SC} deflection induced by basolateral application of DA is concentration-dependent (Fig. 4.3b) (Feng et al. 2013).

DA-induced I_{SC} downward deflection is dose-dependently inhibited by basolateral addition of the D1-like receptor antagonist, SCH-23390, while the D2-like receptor antagonist, sulpride, has no effect (Table 4.5). Basolateral application of the D1-like receptor agonist, SKF38393, produces a dose-dependent inhibition in I_{SC}, which mimics the effect of DA (Feng et al. 2013). The clear expression of D_5R in the basolateral side of the duodenal epithelial cells and Brunner's glands (Feng et al. 2013), and DA binding D_5R with much higher affinity than D_1 among the D1-like receptors (Barry et al. 1995), suggests that the DA-induced duodenal I_{SC} response is mostly mediated by D_5R. In addition to DARs, adrenoceptors are also involved in the mediation of DA-induced intestinal ion transport. The antagonists of β_1- and β_2-adrenoceptors have no effect on DA-induced ΔI_{SC}, but the antagonists of α- and β_3-adrenoceptors significantly reduce DA-induced ΔI_{SC} (Feng et al. 2013) (Table 4.5).

DA-induced ΔI_{SC} in the rat duodenum is mediated by the cAMP signaling pathway. MDL12330A, an inhibitor of adenylate cyclase, significantly reduces

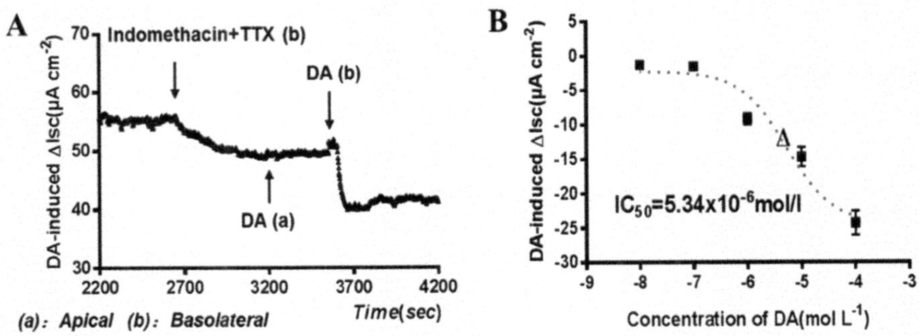

Fig. 4.3 DA-induced I_{SC} response in rat duodenal mucosa. (Feng et al., Transl Res 2013, 161(6), 486–94). (**a**) Original recording showing apical and basolateral application of DA (10 μmol/L) evoked I_{SC} response in rat duodenal mucosa pretreated with indomethacin (10 μmol/L) and tetrodotoxin (1 μmol/L). (**b**) Concentration-response curve of DA-induced decrease of I_{SC} response in rat duodenum

Table 4.5 Effects of different receptor antagonists on DA-induced ΔISC in rat duodenum

Antagonist	Concentration (μmol/L)	*N*	ΔI_{SC} (μA/cm) DA (10 μmol/L)	DA (10 μmol/L) with antagonist
D1-like receptor antagonist, SCH23390	1	8	−11.45 ± 1.12	−5.28 ± 0.94**
	10	8	−11.76 ± 1.34	−2.25 ± 0.27***
D2-like receptor antagonist, sulpride	10	8	−10.41 ± 1.65	−11.20 ± 2.10
α-Adrenoceptor antagonist, phentolamine	1	8	−12.07 ± 1.81	−3.03 ± 1.07**
	10	8	−11.29 ± 1.18	−0.86 ± 0.85***
β_1-Adrenoceptor antagonist, atenolol	100	8	−12.47 ± 1.05	−11.45 ± 1.95
β_2-Adrenoceptor antagonist, ICI118551	10	8	−12.05 ± 1.16	−11.26 ± 1.61
β_3-Adrenoceptor antagonist, SR59230A	1	8	−11.89 ± 1.07	−10.29 ± 1.46
	10	8	−12.47 ± 1.34	−8.38 ± 1.08**

N Number of rats; $^{**}p < 0.01$ $^{***}p < 0.001$

DA-induced ΔI_{SC} (Fig. 4.4a). Both DA and SKF38393 increase intracellular cAMP levels in duodenal mucosa (Fig. 4.4b), suggesting that the cAMP signaling pathway is involved in the DA-induced and D_5R-mediated ΔI_{SC} in the rat duodenum (Feng et al. 2013).

The possible reasons for DA-induced downward deflection in I_{SC} are the net electrogenic cation secretion (such as K^+), anion absorption (such as Cl^- and HCO_3^-), and/or reduction of anion secretion (Zhang et al. 2007). Basolateral addition of DA and SKF38393 fails to induce duodenal bicarbonate secretion (DBS), despite forskolin, a well-known adenylyl cyclase agonist, markedly increases DBS. DA-induced downward ΔI_{SC} in the rat duodenum is not affected by apical

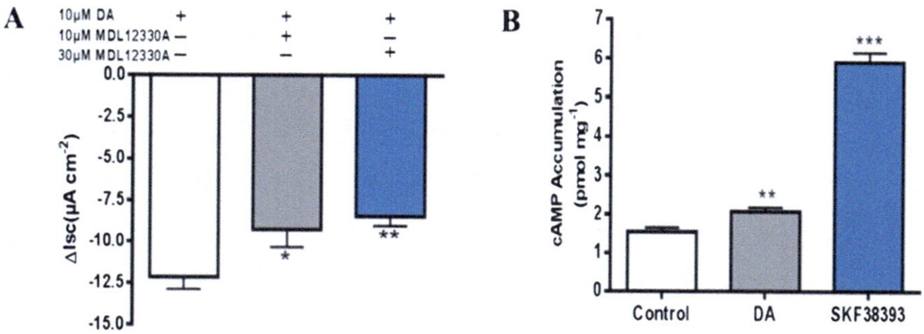

Fig. 4.4 DA-induced I_{SC} response in rat duodenum is mediated by the cAMP pathway. (Feng et al., Transl Res 2013, 161(6), 486–494.) (**a**) The effects of pretreatment with MDL12330A on DA-induced I_{SC} response. (**b**) DA and SKF38393-stimulated intracellular cAMP production in rat duodenal mucosa. $*p < 0.05$, $**p < 0.01$, $***p < 0.001$

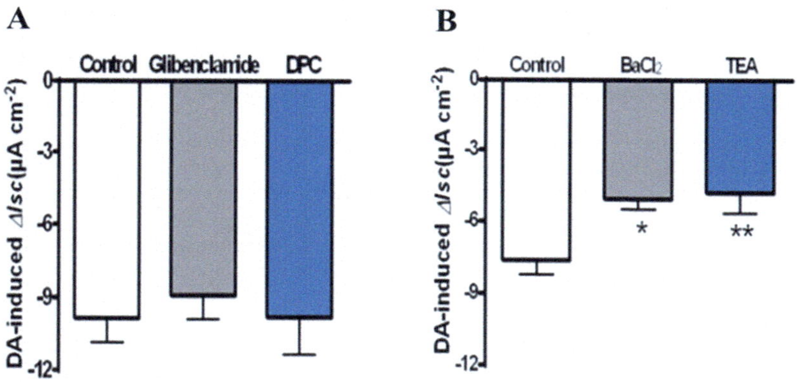

Fig. 4.5 The effects of Cl^- and K^+ channel blockers on the DA-induced I_{SC} response in rat duodenal mucosa. (Feng et al., Transl Res 161(6), 486–494.) (**a**) Effect of DPC (1 mmol/L) and glibenclamide (1 mmol/L) on DA-induced I_{SC} response. (**b**) Effect of $BaCl_2$ (5 mmol/L) and TEA (5 mmol/L) on DA-induced I_{SC} response. $*p < 0.05$, $**p < 0.01$

addition of nonspecific Cl^- channel inhibitors, DPC and glibenclamide (Fig. 4.5a), but is remarkably reduced by apical application of K^+ channel blockers, Ba^{2+} and TEA (Fig. 4.5b), suggesting that DA-induced duodenal downward ΔI_{SC} is mediated by K^+ secretion.

In jejunum, DA (0.1–100 μM) also produces a concentration-dependent decrease in *Isc* with an EC_{50} value of 1.0 μM in young rats and 7.0 μM in adult rats (Fig. 4.6a, b). In the presence of α-adrenoceptor antagonist phentolamine, DA-induced decrease in *Isc* in adult rats is significantly reduced, suggesting the involvement of the α adrenergic receptors in DA-induced changes in *Isc*. Interestingly, in the presence of phentolamine, DA at lower concentrations increases the *Isc* in the jejunum of young rats but not adult rats (Vieira-Coelho and Soares-da-Silva 2000)

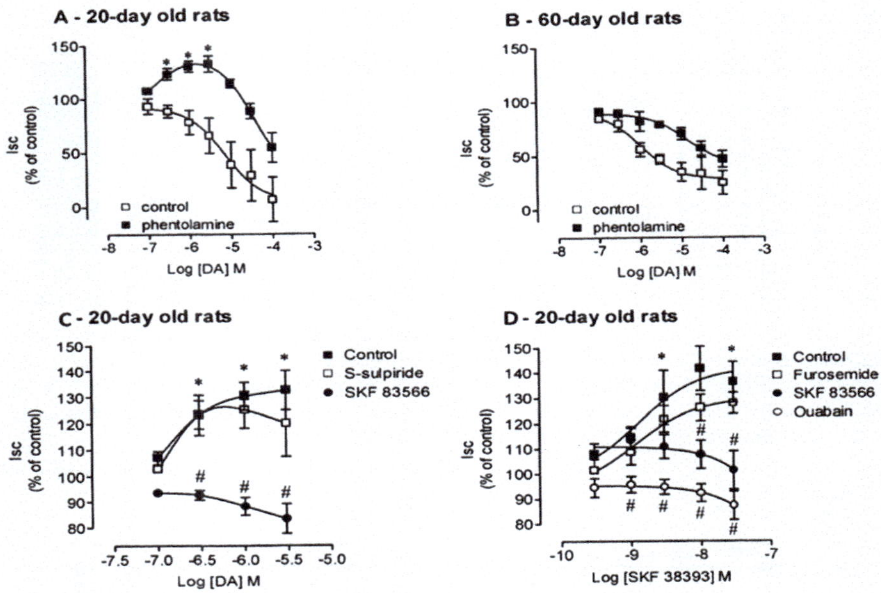

Fig. 4.6 Dose–response curves to DA and SKF 38393 on *Isc* in rat jejunum. (Vieira-Coelho and Soares-da-Silva, BJP, 2000,129, 573–581.) (**a**) DA-induced *Isc* in the presence of phentolamine (0.2 μM) in 20-day-old rats. (**b**) DA-induced *Isc* in the presence of phentolamine in 60-day-old rats. (**c**) DA-induced *Isc* in the presence of SKF83566 (1 mM) or sulpiride (1 μM) in 20-day-old rats. (**d**) SKF38393-induced *Isc* in the presence of SKF83566, ouabain, or furosemide (1 mM) in 20-day-old rats. $^{*}p < 0.05$

(Fig. 4.6a, b). The DA-induced increases in jejunal *Isc* are completely abolished by the D1-like receptor antagonist SKF83566, but not affected by the D2-like receptor antagonist S-sulpiride (Fig. 4.6c, d). The D1-like receptor agonist SKF38393 mimics the effect of DA and induces an increase in *Isc*. SKF38393-induced *Isc* increase is abolished by a Na^+-K^+-ATPase pump inhibitor, ouabain, but not by the Na^+-K^+-$2Cl^-$ transporter blocker, furosemide (Fig. 4.6d). In young rats, DA produces a decrease in cation secretion, and this effect is most probably due to D_1R-mediated inhibition of Na^+-K^+-ATPase (Vieira-Coelho and Soares-da-Silva 2000).

DA is involved in the physiologic regulation of intestinal electrolyte absorption. DA increases net Na^+ and net Cl^- absorption in the rabbit ileum, but has no effect on the residual ion flux (Donowitz et al. 1982). Basolateral application of DA causes a dose-dependent decrease in *Isc*, which is inhibited by the DAR antagonists haloperidol and domperidone, and the α_2-adrenergic receptor antagonist yohimbine, but not altered by the α_1-adrenergic receptor antagonist prazosin and the β-antagonist propranolol. In addition, the α_2-adrenergic agonist clonidine, but not the α_1-agonist methoxamine, causes a dose-dependent decrease in *Isc* (Donowitz et al. 1982).

In in vitro experiments, DA also induces a dose-dependent decrease in *Isc* across the rat distal colon. The α-adrenoceptor antagonist, phentolamine, and the D2-like receptor antagonists, L-741,626 and L-745,870, inhibit DA-induced decrease in *Isc*, suggesting the contribution of both adrenergic and dopaminergic receptors in DA-induced colonic ion transport. The inhibitor of basolateral Na^+-K^+-$2Cl^-$ cotransporter, bumetanide, and the blocker of apical K^+ channels, quinine, inhibit the DA-induced decrease in *Isc*, indicating that stimulation of K^+ secretion contributes to the measured change in *Isc*. DA hyperpolarizes the membrane of colonic epithelial cell and increases cellular K^+ current in patch-clamp experiments. These results demonstrate that DA stimulates colonic K^+ secretion via α-adrenoceptor and D2-like receptor (Al-Jahmany et al. 2004). The freshly isolated rat colonic mucosa close to anus is used in the in vitro experiment of Dr. Zhu JX's lab. The results also show that basolateral addition of DA produces a dose-dependent downward deflection in *Isc* in the late distal colon. However, the DA-induced decrease in *Isc* is not significantly affected by the putative K^+ blockers such as Ba^{2+}, tetraethylammonium (TEA), or 293B (trans-6-cyano-4-(*N*-ethylsulfonyl-*N*-methylamino)-3-hydroxy-2,2-dimethyl-chromane). The response to DA in *Isc* is induced by Cl^- absorption coupled with HCO_3^- secretion in the rat late distal colon (Zhang et al. 2007). Apical addition of nonspecific Cl^- channel/transporter blockers DPC and glibenclamide inhibit basal *Isc* (Fig. 4.7a, b). In some cases, glibenclamide induces an increase in basal *Isc* (Fig. 4.8b, left), suggesting that both anion secretion and absorption exist in basal condition. Furthermore, DPC and glibenclamide significantly inhibit subsequently DA-induced *Isc* (Fig. 4.7c). When glibenclamide is applied after DA treatment, the DA-induced *Isc* is partly reversed with an upward deflection (Fig. 4.7d). The results exclude the possibility that the effect of DA is anti-secretory (Zhang et al. 2007).

Replacing bilateral HCO_3^- completely blocks the DA-evoked ΔI_{SC} (Fig. 4.9a). Reducing basolateral HCO_3^- concentration (from 24 to 3 mM) significantly inhibits the DA-evoked I_{SC} response (Fig. 4.8a, b), implying that DA-stimulated Cl^- absorption depends on basolateral HCO_3^- (Zhang et al. 2007). Taken together, although DA induces a downward *Isc* in rat distal colon, different ion transports contribute to the changes in *Isc*. The possible reasons for that are segmental difference in tissue samples, different techniques used to examine ion transport, and the different concentration of DA used in the studies.

Another study from Dr. Zhu JX's lab shows that β-adrenoceptors (mainly β_2-adrenoceptors), but not DARs, mediate DA-induced ion transport in the rat late distal colon (Zhang et al. 2008). Pretreatment with β-adrenoceptor inhibitor propranolol, not α-adrenoceptor antagonist phentolamine in the serosal side, reduces DA-evoked I_{SC} (Fig. 4.9a), which indicates that the action of DA in the last distal colon is mediated by β-adrenoceptors. The highly selective β_1-adrenoceptor antagonist CGP-20712A and β_2-adrenoceptor antagonist ICI 118,551 cause concentration-dependent inhibition of the DA-induced ΔI_{SC} (Fig. 4.9b–d) (Zhang et al. 2008). Furthermore, the expression levels of β_1- and β_2-adrenoceptors are higher in the mucosa of the colorectum region than the regions away from the anus of rats and humans (Zhang et al. 2010).

In conclusion, both the in vitro and in vivo studies demonstrate that DA promotes intestinal K^+, Cl^-, as well as HCO_3^- transport mainly via D1-like receptors and α_2

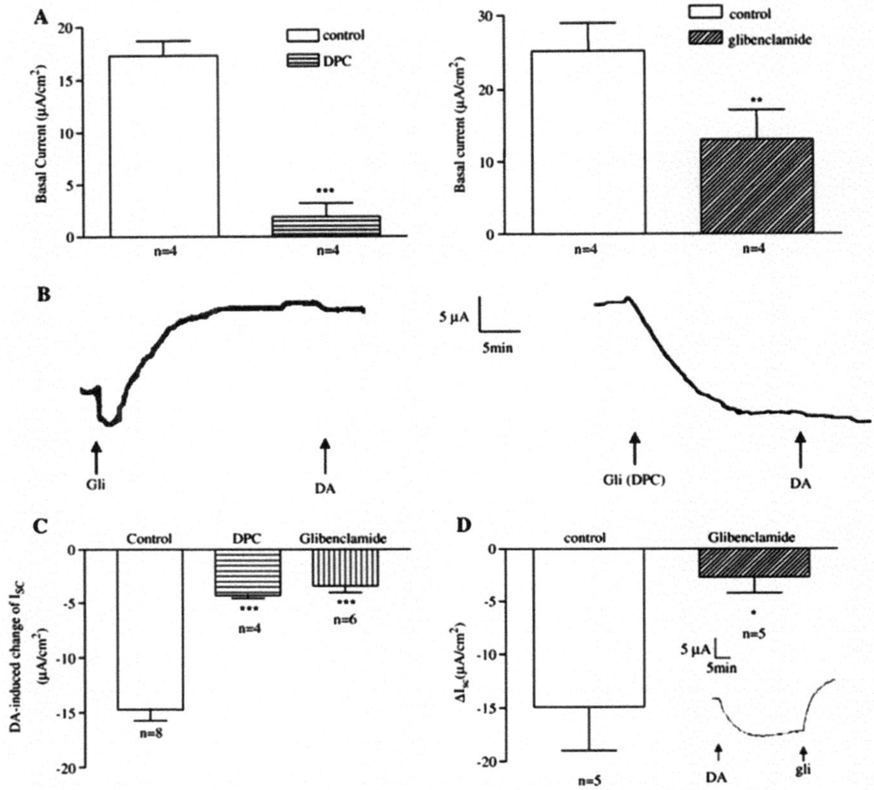

Fig. 4.7 Effect of Cl^- channel blockers/transporter inhibitors on the DA-induced I_{SC} in rat late distal colonic mucosa. (Zhang et al., Eur J Pharmacol 2007, 570:188–195.) (**a**) Effect of glibenclamide and DPC on basal I_{SC} current. (**b**) Representative I_{SC} recordings in response to DPC/glibenclamide (1 mmol/L) and DA (20 µmol/L). (**c**) Effect of glibenclamide and DPC on DA-induced I_{SC} response. (**d**) The DA-induced I_{SC} decrease is reversed by subsequent addition of glibenclamide (1 mmol/L)

adrenergic receptors. The study about colonic ion transport from Dr. Zhu JX demonstrated that absorptive Na^+-K^+-$2Cl^-$ cotransporter (NKCC) isoform 2 expressed in an apical membrane of the colonic epithelia in the rat distal colon and human sigmoid involving in the process of colonic Cl^- absorption coupled with HCO_3^- secretion (Fig. 4.10a) (Zhu et al. 2011). 5-HT, as a key regulator of the gastrointestinal system, induced ion secretion by releasing somatostatin from submucosal neurons (Fig. 4.10b) (Yang et al. 2010).

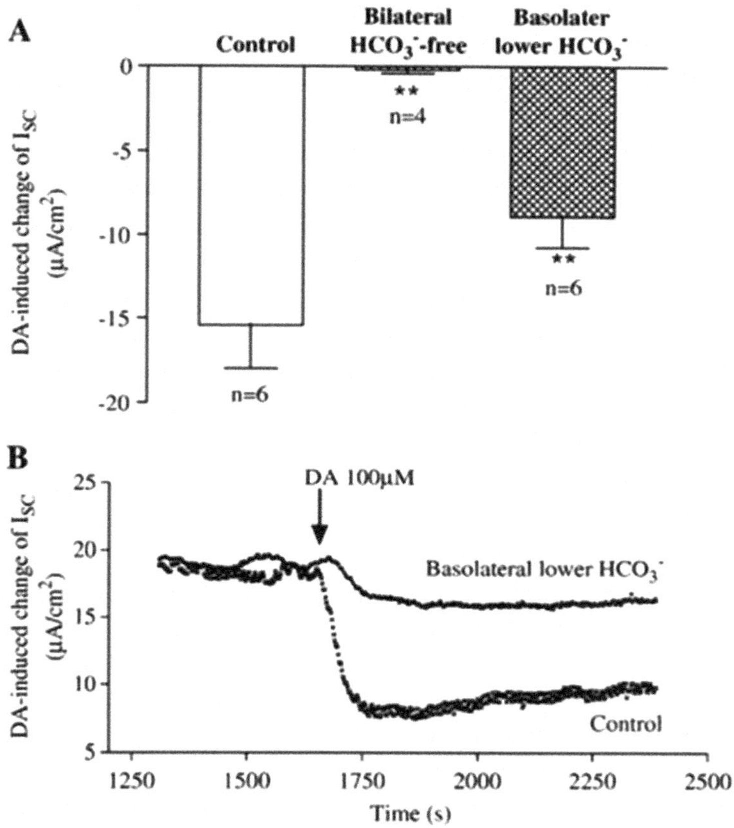

Fig. 4.8 Effect of bicarbonate replacement on the DA-induced *Isc* in rat late distal colonic mucosa. (Zhang et al., Eur J Pharmacol 2007, 570:188–195.) (**a**) Summary of the effects of replacing bilateral bicarbonate or reducing basolateral bicarbonate concentration on DA-induced I_{SC}. (**b**) Representative I_{SC} recording with arrows indicating the time for basolateral addition of DA when the basolateral bicarbonate concentration reducing from 24 to 3 mmol/L. ** $p < 0.01$

4.2.3 Dopamine and Intestinal Absorption

GI epithelium provides a surface for nutrient, water, and drug absorption. Intraperitoneal injection of DA causes stimulation of ileal and colonic water absorption using the single-pass perfusion technique. The effect of DA on the ileum water absorption is antagonized by haloperidol, a DAR antagonist, and yohimbine, a specific α_2-adrenergic antagonist. Intravenous or intraluminal injection of bromocriptine stimulates ileal and colonic water absorption, which is inhibited by haloperidol and yohimbine. Bromocriptine also reverses cholera toxin-induced ileal secretion (Donowitz et al. 1983).

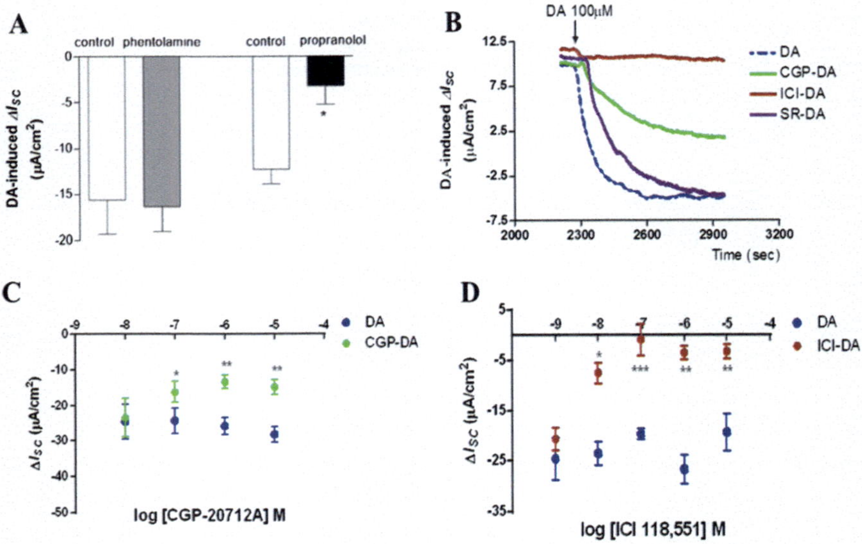

Fig. 4.9 The effect of selective adrenoceptor antagonists on DA-induced ΔI_{SC}. (Zhang et al., Cell tissue Res 2008, 334(1):25–35.) (**a**) Effect of nonselective adrenoceptor antagonists on DA-induced I_{SC}. (**b**) Both the β_1-adrenoceptor antagonist CGP-20712A (CGP, 1 μmol/L) and the β_2-adrenoceptor antagonist ICI 118,551 (ICI, 0.1 μmol/L) inhibit the effect of DA on I_{SC}. (**c**) The selective β_1-adrenoceptor antagonist CGP-20712A (0.01–10 μmol/L) concentration dependently inhibit the effect of DA on I_{SC}. (**d**) The selective β_2-adrenoceptor antagonist ICI 118,551 (0.001–10 μmol/L) concentration dependently inhibit the effect of DA on I_{SC}. $^*p < 0.05$, $^{**}p < 0.01$, $^{***}p < 0.001$

Luminal DA causes significant increase water and electrolyte absorption in canine ileum measured by perfusion with [^{14}C] PEG in in vivo experiment. The α_1 antagonist terazosin and α_2 antagonist yohimbine prevent the luminal DA-induced proabsorptive response (Barry et al. 1995). Luminal DA may serve as a proabsorptive modulator of ileal transport, acting via α_1, α_2, and dopaminergic receptors. The development of similar DA analogs, which maintain the ability to activate mucosal receptors, may be useful in the management of diarrhea associated with some clinical situations as diabetes, short gut syndrome, and small bowel transplantation.

Report about the effect of DA on nutrient absorption in the intestine is very rare. This field requires urgent attention and should be explored in the near future.

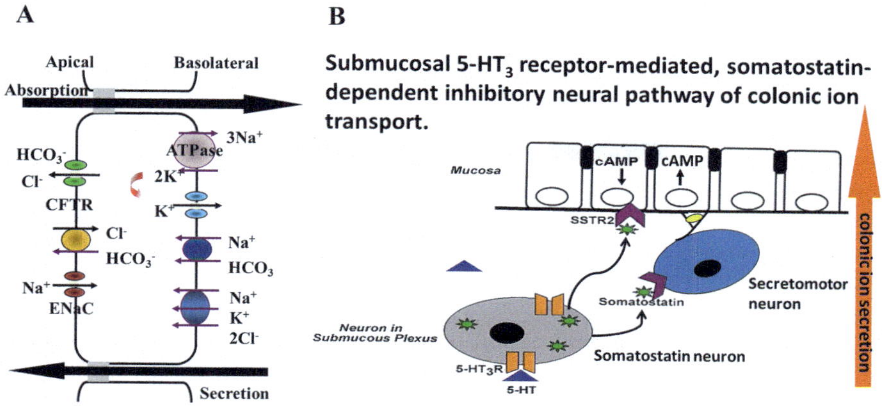

Fig. 4.10 The mechanism of intestinal ion transport. (Zhu et al. Transl Res 2011; 158:146–154; Yang et al. BJP 2010; 159, 1623–1635.) (**a**) A large number of chemical messengers released from enteric nerves, endocrine cells, or immune cells regulate transport proteins in luminal (CFTR Cl^- channels, K^+ channels) and basolateral (Na^+-$2Cl^-$-K^+ cotransporter, K^+ channels, Na^+-K^+-ATPase) membranes. (**b**) Activation of 5-HT_3 receptors in the submucosal plexus leads to the release of somatostatin. Somatostatin binds with to the sst2 receptors on the colonic epithelium and reduces the intracellular cAMP concentration, which leads to a decrease of ion secretion in the rat colon. Somatostatin also inhibits secretomotor neuron activity, which leads to a decrease of ion secretion

4.3 Dopamine and Gastrointestinal Mucosal Barrier

The GI mucosal barrier plays a pivotal role in the uptake and transport of dietary nutrients while simultaneously acting as physical and biochemical barrier to separate the inside of the body from the outside environment (Keely and Talley 2020). The GI epithelium forms a dynamic barrier involving various elements, both intra- and extracellular, which work in a coordinated way to impede the passage of antigens, toxins, and microbial byproducts (Camilleri et al. 2019).

The GI mucosal barrier can be mainly divided into three interacting components: chemical barrier, mechanical barrier, and immune barrier. The chemical barrier is mainly formed by the bicarbonate and mucus secretion, which is the first-line barrier of the mucosal surface. The epithelial cells and intercellular junctional complexes make up the most important physical barrier at the intestinal mucosal surface. The immunological barrier of the intestinal mucosa is located primarily below the intestinal epithelium, which is known as the mucosal immune system. DARs are widely distributed in the intestinal epithelium and almost all immune cell subpopulations. DA plays an important role in the protection of the intestinal mucosal barrier through binding with these DARs.

4.3.1 Dopamine and Gastrointestinal Chemical Barrier

The entire GI mucosal surface is coated with mucus, which serves as the chemical barrier and the first-line of defense (Li et al. 2019). The mucus is a gel layer containing bicarbonate and antimicrobial peptides such as α- and β-defensins. It acts as a lubricant and protective agent against the myriad of damaging agents within the gut lumen. Only one mucus layer is in the small intestinal. However, there are two mucus layers in the colon, a loose outer layer that is inhabited by the gut microbiota and a dense inner layer that is largely devoid of bacteria (Johansson et al. 2011).

Although the mucus layer in the small intestine is thinner than the colon, a large amount of bicarbonate and antimicrobial peptides are secreted to against a heavy load of potential mucosal damage induced by acid, pepsin, certain drugs, and bacterial infection (Bergstrom et al. 2010). The duodenal epithelium secretes bicarbonate at the highest rate comparing to other regions of the small intestine, and the pH of mucus gel is maintained at neutrality in healthy duodenum (Flemstrom and Isenberg 2001).

DBS is currently accepted as the most important defense mechanism against gastric acid-peptic injury (Yin et al. 2018). Multiple bioactive substances, including ACh, 5-HT, and estrogen, are involvement in DBS (Tuo et al. 2011; Yin et al. 2018; Safsten et al. 2006). DA also contributes to the DBS and preservation of the duodenal mucosa and inhibits the formation of duodenal ulcers (Desai et al. 2008; Feng et al. 2020).

Continuous intravenous infusion of DA or the D1-like receptor agonist SKF-38393 causes a dose-dependent increase in DBS in rats; while bolus intravenous injection of different dosage of DA manifests a different pattern of DBS with lower doses of DA induce a dose-dependent increase in DBS, but higher dose of DA causes a transient, small decrease in DBS. Although peripheral application of D2-like receptor antagonist, domperidone (i.v.), has no significant effect on basal DBS, it inhibits SKF-38393-induced increase in DBS (Flemstrom et al. 1993) (Table 4.6).

In the latest study of Dr. Zhu JX's lab, apical but not basolateral addition of DA to the rat duodenum produces a concentration-dependent upward DBS by in vitro pH titration experiment (Fig. 4.11A, B). DA also significantly increases the steady-state pHi in the isolated duodenal upper villous epithelium (Fig. 4.11C, D) (Feng et al. 2020), indicating that the intracellular HCO_3^- production is increased and helpful in increasing HCO_3^- efflux. The mechanism of HCO_3^- transport in duodenal enterocyte has previously been confirmed (Parker and Boron 2013). HCO_3^- inside the enterocytes may come from two sources, the HCO_3^- produced by the carbonic anhydrase reaction inside the enterocytes and the HCO_3^- imported by the basolateral Na^+-HCO_3^- cotransporter (NBC). Secretion of HCO_3^- to the lumen of the duodenum mainly involves the contributions of apical membrane Cl^-/HCO_3^- exchangers and cystic fibrosis transmembrane conductance regulators (CFTR) (Flemstrom and Isenberg 2001).

Table 4.6 The effects of DA and its related receptor agonist/antagonists on duodenal bicarbonate secretion

Compound	Route	Dose of drugs	DBS	Species	Reference
DA	Continuous IV infusion	50 ~ 250 μg/kg/h	Increase DBS	Rat	Flemstrom et al. (1993)
	Bolus IV injection	5 ~ 50 μg/kg	Increase DBS		
	Bolus IV injection	500 μg/kg/h	Decrease DBS		
D1-like receptor agonist SKF-38393	Bolus IV injection	10–200 μg/kg	Increase DBS		
D2-like receptor antagonist domperidone	Bolus IV injection	5, 10, 20 μg/kg	No effect on basal DBS, but inhibits SKF-38393-induced DBS		
DA	Apical addition	0.01–100 μM	A dose-dependent increase of DBS	Rat	Feng et al. (2020)
D2-like receptor agonist quinpirole	Apical addition	0.01–100 nM	A dose-dependent increase of DBS		
Non-selective D2-like receptor antagonist sulpiride	Apical pretreatment	10 μM	Inhibits DA-induced DBS		
Specific D_2R antagonist L741,626	Apical pretreatment	10 μM	Inhibits DA-induced DBS		
Nonselective D1-like receptor antagonist SCH-23390	Apical pretreatment	10 μM	No effect on DA-induced DBS		
α_1-Adrenocepter agonist phenylephrine	Continuous IV infusion	100, 500 μg/kg/h	A dose-dependent increase of DBS	Rat	Nylander and Flemstrom (1986)
α_1-Adrenocepter antagonist prazosin	Bolus IV injection	0.5 mg/kg	Inhibits phenyleph-rine- induced DBS		
α_2-Adrenocepter agonist clonidine	Continuous IV infusion	0.7, 15 μg/kg	Inhibits basal and phenylephrine-induced DBS		
α_2-Adrenocepter agonist clonidine	Continuous IV infusion	150 μg/kg	Inhibits acid-induced DBS	Human	Knutson et al. (1989)

DBS duodenal bicarbonate secretion, *IV* Intravenous

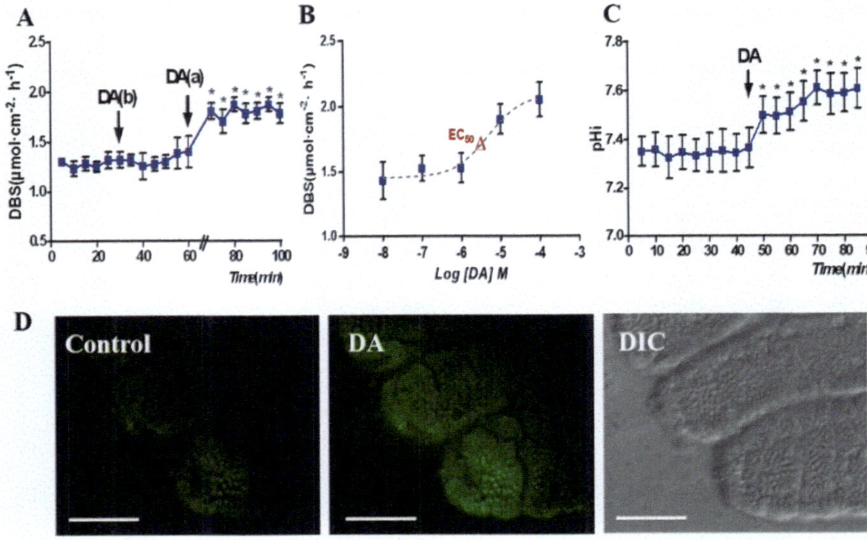

Fig. 4.11 DA increases duodenal HCO_3^- secretion (DBS) in rats. (Feng et al., BJP 2020, 177:3258–3272.) (**A**) Apical (a) and basolateral (b) addition of DA increases DBS. (**B**) Apical addition of DA causes a dose-dependent DBS. (**C**) DA increases intracellular pH in the chorionic epithelial cells of the rat duodenum. (**D**) The distribution of pH-sensitive microfluorometry probe BCECF-AM after addition of DA. Differential interference contrast (DIC) microscopic image showing the three-dimensional morphology of a duodenal villus. Scale bars: 100 μm. $^*p < 0.05$

DARs are widely present in the rat duodenal mucosa. D_2R has the most abundant level among all five subtypes of DARs (Fig. 4.12a). D_2R immunoreactivity is principally distributed on the apical sides of intestinal crypts and Brunner glands (Fig. 4.12b) (Feng et al. 2020). Apical DA-induced DBS is completely inhibited by apical application of D2-like receptor antagonist, sulpiride, and D_2R-specific antagonist, L741, 626, while D1-like receptor antagonist, SCH-23390, has no effect (Fig. 4.12c). In addition, apical addition of D2-like receptor agonist, quinpirode, mimics the effect of DA in DBS, but fails to enhance DBS and pHi in D_2R knockdown mice and in $D_2R^{-/-}$ mice (Fig. 4.12d–f) (Feng et al. 2020).

DBS is prominently controlled by cytosolic cAMP and Ca^{2+} signaling pathways (Tuo et al. 2009; Tuo et al. 2011). MDL12330A, an inhibitor of adenylate cyclase, has no effect on DA- or quinpirole-induced DBS. However, DA-induced DBS is completely blocked by intracellular calcium chelator BAPTA-AM (Fig. 4.13a). Apical addition of quinpirole significantly increases the protein levels of p-PI3K and p-AKT in WT mice but not in $D_2R^{-/-}$ mice (Fig. 4.13b, c). Both BAPTA-AM and the selective PI3K inhibitor LY294002 inhibit quinpirole-stimulated duodenal phosphorylation of Akt in mice (Fig. 4.13d) (Feng et al. 2020), suggesting that DA increases in vitro DBS via apical D_2R and Ca^{2+}/PI3K/Akt pathway in the rodent duodenum (Fig. 4.14).

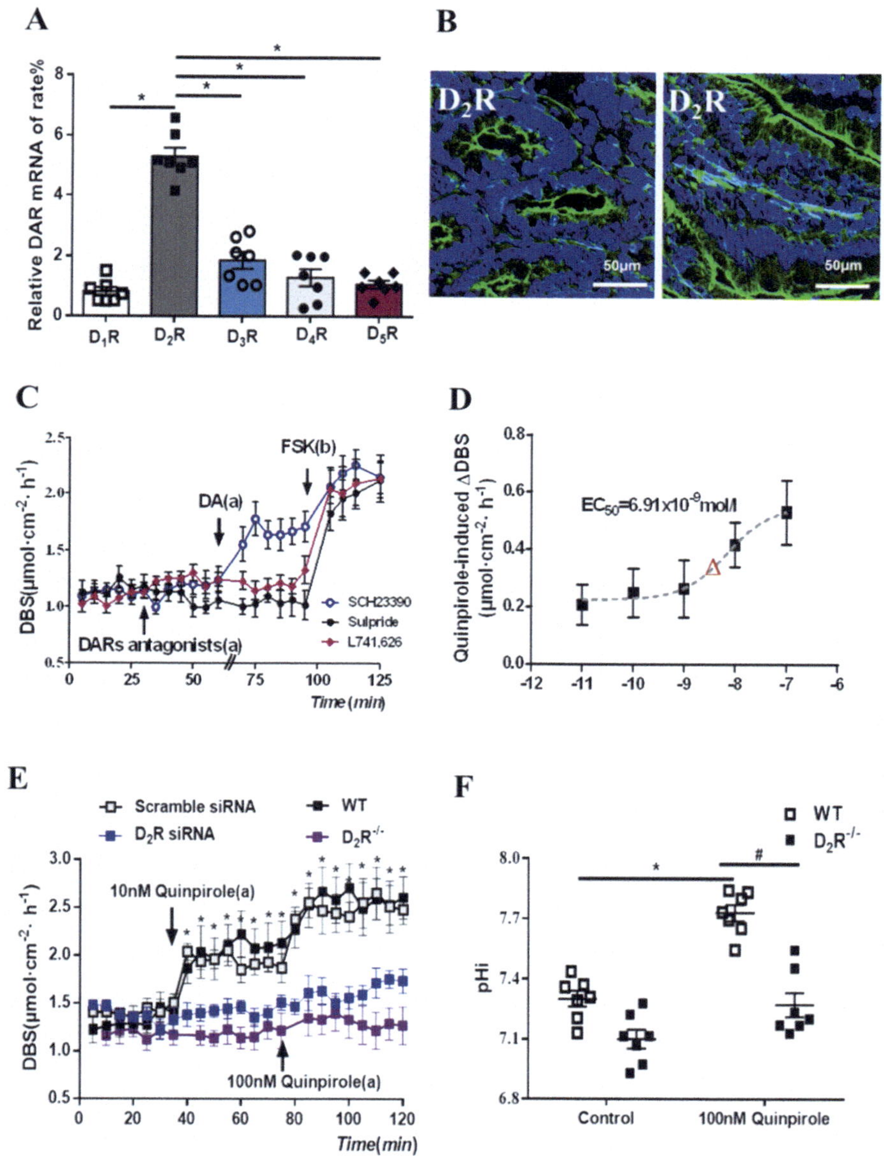

Fig. 4.12 DA increases duodenal HCO_3^- secretion (DBS) via apical D_2 receptors. (Feng et al., BJP 2020, 177:3258–3272.) (**a**) The mRNA expression of DARs in the duodenal epithelium as analyzed by real-time RT-PCR. β-Actin was used as an internal control. (**b**) Immunofluorescence of D_2R in the rat duodenum. (**c**) Effects of DAR antagonists on the response to apical addition of DA-induced DBS. (**d**) Changes in DBS after apical addition of the D2-like receptor agonist quinpirole. (**e**) DBS in the control (scramble siRNA), in vivo D_2R siRNA knockdown mice and the wild-type (WT), $D_2R^{-/-}$ mice after apical addition of quinpirole. (**f**) Intracellular pH in the chorionic epithelial cells of $D_2R^{-/-}$ mice after addition of quinpirole. $^*p < 0.05$ vs. control or WT mice, $^\#p < 0.05$ vs. $D_2R^{-/-}$ mice

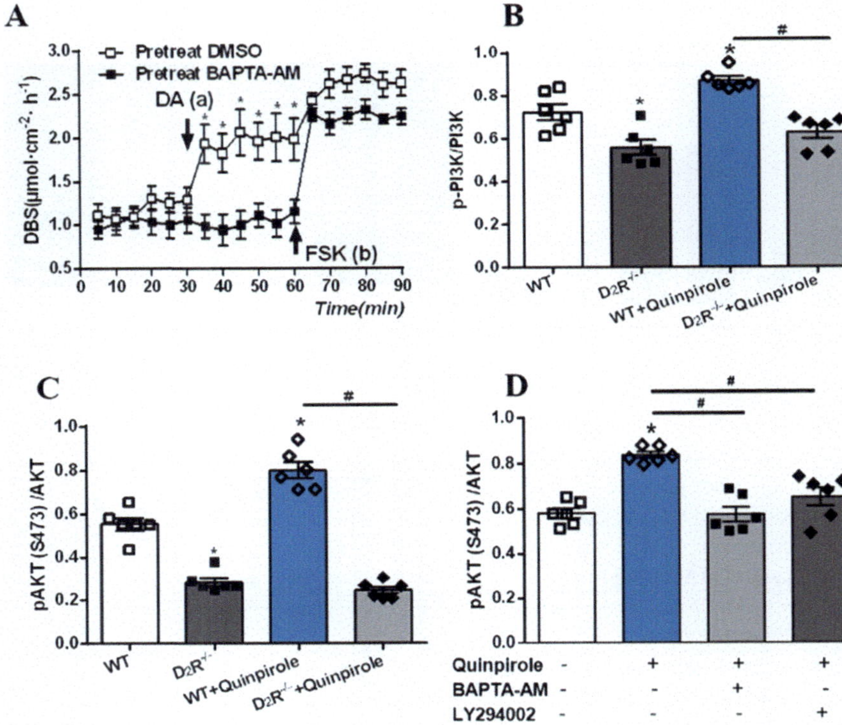

Fig. 4.13 The calcium/PI3K/Akt pathway is involved in DA-induced DBS. (Feng et al., BJP 2020, 177:3258–3272.) (**a**) Effects of the intracellular calcium chelator BAPTA-AM on DA-induced DBS. (**b**) Protein levels of PI3K and p-PI3K in the duodenal mucosa preparations of WT and $D_2R^{-/-}$ mice after treatment with quinpirole. (**c**) The ratio of phosphorylated AKT (S473) to AKT in the duodenal mucosa preparations of WT and $D_2R^{-/-}$ mice after treatment with quinpirole. (**d**) The ratio of phosphorylated AKT (S473) to AKT in the duodenal mucosa preparations after pretreatment with BAPTA-AM and LY294002. $^{*}p < 0.05$, $^{\#}p < 0.05$

Besides DARs, activation of adrenergic receptors also modulates DBS. The α_1-adrenocepter agonist phenylephrine stimulates DBS in rat (Nylander and Flemstrom 1986); however, peripheral α_2-adrenocepter stimulation results in a strong inhibition of DBS in rat, cat, and human (Nylander and Flemstrom 1986; Knutson et al. 1989). The effects of DA and its related receptors on DBS are quite different in different species or drug administration routes (Table 4.6). DA also reduces indomethacin-induced small intestinal lesions, and this effect is abrogated by D_2R antagonist, domperidone (Yasuda et al. 2011).

In colon, the latest study of Dr. Zhu JX's lab demonstrates that DA increases rat distal colonic mucus secretion in a dose-dependent manner (Fig. 4.15a). Pretreatment with the D1-like receptor antagonist SCH23390, not the D2-like receptor antagonist sulpiride, inhibits DA-induced mucus secretion in distal colon (Fig. 4.15b). Semiquantitative PCR and ELISA results show that DA significantly

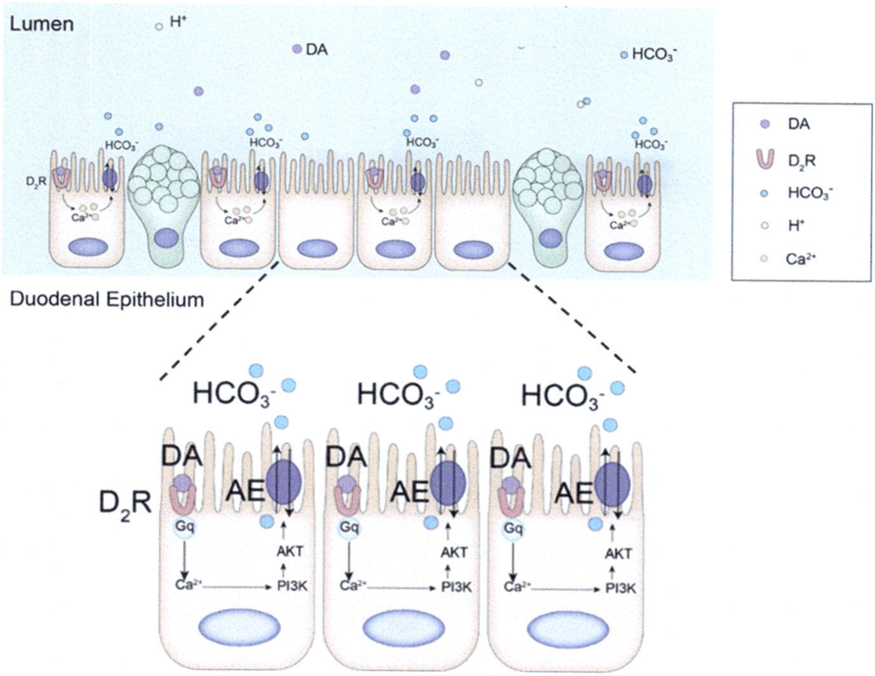

Fig. 4.14 DA-induced DBS via an apical D_2R- and calcium-dependent pathway. DA in the duodenal lumen binds to D_2Rs on the duodenal apical epithelium, thereby increasing bicarbonate secretion through a calcium/PI3K/AKT-dependent pathway

elevates the mRNA level of MUC2 in rat colonic mucosa tissue and the MUC2 content in (Camilleri 2019) incubation medium (Fig. 4.15c, d). D1-like receptor agonist SKF38393 significantly elevates both the MUC2 content in incubation medium and the intracellular cAMP level of colonic mucosa (Fig. 4.15d) but decreases the MUC2 content in colonic mucosa (Fig. 4.15f). Alcian blue-Periodic Acid-Schiff staining (AB-PAS) is a well-known method to observe the expression of mucus in the colon. The number of AB-PAS-positive goblet cells in crypts of the colonic mucosa preparations is significantly reduced after treatment with SKF38393 (Fig. 4.15e) (Li et al. 2019).

D_5R is the only DAR subtype found in the crypts of the rat colonic mucosa, where a lot of goblet cells are located, and D_5R-immunoreactivity positive cells are almost 100% overlapped with the MUC2-marked goblet cells along each crypt (Li et al. 2019). Yang et al. developed several human D_5R transgenic mice ($C_{57}BL/6J$ background), including hD_5^{WT}, D_5^{F173L}, and D_5^{S390G}. MUC2 content in the colonic mucosa of D_5R knocked down mice (hD_5^{F173L} mice) significantly decreases compared with human wild-type (hD_5^{WT}) transgenic mice by ELISA. The number of AB-PAS-positive goblet cells along the whole colonic crypts is significantly decreased in hD_5^{F173L} mice (Fig. 4.16a, b). These results reveal that DA promotes

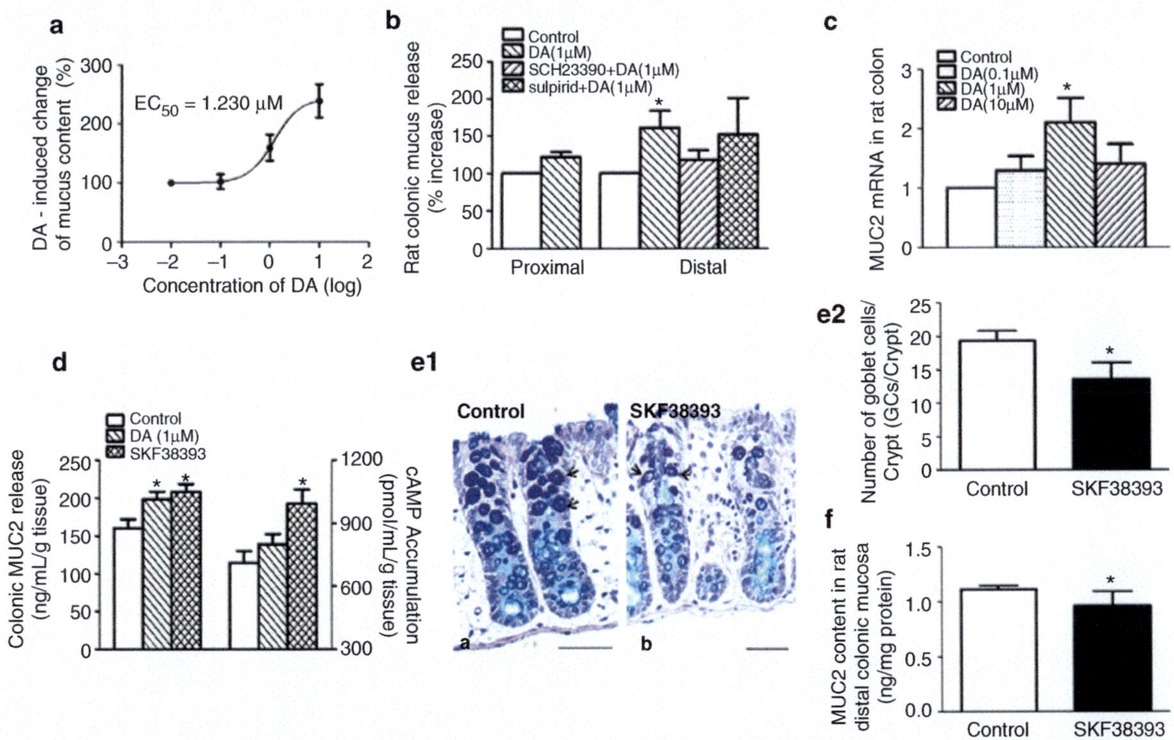

Fig. 4.15 Effects of DA on mucus secretion in the rat colonic mucosa. (Li Y et al. *Am J Physiol Cell Physiol 2019, 316: 393–403.*) (***a***) The dose-responsive curve for DA-induced increase in distal colonic mucus content in the incubation medium. (***b***) Effect of DA on colonic mucus secretion between proximal and distal colon of rat and the effect of DAR antagonist on DA-induced change of distal colonic mucus secretion. (***c***) The mRNA level of MUC2 in DA-stimulated distal colonic mucosa detected by RT-PCR. (***d***) MUC2 content in the incubation medium and intracellular cAMP level in the distal colon. (***e***) AB-PAS-positive goblet cells in the SKF38393-stimulated colonic mucosa. Scale bars, 50 μm. (***f***) MUC2 content in the colonic mucosa detected by ELISA assay. $^{*}p < 0.05$

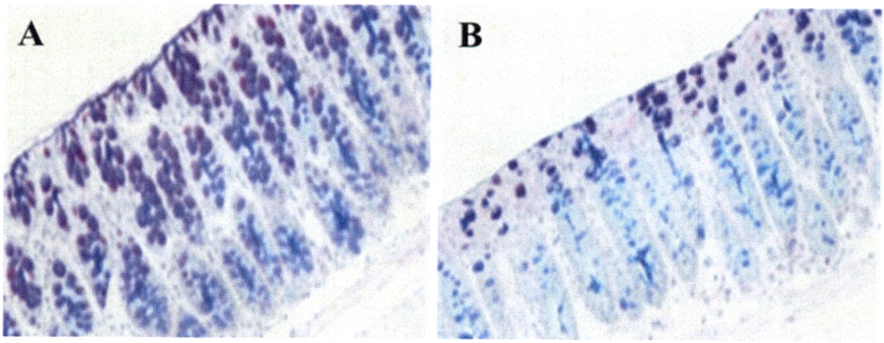

Fig. 4.16 Distal colonic mucus in D_5R-downregulated transgenic mice. (***a***) AB-PAS-positive goblet cells of WT (hD_5^{WT}). (***b***) AB-PAS-positive goblet cells of D_5-downregulated transgenic mice (hD_5^{F173L})

distal colonic mucus secretion via the D_5R. These findings provide new insight into the gut modulation and protection by DA and potential drug targets for the improvement of gut biochemical barrier function.

4.3.2 Dopamine and Gastrointestinal Mechanical Barrier

The intestinal epithelium forms a continuous and polarized mechanical barrier with several types of epithelial cells: absorptive cells, goblet cells, enteroendocrine cells, tuft cells, and Paneth cells. The epithelial cells and intercellular junctional complexes make up the most effective barrier to separate the inner and the outer environments. The mechanical barrier regulates the movement of ions, solutes, and macromolecules through the transepithelial and paracellular pathways to maintain homeostasis (Turner 2009; Neunlist et al. 2013). The transepithelial resistance (TER) measured by the Ussing chamber system and epithelial permeability to FITC-dextran are the common physiological indexes to evaluate paracellular permeability (Tolstanova et al. 2015; Wallon et al. 2005). The horseradish peroxidase (HRP) flux across the epithelium by measurement HRP probes in Ussing chamber is usually used to evaluate transcellular permeability (Hamilton et al. 2015).

The results of Dr. Zhu JX's lab indicate that DA decreases duodenal TER and increases FITC-dextran permeability, which are reversed by treatment of D1-like receptor antagonist SCH-23390, but not by the D2-like receptor antagonist sulpride (Fig. 4.17a, b) (Feng et al. 2017). A strong D_5R-immunoreactive signal is observed in the basolateral side of the duodenal villus (Feng et al. 2017), and DA has much higher binding affinity of D_5R than other DARs (Turner 2009), which suggest that D_5R mediates DA-induced increase in the duodenal epithelial permeability of rats.

The human D_5R transgenic mice developed by Yang et al. are used to confirm the role of D_5R on DA-induced increase in duodenal epithelial permeability (Yang et al. 2015). Compared with D_5^{WT} mice, duodenal mucosa from D_5^{S390G} mice manifests

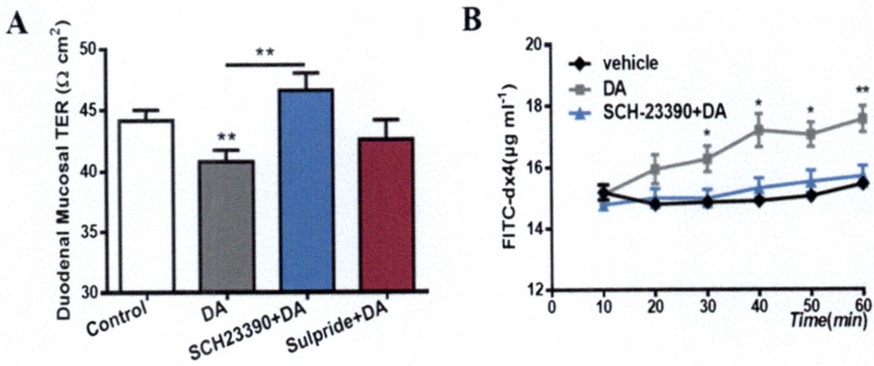

Fig. 4.17 Involvement of DAR in rat duodenal epithelial permeability. (Feng et al., Acta Physiol, 2017, 220, 113–123.) (**a**) The duodenal TER after basolateral addition of DA and pretreatment with either SCH-23390 or sulpride. (**b**) The duodenal FITC-dextran permeability after basolateral addition of DA and pretreatment with SCH-23390. *$p < 0.05$, **$p < 0.01$

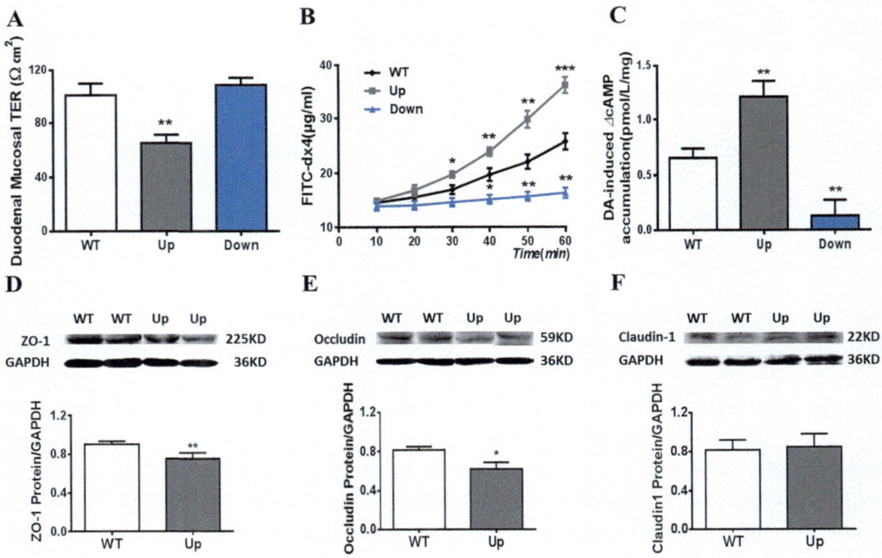

Fig. 4.18 Duodenal epithelial permeability in human D_5R transgenic mice. (Feng et al., Acta Physiol, 2017, 220, 113–123.) (**a**) Duodenal mucosal TER of transgenic mice. (**b**) Duodenal FITC-dextran permeability of transgenic mice. (**c**) Effects of DA-induced intracellular cAMP level in the duodenum of transgenic mice. (**d–f**) Protein expressions of ZO-1, occludin and claudin1 in the duodenum of transgenic mice. *$p < 0.05$, **$p < 0.01$, ***$p < 0.001$

up-regulated D_5R but D_5R in D_5^{F173L} mice show down-regulated D_5R (Yang et al. 2015; Liu et al. 2015). It is worth to emphasize that D_5^{S390G} mice have an increased FITC-dextran permeability (Fig. 4.18a) and remarkably decreased TER (Fig. 4.18b) in the duodenal mucosa, indicating attenuated duodenal mucosa barrier function. In contrast, D_5^{F173L} mice manifest depressant FITC-dextran permeability without significant alteration in TER (Fig. 4.18a, b). Basolateral addition of DA produces a higher intracellular cAMP level of duodenal mucosa in D_5^{S390G} mice than that of D_5^{WT} mice, while the opposite result is observed in D_5^{F173L} mice (Fig. 4.18c). Several studies have reported the correlation of DA or intracellular cAMP pathway with the mucosa barrier. Increase in the intracellular cAMP has been found to inhibit barrier restoration by mean of TER monitor (Zimmerman et al. 2012). Stimulating protein kinase A (PKA) strikingly increases paracellular permeability, and conversely inhibiting PKA diminishes paracellular permeability (Kovbasnjuk et al. 1995). DA has also been reported to increase the conductance of tight junction by binding with D1-like receptor in a cAMP- and PKA-dependent manner (Hu et al. 2010; Piccolino et al. 1984).

Tight junctions are the most important component for the constitutive barrier of epithelial cells, which regulate the permeability of the barrier by tightly sealing the paracellular junctions (Oshima and Miwa 2016). Tight junction proteins include claudins, occludin, and junctional adhesion molecules (JAMs). These molecules bind directly to the scaffold protein zonula occludens (ZOs). The tight junction proteins play primary roles in the membrane barrier function through interaction of ZO proteins with the membrane-spanning proteins claudin and occludin (Suzuki 2013). ZO-1 and occludin, as markers of tight junction integrity, are often detected in epithelial barrier (Rao et al. 2002; Blasig et al. 2011). They are mainly responsible for regulating permeability to larger molecules. Different claudin molecules play different roles in paracellular permeability. Some claudins (1, 3, 5, 9, 11) form sealing and decrease paracellular permeability. Other claudins (2 ,7, 10, 15, 16) form pores and enhance paracellular permeability in a size and charge-selective manner (Sharkey and Savidge 2014). The high duodenal permeability in D_5^{S390G} mice is mainly due to the down-regulation of tight junction proteins ZO-1 and occluding (Fig. 4.18d–f).

6-Hydroxydopamine (6-OHDA) rats, a kind of Parkinson's disease model, have enhanced expression of dopaminergic markers and significantly high DA content in the gut (Tian et al. 2008; Zheng et al. 2014; Zhang et al. 2015). 6-OHDA rats also show an increased FITC-dextran permeability (Fig. 4.19a) and decreased TER (Fig. 4.19b) in the duodenal mucosa. D_5R, but not D_1R is significantly increased in the duodenal mucosa of 6-OHDA rats (Fig. 4.19c–e). Moreover, a much lower expression of ZO-1 and occludin, but not claudin1 is detected in duodenal mucosa of 6-OHDA rats (Fig. 4.19f–g). The impaired duodenal mucosal barrier of 6-OHDA rats is similar to that of D_5^{S390G} mice. Taken together, DA decreases tight junction proteins ZO-1 and occludin, which resulted in elevated duodenal paracellular permeability, via a D_5R-mediated and cAMP-dependent pathway (Fig. 4.20).

The whole GI permeability can be assessed in vivo by determining the permeability of FITC-dextran with a defined molecular size in the blood plasma. TER is also a common physiological index used to evaluate the mucosal mechanical barrier. The results of Dr. Zhu JX's show that 6-OHDA rats have the increased FITC-dextran

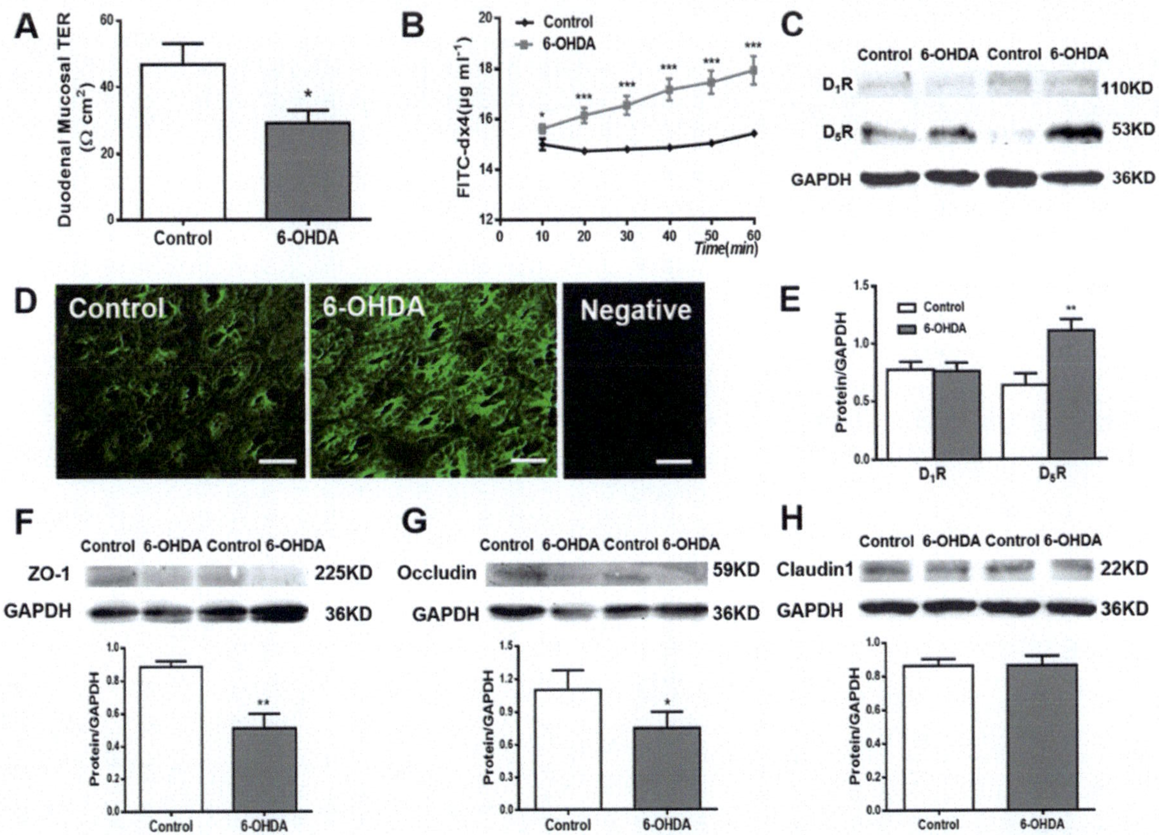

Fig. 4.19 Duodenal epithelial permeability of 6-OHDA rat. (Feng et al., acta physiol, 2017, 220, 113–123.) (**a**) Duodenal TER of control and 6-OHDA rats. (**b**) Duodenal FITC-dextran permeability of control and 6-OHDA rats. (**c**) Representative protein band of D_1R and D_5R in the duodenum of control and 6-OHDA rats (GAPDH, D-glyceraldehyde-3-phosphate dehydrogenase). (**d**) The distribution of D_5R in the duodenum. Scale bar: 50 μm. (**e**) Protein expression of D_1R and D_5R in the duodenum of control and 6-OHDA rats. (**f–h**) Protein expression of ZO-1, occludin, and claudin1 in the duodenal mucosa of control and 6-OHDA rats. $^*p < 0.05$, $^{**}p < 0.01$, $^{***}p < 0.001$

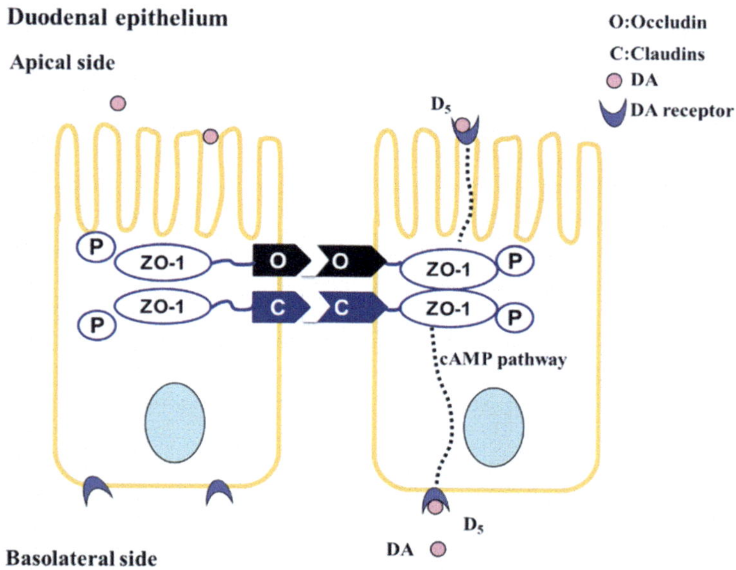

Fig. 4.20 DA enhances duodenal epithelial permeability via a D_5R mediated, cAMP-dependent pathway. In basic condition, DA binds with D_5R in the duodenum to upregulate intracellular cAMP and down-express ZO-1 and occluding, which lead to a higher duodenal FITC-dextran permeability. When D_5R is upregulated in D_5^{S390G} mice and 6-OHDA rats, DA binds with upregulated D_5R to get a lower expression of ZO-1 and occluding, which show an even higher duodenal permeability in D_5^{S390G} mice and 6-OHDA rats

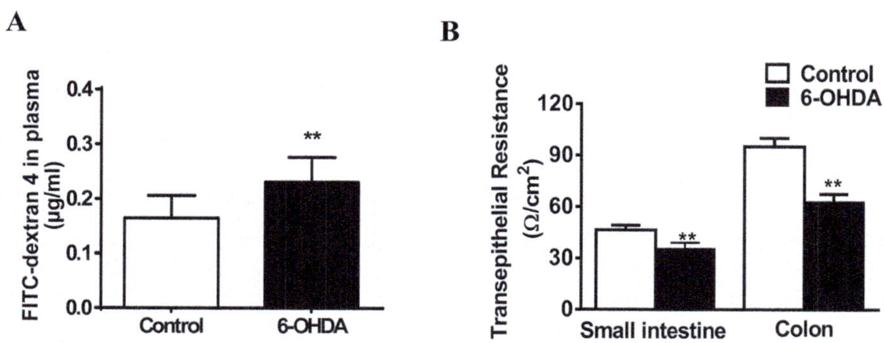

Fig. 4.21 The gastrointestinal mechanical barrier of the 6-OHDA rats. (Feng et al., Physiol Res, 2019, 68: 295–303.) (**a**) The FITC-dextran concentration in the plasma of control and 6-OHDA rats. (**b**) The TERs of small intestinal and colonic preparations in the control and 6-OHDA rats

permeability through the whole GI tract (Fig. 4.21a) and decreased TER (Fig. 4.21b) in the small intestine and colon, suggesting attenuated mucosal integrity alone the whole GI tract (Feng et al. 2019).

4.3.3 Dopamine and Gastrointestinal Immunological Barrier

The immune factors play an important role in mucosa protection (Beck et al. 2004). The GI barrier can provide an immune sentinel function by secreting various cytokines under bacterial stimulation, including the IL-1 family (e.g., IL-1β, IL-18, and IL-33), IL-6, IL-8, and some anti-inflammatory cytokines (e.g., IL-10 and IL-25). Studies have shown that IL-1β participates in the inflammatory responses by augmenting the infiltration of neutrophils via the activation T cells and innate lymphoid cells (Sun et al. 2015a).

DA is synthesized by various immune cells, which widely express DARs (Nakano et al. 2009; Arreola et al. 2016). Immune cells also express the plasma membrane- localized DA reuptake transporter and harbor resident DA uptake transporters (Amenta et al. 2001), suggesting that immune cells utilize DA to regulate immune responses.

Various studies have suggested that DA plays a role in regulating innate and adaptive responses. During innate immune responses, DA regulates cytokine secretion, for example, DA suppresses secretion of IFN-γ, TNF-α, and IL-1β, but promotes secretion of IL-10 and CXCL1 (Kawano et al. 2018). DA suppresses T-cell activation (Basu et al. 2010; Bergquist et al. 1994), which affects adaptive responses through a shift in the Th cell balance. The effects of DAR agonists and antagonists are also examined during innate and adaptive immune responses (Arreola et al. 2016). Activating D_2R affects the transcription of pro-inflammatory cytokines, including IL-1β and TNF-α (Shao et al. 2013).

Moreover, DA prevents nucleotide-binding oligomerization domain-like receptor 3 (NLRP3) inflammasome-dependent inflammations from occurring via activation of D_1R in in vivo studies (Yan et al. 2015). DA inhibits toll-like receptor 2 (TLR2)-induced NF-kB activation and inflammation via the D_5R (Wu et al. 2020). In fact, DA modulates the activation and chemotaxis of different immune cells by acting on its receptors and act as an important regulator of the immune system (Pacheco et al. 2014).

The results of Dr. Zhu JX's lab indicate that 6-OHDA rats display a chronic intestinal disorder with impaired GI transit time and attenuated mucosal barrier, which is correlated with exaggerated immune responses with increased pro-inflammatory factors IL-1β and IL-8 and decreased anti-inflammatory factor IL-10 (Fig. 4.22). Similar results have also been reported in the colonic walls isolated from the 6-OHDA rats with increased levels of malondialdehyde (MDA), TNF, and IL-1β, suggesting the presence of gut inflammation in the colon of 6-OHDA rats (Pellegrini et al. 2016). However, the mechanism underlying the effects of DA on GI immunological barrier as an anti-inflammatory factor is unclear.

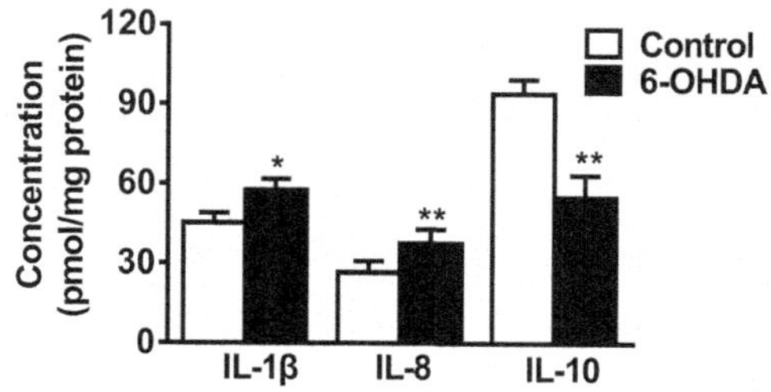

Fig. 4.22 The expression of inflammatory cytokines in the 6-OHDA rats. (Feng et al., Physiol Res, 2019, 68: 295–303.) The expression of pro-inflammatory factors IL-1β and IL-8 and anti-inflammatory factor IL-10 in the control and 6-OHDA rats

4.4 Dopamine and Gastrointestinal Mucosal Blood Flow

Blood flow and oxygen delivery to the GI mucosa are also involved in the regulation of mucosal functions. DA (10 μg/kg/min) increases the bethanechol-stimulated gastric mucosal blood flow. The α-adrenoceptor antagonist phentolamine further increases DA-elevated blood flow, indicating a primary action of the α-adrenoceptors on blood flow (Hovendal et al. 1982). DA induces a pronounced jejunal mucosal vasodilatation up to 400% of control values in anesthetized cats, while the net fluid and sodium absorption increases by 50%. The effect of DA on isolated jejunal blood flow is unaffected by phentolamine, whereas the absorptive response is abolished. The findings indicate that DA probably induces a mucosal vasodilation via dopaminergic mechanism and enhances fluid transport via another α-adrenergic mechanism (Sjovall et al. 1984). DA also increases small intestinal blood flow in full-term and preterm piglets (Ferrara et al. 1995). Although DA (3 μg/kg/min) does not prevent endotoxemia, it delays and attenuates the decrease in the intestinal villus microcirculation during endotoxemia (Schmidt et al. 1996). DA increases oxygen delivery to the intestinal mucosa and counteracts hemorrhage-induced mucosal hypoxia. Intravenous infusion of DA (16 μg/kg/min) improves tissue oxygenation of the small intestinal mucosa during moderate hemorrhage and subsequent resuscitation (Germann et al. 1997).

4.5 Dopamine and Gastrointestinal Mucosal Regulation

The GI tract is highly innervated by extrinsic and intrinsic nerves. The extrinsic innervation mainly includes the classic vagal parasympathetic and sympathetic components. The intrinsic innervation is represented by the enteric nervous system (ENS), a complex neural network controlling multifarious cell populations, including smooth muscle cells, mucosal secretory cells, endocrine cells, microvasculature, immune and inflammatory cells (Veiga-Fernandes and Mucida 2016).

DA, as a neurotransmitter in the central nervous system as well as a neurotransmitter/modulator in the peripheral tissues, has received so far extensive attention in GI mucosal regulation. DA plays a role in health and diseases such as peptic ulcer, inflammatory bowel disease (IBD), cancer, and Parkinson's disease. However, relatively little information about how DA modulates GI mucosal function is reported, although DA may affect multiple systems such as nervous system, endocrine system, immune system, and local vascular system. A complete and exhaustive summary of currently available evidence about DA and GI mucosal regulation is provided here.

4.5.1 Dopamine and Gastrointestinal Neural Regulation

A large variety of neurons and neurotransmitters take part in the ENS. However, the nature of these neurons and their role in the regulation of GI functions are debatable. In particular, an emphasis is posed on neurodegenerative disorders, such as Parkinson's disease, which is associated with the loss of catecholamine neurons. Natale et al. base on careful morphological studies showing that the only catecholamine-containing neurons within ENS are dopaminergic. This means that DA neurons are the sole catecholamine cell within intrinsic circuitries affecting gut function (Natale et al. 2017).

The ENS of the gut is arranged in two main ganglionated layers, submucosal plexus and myenteric plexus. Dopaminergic neurons are present in both plexuses and are upregulated after extrinsic denervation. The submucosal plexus mainly innervates the mucosa layer and regulates the GI epithelial cell functions and blood flow.

The pylorus, an important part of the digestive tract controlling the flow of chyme between the stomach and the duodenum, is widely innervated by intrinsic and extrinsic nerves. Double immunocytochemistry reveals that the tyrosine hydroxylase (TH, dopaminergic maker)-positive cells also contain vasoactive intestinal peptide (VIP), choline acetyltransferase (ChAT), and neuronal nitric oxide synthase (nNOS) in the innervate pylorus of the domestic pig (Zalecki 2012). In situ hybridization reveals that subsets of submucosal neurons contain mRNA encoding D_2R or D_3R. Subsets of submucosal neurons are also D_1R, D_2R, or D_3R immunoreactive from stomach through distal colon (Nasser et al. 2006). The results of Dr. Zhu JX's lab also indicate that D_1R, D_2R and D_5R are expressed in the colonic submucosal plexus of rat by real time-PCR and Western blot. And these receptors are located on VIP-positive neurons and cholinergic neurons in the submucosal plexus by double-labeling immunofluorescence (Zhang et al. 2017). Intestinal type-3 innate lymphoid cells (ILC3s) have high expression of the G protein-coupled receptor VIP receptor 2 (VIPR2), and activation by VIP markedly enhances the production of IL-22 and the barrier function of the epithelium (Seillet et al. 2020). Moreover, infection-induced ACh release from neurons stimulates muscarinic signaling in the intestinal epithelium, driving the downstream Wnt pathway and resulting in the expression of C-type lectin and lysozyme genes that enhance host defense (Labed et al. 2018). These results reveal a tight connection between the nervous system and the intestinal epithelium, with important implications for host defense, immune homeostasis, and cancer. This provides a novel therapeutic target for functional GI disorders.

4.5.2 Dopamine and Gastrointestinal Endocrine Regulation

Numerous GI endocrine cells produce and secrete satiety hormones in response to food consumption and digestion. DAR subtypes play critical roles in the regulation of hormone secretion, which are considered important pharmacological targets to inhibit hormone over secretion. Somatostatin, a peptide hormone secreted by D cells, plays an important role in regulating intestinal ion secretion and absorption. It has been reported that there is an association between the response to somatostatin analog treatment and D_4R and D_5R expression (Venegas-Moreno et al. 2018). The results of Dr. Zhu JX's lab also indicate that D_2Rs are distributed on the somatostatin-secreting D cells in human, rat, or mice gastric corpus, which might mediate the regulation of DA on the somatostatin secretion (Sun et al. 2015b). Somatostatin stimulates intestinal Na^+ absorption by increasing intestinal NHE8 expression through the SSTR2-p38 MAPK pathway (Wang et al. 2011). Moreover, exogenous somatostatin (octreotide) administration effectively stimulates mice in vivo colonic MUC2 expression and mucus secretion. In human goblet-like LS174T cells, somatostatin exposure also significantly stimulates MUC2 expression and mucus secretion. Further studies indicate that somatostatin participates in colonic mucus barrier regulation through somatostatin receptor 5-Notch-Hes1-MUC2 signaling pathway (Song et al. 2020).

A significant sex difference is observed for DA-associated food addiction. Plasma DA levels correlate positively with disordered eating behaviors in females and negatively in males. The results provide evidence that depressogenic excess eating and weight gain are associated with peripheral DA levels (Mills et al. 2020). In addition, estrogen stimulates DBS in human and the expression of estrogen receptors in duodenal mucosa has no difference between men and women. The basal DBS in adults is significantly higher in women than in men, while the time course and the net peak of 17β-estradiol-stimulated DBS are similar between men and women (Tuo et al. 2011).

Multiple enteroendocrine cells are located in GI mucosal epithelium and also involved in GI mucosal regulation. After administration of L-DOPA, the enterochromaffin-like cells in the oxyntic gland area of the rat stomach produce and temporarily store DA. These cells have the appearance of polypeptide hormone-secreting cells, comprise two main cell types, the most predominant one having vesicular type granules (EGL cells), the second most predominant one having smaller, uniformly electron dense granules (A-like cells). Ghrelin is first described by Kojima et al., which is a stomach-derived 28-amino acid peptide hormone with a unique modification of acylation. DA stimulates ghrelin secretion from a ghrelin-producing cell line, MGN (mouse ghrelinoma) 3–1 cells. And the secretion manner of that is similar to the previously known ghrelin stimulators, epinephrine, and norepinephrine. MGN3-1 cells express mRNA encoding D_1R and D_2R. The D_1R agonist fenoldopam, not D_2R, D_3R agonist bromocriptine, stimulates ghrelin secretion, while D_1R antagonist SKF83566 attenuates the stimulatory effect of DA. These results indicate that D_1R mediates the stimulatory effect of DA on ghrelin secretion (Iwakura et al. 2011). Ghrelin markedly attenuates intestinal mucosal injury at both histomorphometric and ultrastructural levels post-intracerebral hemorrhage (ICH). Ghrelin reduced ICH-induced intestinal permeability according to fluorescein isothiocyanate conjugated-dextran (FITC-D) and Evans blue extravasation assays. Concomitantly, the intestinal tight junction-related protein markers, ZO-1 and claudin-5, are upregulated by ghrelin post-ICH. Ghrelin reduces intestinal barrier dysfunction, thereby reducing mortality and weight loss (Cheng et al. 2016).

Moreover, DARs are present in rodent and human pancreas, which provides a morphological basis for studying the pancreatic endocrine effects of DA and suggests a new target for the treatment of diabetes. This part will be discussed in Chap. 6.

4.6 Dopamine and Gastrointestinal Mucosal Disorders

A wide array of dopaminergic drugs is used in therapeutics for clinical indications, such as Parkinson's disease, gastric cancer, IBD, diabetes, and traumatic brain injury; with usually favorable therapeutic index and innovative treatment at low price. Dopaminergic drugs are benefit for patients as well as for the healthcare systems (Natale et al. 2017).

A high incidence of gastric ulcer has been noted in the patients of a well-known dopaminergic neuronal degenerative disease, Parkinson disease (Ozdemir et al. 2007). Antidopaminergic drugs induce ulcers, and DA agonists protect against peptic ulcer formation (Rasheed et al. 2010). Mice treated with intraperitoneal injection of 1-methyl-4-phenyl-1,2,3,6-tetrahydropyridine (MPTP) always manifest gastric ulcer, therefore act as gastric ulcer mouse model, which are also called MPTP mice. The MPTP mice show a significant decrease in dopaminergic markers, TH and DA transporter (DAT), as well as DA content in the gut (Tian et al. 2008; Sikiric et al. 1999). These studies suggest that DA has an important protective effect on the gastric mucosa.

A network meta-analysis shows that metoclopramide, trimebutine, mosapride, and domperidone have better efficacy for the treatment of functional dyspepsia than itopride or acotiamide. The short-term use of metoclopramide and domperidone, or the alternative use of trimebutine and mosapride is recommended for the symptomatic relief of functional dyspepsia (Yang et al. 2017). Although domperidone has no effect on inhibiting gastric acid secretion and protecting gastric mucosa, omeprazole (a proton pump inhibitor) combined with domperidone achieves an ideal effect in the treatment of chronic superficial gastritis (Wang et al. 2017). Omeprazole gives full play to the acid and helicobacter pylori inhibition, while domperidone has the emesis resistance and gastric emptying effect. These two drugs act on the different clinical

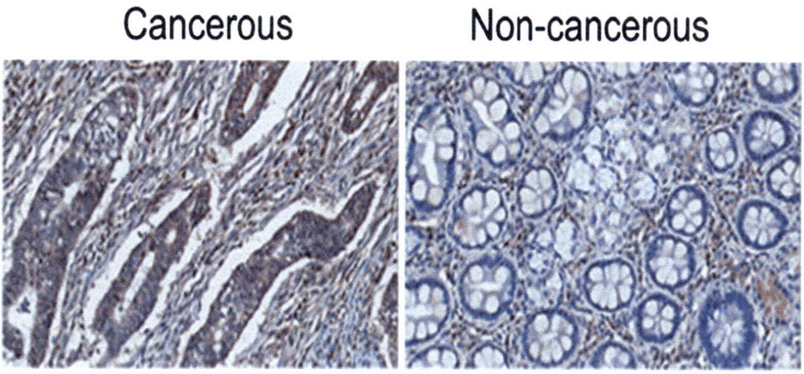

Fig. 4.23 The protein expression of D_2R in gastric cancer and non-cancerous tissues. (Mu, et al. Oncol Lett. 2017, 13(3):1223–1227.) Left: The expression of D_2R in gastric cancer; Right: The expression of D_2R in non-cancerous tissues

symptoms of chronic superficial gastritis, with remarkable effects and less adverse reactions (Wang et al. 2017).

The expression of DARs has been reported to be correlated with tumors (Clague et al. 2010; Li et al. 2014). The expression of D_2R decreases in gastric cancer cells and DA suppresses gastric cancer cell invasion and migration via D_2R to inhibit the EGFR/AKT/MMP-13 pathway (Huang et al. 2016). D_2R is also involved in inhibiting IGF-I-induced gastric cancer cell growth (Ganguly et al. 2010). However, the percentage of gastric cancer cases with a high expression level of D_2R (51.2%) is higher than that of cases with a low expression level of D_2R (39.3%) in 84 paired gastric cancer tissues and respective adjacent non-cancerous tissues (Fig. 4.23). Gastric cancer patients with a higher expression of D_2R have shorter survival durations (Fig. 4.24) (Mu et al. 2017). D_5R is a negative regulator of tumor cell sensitivity to D_2R antagonism in glioblastoma cell lines (Prabhu et al. 2019). Overall, overexpression of D_2R is associated with a poor clinical prognosis in cancer patients. DARs may be considered as a prognosis marker for the cancer patients.

Cancer stem cells (CSCs) lead to drug resistance and eradication of CSCs is an effective therapeutic strategy. A group of antipsychotic DAR antagonists restrain and sensitize CSCs to existing chemotherapeutics by a process called differentiation approach (Roney and Park 2018). Thioridazine, an antipsychotic D_2R antagonist used in the treatment of schizophrenia and psychosis, demonstrates anti-CSC property without affecting normal stem cells (Sachlos et al. 2012). Thioridazine inhibits CSCs in various cancer types, including gastric, liver, glioblastoma, and ovarian cancers. Thioridazine is also a potential drug candidate against colorectal CSCs by suppressing the proliferation and inducing apoptosis (Strobl et al. 1990; Gil-Ad et al. 2004; Kang et al. 2012). However, most studies are in cell lines, no large scale clinical data is found.

IBD is a chronic inflammatory condition of the GI tract encompassing two main clinical entities: Crohn's disease (CD) and ulcerative colitis (UC). The levels of

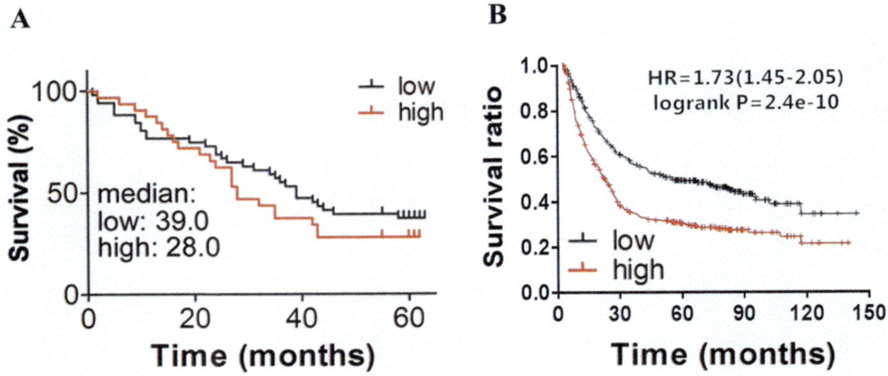

Fig. 4.24 The survival duration of gastric cancer patients with higher expression levels of D_2R. (Mu, et al. Oncol Lett. 2017, 13(3):1223–1227.) (**a**) Correlation between the expression levels of D_2R and the survival durations of patients with gastric cancer. (**b**) Analysis of the correlation between the survival durations of patients with gastric cancer and the expression levels of D_2R

L-DOPA, a source for DA, in the inflamed colon mucosa of CD and UC patients are twice than those in controls (Fig. 4.25a), whereas DA levels in the inflamed mucosa of CD and UC patients are markedly lower than in controls, resulting in significant reductions in DA/L-DOPA ratios in the tissue (Fig. 4.25b, c). However, no significant differences are observed between levels of DA in the non-inflamed mucosa of CD patients and control subjects (Fig. 4.25b). Levels of DOPAC, the major metabolite of DA, in the mucosa of CD patients do not differ from those in control subjects, resulting in a slight increase in DOPAC/DA ratios in the tissue (Fig. 4.25d, e). The finding that levels of DA are markedly reduced in the inflamed mucosa of CD patients, with no changes in the non-inflamed mucosa, suggests that the inflamed tissue is endowed with a reduced capacity to synthesize the amine, which is related to changes in colonic mucus secretion (Magro et al. 2002). Similarly, DA levels are significantly lower in the inflamed mucosa of the distal colon in the trinitrobenzene sulphonic acid-induced colitis model. DA lightens colitis by attenuating vascular permeability via the D_2R (Tolstanova et al. 2015) and promotes UC by triggering inflammatory responses through DAR on the colonic immune cells (van de Kerkhof et al. 2006).

Numerous reports have reported that DA plays pivotal roles in the pathological mechanisms of IBD. The impaired colonic barrier is crucial in the pathogenesis of IBD (Michielan and D'Inca 2015), and the DARs, especially D_2R, is involved in the development of IBD (Tolstanova et al. 2015).

Although dopaminergic drugs are used in clinical therapeutics, the mechanisms and pathways are not clear, which will be further studied to verify the potential targets.

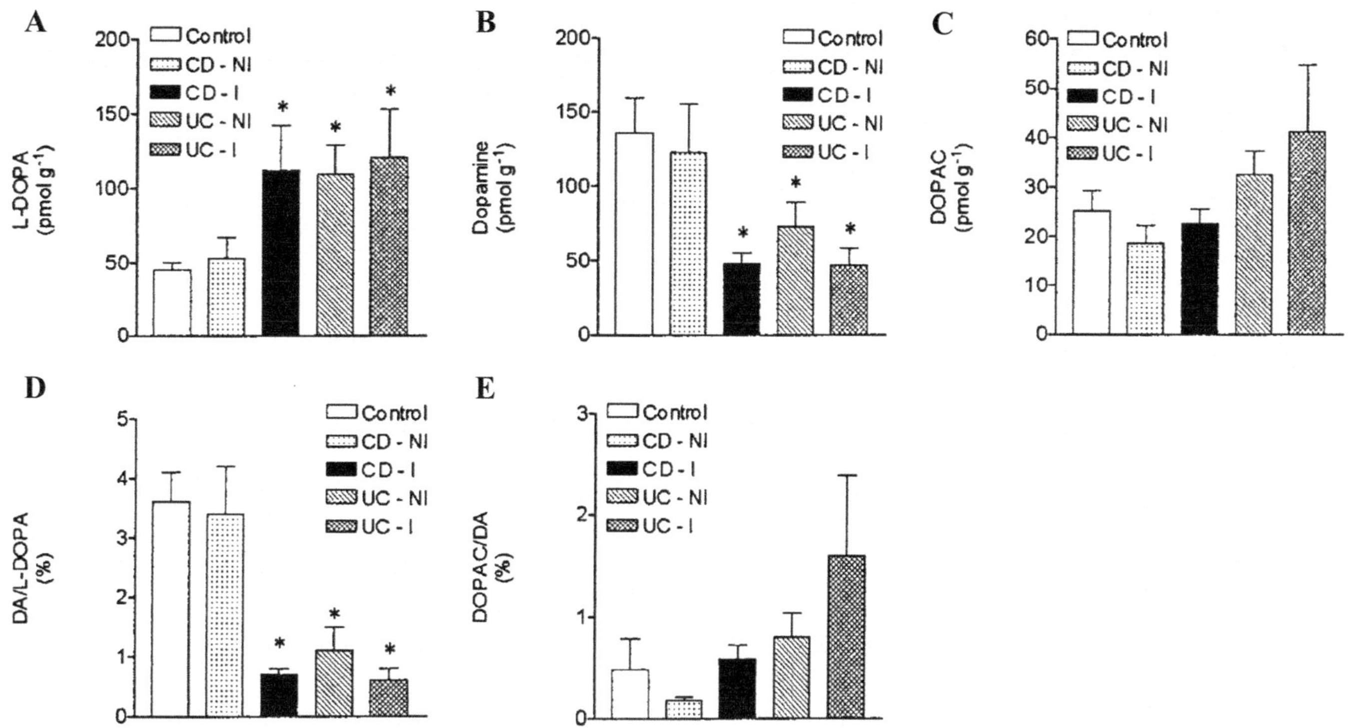

Fig. 4.25 The levels of L-DOPA, DA, and DOPAC in biopsy specimens from colonic mucosa of IBD patients. (Magro et al., Dig Dis Sci, 2002, 47 (1): 216–224.) (**a**) Absolute and relative tissue levels of L-DOPA in biopsy specimens of colonic mucosa from control and the patients afflicted with CD and UC. (**b**) Absolute and relative tissue levels of DA in biopsy specimens of colonic mucosa from control and the patients afflicted with CD and UC. (**c**) The ratio of L-DOPA and DA in biopsy specimens of colonic mucosa from control and the patients afflicted with CD and UC. (**d**) Absolute and relative tissue levels of DOPAC in biopsy specimens of colonic mucosa from control and the patients afflicted with CD and UC. (**e**) The ratio of DOPAC and DA in biopsy specimens of colonic mucosa from control and the patients afflicted with CD and UC. *$p < 0.05$. *CD* Crohn's disease, *UC* ulcerative colitis, *NI* noninflamed colonic mucosa, *I* inflamed colonic mucosa

4.7 Perspectives

The GI mucosa is covered by epithelial cells joined together by intercellular junction complexes, which regulate the internal balance of ions, fluids, and nutrients, and defends against invasion of food antigens, microorganisms, and toxins (Neunlist et al. 2007; Zimmerman et al. 2012). DA and DARs are found in significant quantities in the intestine with high concentrations in the mucosal layer (Feng et al. 2020; Tian et al. 2008). DA performs important regulatory effects on GI mucosal function, such as stimulating epithelial ion transport (Feng et al. 2013; Zhang et al. 2007), evoking fluid absorption (Zhang et al. 2007), regulating barrier function (Feng et al. 2017; Li et al. 2019), and modulating mucosal immune system including cytokine secretion, cell adhesion, cytotoxicity, and chemotaxis (Shao et al. 2013; Wu et al. 2020).

Based on the different concentration of DA and the differential distribution of DARs in different segment of the GI mucosa, variations of responses are observed. It is essential to pay more attention to the roles of DA in gut functional regulation and functional disorders to verify the potential targets for novel therapeutics.

References

Al-Jahmany AA, Schultheiss G, Diener M (2004) Effects of dopamine on ion transport across the rat distal colon. Pflugers Arch 448:605–612

Amenta F, Bronzetti E, Cantalamessa F, El-Assouad D, Felici L, Ricci A, Tayebati SK (2001) Identification of dopamine plasma membrane and vesicular transporters in human peripheral blood lymphocytes. J Neuroimmunol 117:133–142

Arreola R, Alvarez-Herrera S, Perez-Sanchez G, Becerril-Villanueva E, Cruz-Fuentes C, Flores-Gutierrez EO, Garces-Alvarez ME, de la Cruz-Aguilera DL, Medina-Rivero E, Hurtado-Alvarado G, Quintero-Fabian S, Pavon L (2016) Immunomodulatory effects mediated by dopamine. J Immunol Res 2016:3160486

Barry MK, Maher MM, Gontarek JD, Jimenez RE, Yeo CJ (1995) Luminal dopamine modulates canine ileal water and electrolyte transport. Dig Dis Sci 40:1738–1743

Basu B, Sarkar C, Chakroborty D, Ganguly S, Shome S, Dasgupta PS, Basu S (2010) D1 and D2 dopamine receptor-mediated inhibition of activated normal T cell proliferation is lost in jurkat T leukemic cells. J Biol Chem 285:27026–27032

Beck GC, Brinkkoetter P, Hanusch C, Schulte J, van Ackern K, van der Woude FJ, Yard BA (2004) Clinical review: immunomodulatory effects of dopamine in general inflammation. Crit Care 8:485–491

Bergquist J, Tarkowski A, Ekman R, Ewing A (1994) Discovery of endogenous catecholamines in lymphocytes and evidence for catecholamine regulation of lymphocyte function via an autocrine loop. Proc Natl Acad Sci U S A 91:12912–12916

Bergstrom KS, Kissoon-Singh V, Gibson DL, Ma C, Montero M, Sham HP, Ryz N, Huang T, Velcich A, Finlay BB, Chadee K, Vallance BA (2010) Muc2 protects against lethal infectious colitis by disassociating pathogenic and commensal bacteria from the colonic mucosa. PLoS Pathog 6:e1000902

Blasig IE, Bellmann C, Cording J, Del Vecchio G, Zwanziger D, Huber O, Haseloff RF (2011) Occludin protein family: oxidative stress and reducing conditions. Antioxid Redox Signal 15:1195–1219

Caldara R, Ferrari C, Romussi M, Bierti L, Gandini S, Curtarelli G (1978) Effect of dopamine infusion on gastric and pancreatic secretion and on gastrin release in man. Gut 19:724–728
Caldara R, Ferrari C, Barbieri C (1980) Dopamine and stomach. Lancet 1:95–96
Camilleri M (2019) Leaky gut: mechanisms, measurement and clinical implications in humans. Gut 68:1516–1526
Camilleri M, Lyle BJ, Madsen KL, Sonnenburg J, Verbeke K, Wu GD (2019) Role for diet in normal gut barrier function: developing guidance within the framework of food-labeling regulations. Am J Physiol Gastrointest Liver Physiol 317:G17–G39
Cheng Y, Wei Y, Yang W, Cai Y, Chen B, Yang G, Shang H, Zhao W (2016) Ghrelin attenuates intestinal barrier dysfunction following intracerebral hemorrhage in mice. Int J Mol Sci 17 (12):2032
Clague J, Cinciripini P, Blalock J, Wu X, Hudmon KS (2010) The D2 dopamine receptor gene and nicotine dependence among bladder cancer patients and controls. Behav Genet 40:49–58
Desai JK, Parmar NS (1994) Gastric and duodenal anti-ulcer activity of sulpiride, a dopamine D2 receptor antagonist, in rats. Agents Actions 42:149–153
Desai JK, Goyal RK, Parmar NS (1999) Characterization of dopamine receptor subtypes involved in experimentally induced gastric and duodenal ulcers in rats. J Pharm Pharmacol 51:187–192
Desai JC, Sanyal SM, Goo T, Benson AA, Bodian CA, Miller KM, Cohen LB, Aisenberg J (2008) Primary prevention of adverse gastroduodenal effects from short-term use of non-steroidal anti-inflammatory drugs by omeprazole 20 mg in healthy subjects: a randomized, double-blind, placebo-controlled study. Dig Dis Sci 53:2059–2065
Donowitz M, Cusolito S, Battisti L, Fogel R, Sharp GW (1982) Dopamine stimulation of active Na and Cl absorption in rabbit ileum: interaction with alpha 2-adrenergic and specific dopamine receptors. J Clin Invest 69:1008–1016
Donowitz M, Elta G, Battisti L, Fogel R, Label-Schwartz E (1983) Effect of dopamine and bromocriptine on rat ileal and colonic transport. Stimulation of absorption and reversal of cholera toxin-induced secretion. Gastroenterology 84:516–523
Dupuy D, Szabo S (1986) Protection by metals against ethanol-induced gastric mucosal injury in the rat. Comparative biochemical and pharmacologic studies implicate protein sulfhydryls. Gastroenterology 91:966–974
Eliassi A, Aleali F, Ghasemi T (2008) Peripheral dopamine D2-like receptors have a regulatory effect on carbachol-, histamine- and pentagastrin-stimulated gastric acid secretion. Clin Exp Pharmacol Physiol 35:1065–1070
Feng XY, Li Y, Li LS, Li XF, Zheng LF, Zhang XL, Fan RF, Song J, Hong F, Zhang Y, Zhu JX (2013) Dopamine D1 receptors mediate dopamine-induced duodenal epithelial ion transport in rats. Transl Res 161:486–494
Feng XY, Zhang DN, Wang YA, Fan RF, Hong F, Zhang Y, Li Y, Zhu JX (2017) Dopamine enhances duodenal epithelial permeability via the dopamine D5 receptor in rodent. Acta Physiol (Oxf) 220:113–123
Feng XY, Yang J, Zhang X, Zhu J (2019) Gastrointestinal non-motor dysfunction in Parkinson's disease model rats with 6-hydroxydopamine. Physiol Res 68:295–303
Feng XY, Yan JT, Li GW, Liu JH, Fan RF, Li SC, Zheng LF, Zhang Y, Zhu JX (2020) Source of dopamine in gastric juice and luminal dopamine-induced duodenal bicarbonate secretion via apical dopamine D2 receptors. Br J Pharmacol 177:3258–3272
Ferrara JJ, Dyess DL, Peeples GL, Christenberry DP, Roberts WS, Tacchi EJ, Swafford AN, Ardell JL, Powell RW (1995) Effects of dopamine and dobutamine on regional blood flow distribution in the neonatal piglet. Ann Surg 221:531–540. discussion 540–2
Flemstrom G, Isenberg JI (2001) Gastroduodenal mucosal alkaline secretion and mucosal protection. News Physiol Sci 16:23–28
Flemstrom G, Safsten B (1994) Role of dopamine and other stimuli of mucosal bicarbonate secretion in duodenal protection. Dig Dis Sci 39:1839–1842
Flemstrom G, Safsten B, Jedstedt G (1993) Stimulation of mucosal alkaline secretion in rat duodenum by dopamine and dopaminergic compounds. Gastroenterology 104:825–833

Ganguly S, Basu B, Shome S, Jadhav T, Roy S, Majumdar J, Dasgupta PS, Basu S (2010) Dopamine, by acting through its D2 receptor, inhibits insulin-like growth factor-I (IGF-I)-induced gastric cancer cell proliferation via up-regulation of Kruppel-like factor 4 through down-regulation of IGF-IR and AKT phosphorylation. Am J Pathol 177:2701–2707
Germann R, Haisjackl M, Schwarz B, Salak N, Deusch E, Pajk W, Wolf HJ, Riedmann B, Hasibeder W (1997) Dopamine and intestinal mucosal tissue oxygenation in a porcine model of haemorrhage. Br J Anaesth 79:357–362
Gil-Ad I, Shtaif B, Levkovitz Y, Dayag M, Zeldich E, Weizman A (2004) Characterization of phenothiazine-induced apoptosis in neuroblastoma and glioma cell lines: clinical relevance and possible application for brain-derived tumors. J Mol Neurosci 22:189–198
Glavin GB, Hall AM (1994) Clozapine, a dopamine DA4 receptor antagonist, reduces gastric acid secretion and stress-induced gastric mucosal injury. Life Sci 54:PL261–PL264
Glavin GB, Hall AM (1995) Central and peripheral dopamine D1/DA1 receptor modulation of gastric secretion and experimental gastric mucosal injury. Gen Pharmacol 26:1277–1279
Guldvog I, Linnestad P, Schrumpf E, Berstad A (1984) Dopaminergic and adrenergic influence on gastric acid and pepsin secretion stimulated by food. The role of vagal innervation. Scand J Gastroenterol Suppl 89:113–116
Hamilton MK, Boudry G, Lemay DG, Raybould HE (2015) Changes in intestinal barrier function and gut microbiota in high-fat diet-fed rats are dynamic and region dependent. Am J Physiol Gastrointest Liver Physiol 308:G840–G851
Hernandez DE, Mason GA, Walker CH, Valenzuela JE (1987) Dopamine receptors in human gastrointestinal mucosa. Life Sci 41:2717–2723
Hirst BH, Reed JD, Gomez-Pan A, Labib LA (1976) Bromocriptine potentiation of gastric acid secretion in cats. Clin Endocrinol 5:723–729
Hovendal CP, Bech K, Gottrup F, Andersen D (1982) Effect of dopamine on pentagastrin-stimulated gastric acid secretion and mucosal blood flow in dogs with gastric fistula. Scand J Gastroenterol 17:97–102
Hu EH, Pan F, Volgyi B, Bloomfield SA (2010) Light increases the gap junctional coupling of retinal ganglion cells. J Physiol 588:4145–4163
Huang H, Wu K, Ma J, Du Y, Cao C, Nie Y (2016) Dopamine D2 receptor suppresses gastric cancer cell invasion and migration via inhibition of EGFR/AKT/MMP-13 pathway. Int Immunopharmacol 39:113–120
Iwakura H, Ariyasu H, Hosoda H, Yamada G, Hosoda K, Nakao K, Kangawa K, Akamizu T (2011) Oxytocin and dopamine stimulate ghrelin secretion by the ghrelin-producing cell line, MGN3-1 in vitro. Endocrinology 152:2619–2625
Johansson ME, Ambort D, Pelaseyed T, Schutte A, Gustafsson JK, Ermund A, Subramani DB, Holmen-Larsson JM, Thomsson KA, Bergstrom JH, van der Post S, Rodriguez-Pineiro AM, Sjovall H, Backstrom M, Hansson GC (2011) Composition and functional role of the mucus layers in the intestine. Cell Mol Life Sci 68:3635–3641
Kang S, Dong SM, Kim BR, Park MS, Trink B, Byun HJ, Rho SB (2012) Thioridazine induces apoptosis by targeting the PI3K/Akt/mTOR pathway in cervical and endometrial cancer cells. Apoptosis 17:989–997
Kawano M, Takagi R, Saika K, Matsui M, Matsushita S (2018) Dopamine regulates cytokine secretion during innate and adaptive immune responses. Int Immunol 30:591–606
Keely S, Talley NJ (2020) Duodenal bile acids as determinants of intestinal mucosal homeostasis and disease. Neurogastroenterol Motil 32:e13854
Knutson L, Odlind B, Hallgren R (1989) A new technique for segmental jejunal perfusion in man. Am J Gastroenterol 84:1278–1284
Kovbasnjuk O, Chatton JY, Friauf WS, Spring KR (1995) Determination of the Na permeability of the tight junctions of MDCK cells by fluorescence microscopy. J Membr Biol 148:223–232
Kunzelmann K, Mall M (2002) Electrolyte transport in the mammalian colon: mechanisms and implications for disease. Physiol Rev 82:245–289

Labed SA, Wani KA, Jagadeesan S, Hakkim A, Najibi M, Irazoqui JE (2018) Intestinal epithelial Wnt signaling mediates acetylcholine-triggered host defense against infection. Immunity 48:963–978.e3

Li J, Zhu S, Kozono D, Ng K, Futalan D, Shen Y, Akers JC, Steed T, Kushwaha D, Schlabach M, Carter BS, Kwon CH, Furnari F, Cavenee W, Elledge S, Chen CC (2014) Genome-wide shRNA screen revealed integrated mitogenic signaling between dopamine receptor D2 (DRD2) and epidermal growth factor receptor (EGFR) in glioblastoma. Oncotarget 5:882–893

Li Y, Zhang Y, Zhang XL, Feng XY, Liu CZ, Zhang XN, Quan ZS, Yan JT, Zhu JX (2019) Dopamine promotes colonic mucus secretion through dopamine D5 receptor in rats. Am J Physiol Cell Physiol 316:C393–C403

Liu X, Wang W, Chen W, Jiang X, Zhang Y, Wang Z, Yang J, Jones JE, Jose PA, Yang Z (2015) Regulation of blood pressure, oxidative stress and AT1R by high salt diet in mutant human dopamine D5 receptor transgenic mice. Hypertens Res 38:394–399

Maeda-Hagiwara M, Watanabe K (1983) Influence of dopamine receptor agonists on gastric acid secretion induced by intraventricular administration of thyrotropin-releasing hormone in the perfused stomach of anaesthetized rats. Br J Pharmacol 79:297–303

Magro F, Vieira-Coelho MA, Fraga S, Serrao MP, Veloso FT, Ribeiro T, Soares-da-Silva P (2002) Impaired synthesis or cellular storage of norepinephrine, dopamine, and 5-hydroxytryptamine in human inflammatory bowel disease. Dig Dis Sci 47:216–224

Marmon LM, Albrecht F, Canessa LM, Hoy GR, Jose PA (1993) Identification of dopamine1A receptors in the rat small intestine. J Surg Res 54:616–620

Martens EC, Neumann M, Desai MS (2018) Interactions of commensal and pathogenic microorganisms with the intestinal mucosal barrier. Nat Rev Microbiol 16:457–470

Michielan A, D'Inca R (2015) Intestinal permeability in inflammatory bowel disease: pathogenesis, clinical evaluation, and therapy of leaky gut. Mediat Inflamm 2015:628157

Mills JG, Thomas SJ, Larkin TA, Deng C (2020) Overeating and food addiction in major depressive disorder: links to peripheral dopamine. Appetite 148:104586

Mu J, Huang W, Tan Z, Li M, Zhang L, Ding Q, Wu X, Lu J, Liu Y, Dong Q, Xu H (2017) Dopamine receptor D2 is correlated with gastric cancer prognosis. Oncol Lett 13:1223–1227

Nakano K, Higashi T, Takagi R, Hashimoto K, Tanaka Y, Matsushita S (2009) Dopamine released by dendritic cells polarizes Th2 differentiation. Int Immunol 21:645–654

Nasser Y, Ho W, Sharkey KA (2006) Distribution of adrenergic receptors in the enteric nervous system of the guinea pig, mouse, and rat. J Comp Neurol 495:529–553

Natale G, Ryskalin L, Busceti CL, Biagioni F, Fornai F (2017) The nature of catecholamine-containing neurons in the enteric nervous system in relationship with organogenesis, normal human anatomy and neurodegeneration. Arch Ital Biol 155:118–130

Neunlist M, Aubert P, Bonnaud S, Van Landeghem L, Coron E, Wedel T, Naveilhan P, Ruhl A, Lardeux B, Savidge T, Paris F, Galmiche JP (2007) Enteric glia inhibit intestinal epithelial cell proliferation partly through a TGF-beta1-dependent pathway. Am J Physiol Gastrointest Liver Physiol 292:G231–G241

Neunlist M, Van Landeghem L, Mahe MM, Derkinderen P, des Varannes SB, Rolli-Derkinderen M (2013) The digestive neuronal-glial-epithelial unit: a new actor in gut health and disease. Nat Rev Gastroenterol Hepatol 10:90–100

Nylander O, Flemstrom G (1986) Effects of alpha-adrenoceptor agonists and antagonists on duodenal surface epithelial HCO3-secretion in the rat in vivo. Acta Physiol Scand 126:433–441

Oshima T, Miwa H (2016) Gastrointestinal mucosal barrier function and diseases. J Gastroenterol 51:768–778

Ozdemir V, Jamal MM, Osapay K, Jadus MR, Sandor Z, Hashemzadeh M, Szabo S (2007) Cosegregation of gastrointestinal ulcers and schizophrenia in a large national inpatient discharge database: revisiting the "brain-gut axis" hypothesis in ulcer pathogenesis. J Investig Med 55:315–320

Pacheco R, Contreras F, Zouali M (2014) The dopaminergic system in autoimmune diseases. Front Immunol 5:117

Parker MD, Boron WF (2013) The divergence, actions, roles, and relatives of sodium-coupled bicarbonate transporters. Physiol Rev 93:803–959

Pellegrini C, Fornai M, Colucci R, Tirotta E, Blandini F, Levandis G, Cerri S, Segnani C, Ippolito C, Bernardini N, Cseri K, Blandizzi C, Hasko G, Antonioli L (2016) Alteration of colonic excitatory tachykininergic motility and enteric inflammation following dopaminergic nigrostriatal neurodegeneration. J Neuroinflammation 13:146

Piccolino M, Neyton J, Gerschenfeld HM (1984) Decrease of gap junction permeability induced by dopamine and cyclic adenosine 3′:5′-monophosphate in horizontal cells of turtle retina. J Neurosci 4:2477–2488

Prabhu VV, Madhukar NS, Gilvary C, Kline CLB, Oster S, El-Deiry WS, Elemento O, Doherty F, VanEngelenburg A, Durrant J, Tarapore RS, Deacon S, Charter N, Jung J, Park DM, Gilbert MR, Rusert J, Wechsler-Reya R, Arrillaga-Romany I, Batchelor TT, Wen PY, Oster W, Allen JE (2019) Dopamine receptor D5 is a modulator of tumor response to dopamine receptor D2 antagonism. Clin Cancer Res 25:2305–2313

Rao RK, Basuroy S, Rao VU, Karnaky KJ Jr, Gupta A (2002) Tyrosine phosphorylation and dissociation of occludin-ZO-1 and E-cadherin-beta-catenin complexes from the cytoskeleton by oxidative stress. Biochem J 368:471–481

Rasheed N, Ahmad A, Singh N, Singh P, Mishra V, Banu N, Lohani M, Sharma S, Palit G (2010) Differential response of A 68930 and sulpiride in stress-induced gastric ulcers in rats. Eur J Pharmacol 643:121–128

Roney MSI, Park SK (2018) Antipsychotic dopamine receptor antagonists, cancer, and cancer stem cells. Arch Pharm Res 41:384–408

Sachlos E, Risueno RM, Laronde S, Shapovalova Z, Lee JH, Russell J, Malig M, McNicol JD, Fiebig-Comyn A, Graham M, Levadoux-Martin M, Lee JB, Giacomelli AO, Hassell JA, Fischer-Russell D, Trus MR, Foley R, Leber B, Xenocostas A, Brown ED, Collins TJ, Bhatia M (2012) Identification of drugs including a dopamine receptor antagonist that selectively target cancer stem cells. Cell 149:1284–1297

Safsten B, Sjoblom M, Flemstrom G (2006) Serotonin increases protective duodenal bicarbonate secretion via enteric ganglia and a 5-HT4-dependent pathway. Scand J Gastroenterol 41:1279–1289

Schmidt H, Secchi A, Wellmann R, Bohrer H, Bach A, Martin E (1996) Effect of low-dose dopamine on intestinal villus microcirculation during normotensive endotoxaemia in rats. Br J Anaesth 76:707–712

Schrumpf E, Linnestad P (1982) Effect of cholinergic, adrenergic, and dopaminergic blockade on gastrin secretion in healthy subjects. Scand J Gastroenterol 17:29–31

Schubert ML (2009) Gastric exocrine and endocrine secretion. Curr Opin Gastroenterol 25:529–536

Seillet C, Luong K, Tellier J, Jacquelot N, Shen RD, Hickey P, Wimmer VC, Whitehead L, Rogers K, Smyth GK, Garnham AL, Ritchie ME, Belz GT (2020) The neuropeptide VIP confers anticipatory mucosal immunity by regulating ILC3 activity. Nat Immunol 21:168–177

Shao EH, Hayes EM, Khwaja HA, Efthimiou E (2011) The importance of a travel history in the preoperative assessment of an elective surgical patient. BMJ Case Rep 2011:bcr0720114564

Shao W, Zhang SZ, Tang M, Zhang XH, Zhou Z, Yin YQ, Zhou QB, Huang YY, Liu YJ, Wawrousek E, Chen T, Li SB, Xu M, Zhou JN, Hu G, Zhou JW (2013) Suppression of neuroinflammation by astrocytic dopamine D2 receptors via alpha B-crystallin. Nature 494:90–94

Sharkey KA, Savidge TC (2014) Role of enteric neurotransmission in host defense and protection of the gastrointestinal tract. Auton Neurosci 181:94–106

Sikiric P, Marovic A, Matoz W, Anic T, Buljat G, Mikus D, Stancic-Rokotov D, Separovic J, Seiwerth S, Grabarevic Z, Rucman R, Petek M, Ziger T, Sebecic B, Zoricic I, Turkovic B, Aralica G, Perovic D, Duplancic B, Lovric-Bencic M, Rotkvic I, Mise S, Jagic V, Hahn V (1999) A behavioural study of the effect of pentadecapeptide BPC 157 in Parkinson's disease models in mice and gastric lesions induced by 1-methyl-4-phenyl-1,2,3,6-tetrahydrophyridine. J Physiol Paris 93:505–512

Sjovall H, Redfors S, Biber B, Jodal M, Lundgren O (1984) Effect of intravenous dopamine infusion on intramural blood flow distribution and fluid absorption in the feline small intestine. Scand J Gastroenterol 19:411–416

Song S, Li X, Geng C, Li Y, Wang C (2020) Somatostatin stimulates colonic MUC2 expression through SSTR5-notch-Hes1 signaling pathway. Biochem Biophys Res Commun 521:1070–1076

Strobl JS, Kirkwood KL, Lantz TK, Lewine MA, Peterson VA, Worley JF (1990) Inhibition of human breast cancer cell proliferation in tissue culture by the neuroleptic agents pimozide and thioridazine. Cancer Res 50:5399–5405

Sun L, Feng XY, Zheng LF, Zhang XL, Zhang Y, Zhu JX (2015a) Distribution of dopamine receptors in gastric mucosal somatostatin-secreting cell of human and mice. Capit Med Univ 36:432–436

Sun M, He C, Cong Y, Liu Z (2015b) Regulatory immune cells in regulation of intestinal inflammatory response to microbiota. Mucosal Immunol 8:969–978

Suzuki T (2013) Regulation of intestinal epithelial permeability by tight junctions. Cell Mol Life Sci 70:631–659

Szabo S (1979) Dopamine disorder in duodenal ulceration. Lancet 2:880–882

Tian YM, Chen X, Luo DZ, Zhang XH, Xue H, Zheng LF, Yang N, Wang XM, Zhu JX (2008) Alteration of dopaminergic markers in gastrointestinal tract of different rodent models of Parkinson's disease. Neuroscience 153:634–644

Tolstanova G, Deng X, Ahluwalia A, Paunovic B, Prysiazhniuk A, Ostapchenko L, Tarnawski A, Sandor Z, Szabo S (2015) Role of dopamine and D2 dopamine receptor in the pathogenesis of inflammatory bowel disease. Dig Dis Sci 60:2963–2975

Tsai LH, Cheng JT (1990) Neuropeptide Y (NPY) inhibits dimethylphenylpiperazinium (DMPP)-induced gastric acid secretion in isolated rat stomach. Neurosci Res 8:21–28

Tsai LH, Cheng JT (1995) Stimulatory effect of dopamine on acid secretion from the isolated rat stomach. Neurosci Res 21:235–240

Tuo B, Wen G, Zhang Y, Liu X, Wang X, Liu X, Dong H (2009) Involvement of phosphatidylinositol 3-kinase in cAMP- and cGMP-induced duodenal epithelial CFTR activation in mice. Am J Physiol Cell Physiol 297:C503–C515

Tuo B, Wen G, Wei J, Liu X, Wang X, Zhang Y, Wu H, Dong X, Chow JY, Vallon V, Dong H (2011) Estrogen regulation of duodenal bicarbonate secretion and sex-specific protection of human duodenum. Gastroenterology 141:854–863

Turner JR (2009) Intestinal mucosal barrier function in health and disease. Nat Rev Immunol 9:799–809

van de Kerkhof EG, Ungell AL, Sjoberg AK, de Jager MH, Hilgendorf C, de Graaf IA, Groothuis GM (2006) Innovative methods to study human intestinal drug metabolism in vitro: precision-cut slices compared with ussing chamber preparations. Drug Metab Dispos 34:1893–1902

Veiga-Fernandes H, Mucida D (2016) Neuro-immune interactions at barrier surfaces. Cell 165:801–811

Venegas-Moreno E, Vazquez-Borrego MC, Dios E, Gros-Herguido N, Flores-Martinez A, Rivero-Cortes E, Madrazo-Atutxa A, Japon MA, Luque RM, Castano JP, Cano DA, Soto-Moreno A (2018) Association between dopamine and somatostatin receptor expression and pharmacological response to somatostatin analogues in acromegaly. J Cell Mol Med 22:1640–1649

Vieira-Coelho MA, Soares-da-Silva P (2000) Ontogenic aspects of D1 receptor coupling to G proteins and regulation of rat jejunal Na+, K+ ATPase activity and electrolyte transport. Br J Pharmacol 129:573–581

Wallace RA, Wallace L, Harrold M, Miller D, Uretsky NJ (1989) Interaction of permanently charged chlorpromazine and dopamine analogs with the striatal D-1 dopaminergic receptor. Biochem Pharmacol 38:2019–2025

Wallon C, Braaf Y, Wolving M, Olaison G, Soderholm JD (2005) Endoscopic biopsies in Ussing chambers evaluated for studies of macromolecular permeability in the human colon. Scand J Gastroenterol 40:586–595

Wang C, Xu H, Chen H, Li J, Zhang B, Tang C, Ghishan FK (2011) Somatostatin stimulates intestinal NHE8 expression via p38 MAPK pathway. Am J Physiol Cell Physiol 300:C375–C382

Wang Q, Ji T, Zheng LF, Feng XY, Wang ZY, Lian H, Song J, Li XF, Zhang Y, Zhu JX (2012) Cellular localization of dopamine receptors in the gastric mucosa of rats. Biochem Biophys Res Commun 417:197–203

Wang F, Zhang X, Wang J (2017) Effects of domperidone in combination with omeprazole in the treatment of chronic superficial gastritis. Pak J Med Sci 33:306–309

Wu Y, Hu Y, Wang B, Li S, Ma C, Liu X, Moynagh PN, Zhou J, Yang S (2020) Dopamine uses the DRD5-ARRB2-PP2A Signaling Axis to block the TRAF6-mediated NF-kappaB pathway and suppress systemic inflammation. Mol Cell 78:42–56.e6

Yan Y, Jiang W, Liu L, Wang X, Ding C, Tian Z, Zhou R (2015) Dopamine controls systemic inflammation through inhibition of NLRP3 inflammasome. Cell 160:62–73

Yang N, Liu SM, Zheng LF, Ji T, Li Y, Mi XL, Xue H, Ren W, Xu JD, Zhang XH, Li LS, Zhang Y, Zhu JX (2010) Activation of submucosal 5-HT(3) receptors elicits a somatostatin-dependent inhibition of ion secretion in rat colon. Br J Pharmacol 159:1623–1625

Yang S, Yang Y, Yu P, Yang J, Jiang X, Villar VA, Sibley DR, Jose PA, Zeng C (2015) Dopamine D1 and D5 receptors differentially regulate oxidative stress through paraoxonase 2 in kidney cells. Free Radic Res 49:397–410

Yang YJ, Bang CS, Baik GH, Park TY, Shin SP, Suk KT, Kim DJ (2017) Prokinetics for the treatment of functional dyspepsia: Bayesian network meta-analysis. BMC Gastroenterol 17:83

Yasuda M, Kawahara R, Hashimura H, Yamanaka N, Iimori M, Amagase K, Kato S, Takeuchi K (2011) Dopamine D(2)-receptor antagonists ameliorate indomethacin-induced small intestinal ulceration in mice by activating alpha7 nicotinic acetylcholine receptors. J Pharmacol Sci 116:274–282

Yin J, Tse CM, Avula LR, Singh V, Foulke-Abel J, de Jonge HR, Donowitz M (2018) Molecular basis and differentiation-associated alterations of anion secretion in human duodenal enteroid monolayers. Cell Mol Gastroenterol Hepatol 5:591–609

Zalecki M (2012) Localization and neurochemical characteristics of the extrinsic sympathetic neurons projecting to the pylorus in the domestic pig. J Chem Neuroanat 43:1–13

Zhang GH, Zhu JX, Xue H, Fan J, Chen X, Tsang LL, Chung YW, Xing Y, Chan HC (2007) Dopamine stimulates Cl(−) absorption coupled with HCO(3)(−) secretion in rat late distal colon. Eur J Pharmacol 570:188–195

Zhang XH, Zhang XF, Zhang JQ, Tian YM, Xue H, Yang N, Zhu JX (2008) Beta-adrenoceptors, but not dopamine receptors, mediate dopamine-induced ion transport in late distal colon of rats. Cell Tissue Res 334:25–35

Zhang XH, Ji T, Guo H, Liu SM, Li Y, Zheng LF, Zhang Y, Zhang XF, Duan DP (2010) Expression and activation of beta-adrenoceptors in the colorectal mucosa of rat and human. Neurogastroenterol Motil 22:e325–e334

Zhang X, Li Y, Liu C, Fan R, Wang P, Zheng L, Hong F, Feng X, Zhang Y, Li L, Zhu J (2015) Alteration of enteric monoamines with monoamine receptors and colonic dysmotility in 6-hydroxydopamine-induced Parkinson's disease rats. Transl Res 166:152–162

Zhang Y, Li Y, Zhu JX, Zhao WM (2017) Expression and cellular distribution of dopamine receptors in the rat colonic submucosal plexus. Capit Med Univ 38:411–416

Zheng LF, Song J, Fan RF, Chen CL, Ren QZ, Zhang XL, Feng XY, Zhang Y, Li LS, Zhu JX (2014) The role of the vagal pathway and gastric dopamine in the gastroparesis of rats after a 6-hydroxydopamine microinjection in the substantia nigra. Acta Physiol (Oxf) 211:434–446

Zheng LF, Ji T, Guo ZH, Wang T, Xiu XL, Liu XY, Li SC, Sun L, Xue H, Zhang Y, Zhu JX (2020) Na(+)-K(+)-2Cl(−) cotransporter 2 located in the human and murine gastric mucosa is involved in secretagogue-induced gastric acid secretion and is downregulated in lipopolysaccharide-treated mice. Eur J Pharmacol 880:173162

Zhu JX, Xue H, Ji T, Xing Y (2011) Cellular localization of NKCC2 and its possible role in the cl-absorption in the rat and human distal colonic epithelia. Transl Res 158:146–154

Zimmerman NP, Kumar SN, Turner JR, Dwinell MB (2012) Cyclic AMP dysregulates intestinal epithelial cell restitution through PKA and RhoA. Inflamm Bowel Dis 18:1081–1091

Chapter 5
Dopamine and Gastrointestinal Motility

Li-Fei Zheng, Sumei Liu, Li Zhou, Xiao-Li Zhang, Xiao Yu, and Jin-Xia Zhu

Abstract Dopamine (DA) plays an important role in regulating gastrointestinal (GI) motility function, from the esophagus to the colon. In this chapter, we focus on the role of DA in regulating esophageal, gastric, and intestinal motility under physiological processes. We describe the effect of DA on contractile activity of the lower esophageal sphincter (LES), esophageal peristalsis, gastric myoelectricity, intragastric pressure, antral motility, gastric emptying, antroduodenal coordination, intestinal motility, and GI transit time. DA antagonists have prokinetic effect on GI tract. We review the effect of DA antagonists on gastroparesis and gastroesophageal reflux, antroduodenal coordination, and colonic motility disorders. We discuss the possible mode of DA in GI motility regulation, via acting on smooth muscle cells, intrinsic neurons, and extrinsic neurons. In pathological states, such as Parkinson's disease (PD), we summarize the symptoms and underlying mechanism of abnormal GI motility in PD patients. Furthermore, we review GI motility function in several PD animal models.

Keywords Dopamine · Gastrointestinal motility · Dopamine receptor agonist · Dopamine receptor antagonist · Muscularis externa · Enteric nervous system

---

L.-F. Zheng (✉) · X.-L. Zhang · X. Yu · J.-X. Zhu
Department of Physiology and Pathophysiology, School of Basic Medical Science, Capital Medical University, Beijing, China
e-mail: 13581582937@163.com

S. Liu
Department of Biology, College of Science and Health, University of Wisconsin-La Crosse, La Crosse, WI, USA
e-mail: sliu@uwlax.edu

L. Zhou
Department of Physiology and Pathophysiology, School of Basic Medical Science, Capital Medical University, Beijing, China

Department of Human Anatomy, Henan Key Laboratory of Medical Tissue Regeneration, School of Basic Medical Sciences, Xinxiang Medical University, Xinxiang, China

J.-X. Zhu (ed.), *Dopamine in the Gut*, https://doi.org/10.1007/978-981-33-6586-5_5

Abbreviations

5-HIAA	5-Hydroxyindoleaceticacid
5-HT	5-Hydroxytryptamine
6-OHDA	6-Hydroxy-dopamine
ACh	Acetylcholine
AP	Anteroposterior
At2	Angiotensin II type 2 receptor
BAC	Bacterial artificial chromosome
BBB	Blood–brain barrier
ChAT	Choline acetyltransferase
CNS	Central nervous system
cNTS	Subnucleus centralis of the nucleus tractus solitarius
COMT	Catechol-*O*-methyltransferase
COX-2	Cyclooxygenase-2
D_1R	Dopamine D_1 receptor
D_2R	Dopamine D_2 receptor
D_3R	Dopamine D_3 receptor
D_4R	Dopamine D_4 receptor
D_5R	Dopamine D_5 receptor
DA	Dopamine
DARs	Dopamine receptors
DAT	Dopamine transporter
DMSO	Dimethyl sulfoxide
DMV	Dorsal motor nucleus of the vagus
DOPAC	3,4-Dihydroxyphenylacetic acid
DV	Dorsoventral
DVC	Dorsal vagal complex
EFS	Electrical field stimulation
EGG	Electrogastrogram
ENS	Enteric nervous system
FD	Functional dyspepsia
FDA	Food and Drug Administration
FGID	Functional gastrointestinal disorders
FJ-C	Fluorojade-C
GER	Gastroesophageal reflux
GFAP	Glial fibrillary acidic protein
GFP	Green fluorescent protein
GI	Gastrointestinal
hpGRF	Human pancreatic growth hormone-releasing factor
HVA	Homovanillic acid
i.c.	Intracisternal
i.c.v.	Intracerebroventricular
i.m.	Intramuscular

i.p.	Intraperitoneal
i.t.	Intrathecal
i.v.	Intravenous
Iba1	Ionized calcium binding adaptor molecule 1
IBS	Irritable bowel syndrome
ICCs	Interstitial cells of Cajal
IL-17	Interleukin-17
IL-1β	Interleukin-1 beta
IL-6	Interleukin-6
iNOS	Inducible nitric oxide synthase
L-DOPA	Levodopa
LES	Esophageal sphincter
L-NAME	Nitro-L-arginine methyl ester
LPS	Lipopolysaccharides
MDA	Malondialdehyde
MFB	Medial forebrain bundle
ML	Medio-lateral
MMC	Migrating motor (myoelectric) complex
MPTP	1-Methyl 4-phenyl 1,2,3,6- tetrahydropyridine
MRI	Magnetic resonance imaging
NADPH	Nicotinamide adenine dinucleotide phosphate
NAmb	Nucleus ambiguus
NF-κB	Nuclear factor kappa b
NK1	Neurokin 1
nNOS	Neuronal nitric oxide synthase
NOS	Nitric oxide synthase
NT	Neurotensin
p.o.	Per os
PD	Parkinson's disease
PFFs	Preformed fibril α-synuclein
PGP9.5	Protein gene product 9.5
p-α-Syn	Phosphorylated α-synuclein
RAS	Renin-angiotensin system
RC	Rostro-caudal
s.c.	Subcutaneous
SCFA	Short-chain fatty acids
SD	Sprague-Dawley
SIBO	Small intestinal bacterial overgrowth
SMCs	Smooth muscle cells
SN	Substantia nigra
SNpc	Substantia nigra pars compacta
SP	Substance P
t.i.d.	Three times a day
TBK1	TANK-binding kinase 1

TH	Tyrosine hydroxylase
TLR-4	Toll-like receptor 4
TNF	Tumor necrosis factor
TNF-α	Tumor necrosis factor α
TRH	Thyrotropin-releasing hormone
VIP	Vasoactive intestinal peptiepeptide
VMAT2	Type 2 vesicular monoamine transporter
α-Syn	Alpha-synuclein
β1	β1-Adrenoceptor
β3	β3-Adrenoceptor

5.1 Introduction

Gastrointestinal (GI) motility in digestive and interdigestive states is caused by contractions of the gut smooth muscle. Two layers of smooth muscle physically drive motility in the GI tract: an inner circular muscle layer and an outer longitudinal muscle layer. GI motility is traditionally believed to be dictated by coordinated activities from smooth muscle cells (SMCs), pacemaker interstitial cells of Cajal (ICCs), and motor neurons in the enteric nervous system (ENS) (Sanders et al. 2012). Gut smooth muscle contraction is regulated and controlled by intrinsic ENS (the "brain of the gut") and extrinsic autonomic nervous system. It is also modulated by GI innate immune system, enteric endocrine system, and gut microbiota.

The extrinsic innervation includes the classic vagal parasympathetic and sympathetic components, with afferent sensory and efferent secretomotor fibers. The intrinsic innervation is represented by the ENS, which is recognized as a complex neural network controlling a variety of cell populations, including SMCs, mucosal secretory cells, endocrine cells, microvasculature, immune, and inflammatory cells. The ENS is involved in regulating GI secretion, absorption, and motility. In particular, this network is organized in two plexuses, the myenteric (Auerbach) and submucosal (Meissner) plexuses, each one providing quite autonomous control of GI functions. The myenteric plexus is located between the circular and longitudinal muscle layers and regulates contraction and relaxation of the musculature, producing motility which mix and propel the luminal contents. Located between the circular muscle and submucosa, the submucosal plexus regulates GI fluid secretions and blood flow. A large variety of neurons and neurotransmitters take part in the ENS, such as dopamine (DA)-containing neurons.

As a catecholamine neural transmitter, DA is derived not only from the central nervous system (CNS) (such as the substantia nigra (SN), the dorsal motor nucleus of the vagus (DMV)) but also from the peripheral organs (such as gastric parietal cells). DA receptors (DARs) are widely distributed throughout the GI tract, including the ENS, gut innate immune cells, and enteric endocrine cells. They play an important role in regulating GI motility, including esophageal motility, gastric myoelectricity, and postprandial and interdigestive contractions of the stomach, small intestine, and large intestine.

An important role for endogenous DA as an inhibitory neuromodulator of food intake, lower esophageal sphincter (LES) contraction, gastric contractile activity, colonic motility, gastroduodenal motility, and GI transit is suggested. The regulation of gut motility through DA requires binding to DARs. DARs are widely distributed on GI muscularis externa, including enteric neurons and SMCs. DA exerts a direct effect through activation of DARs on GI muscle cells or an indirect effect through inhibition of acetylcholine (ACh) release from intrinsic cholinergic motor neurons via the activation of pre-junctional D_2 receptors (D_2Rs). DAR antagonists (metoclopramide, domperidone, alizapride) may be useful for a number of GI motility disorders such as gastro-esophageal reflux, gastroparesis, intestinal pseudo-obstruction, cystic fibrosis, and constipation (Demol et al. 1989). Among the D_2R antagonists, we will discuss metoclopramide, domperidone, levosulpiride, and itopride in detail. From a therapeutic point of view, metoclopramide and domperidone are used in gastric motility disorders. Domperidone, which is widely used in many countries (Reddymasu et al. 2007), acts as an antiemetic agent via the chemoreceptor trigger zone (CTZ) and a prokinetic agent through its effects on stomach and small intestinal motility. Unlike metoclopramide, domperidone does not cause any adverse neurological symptoms as it has minimal penetration through the blood–brain barrier. It thus provides an excellent safety profile for long-term administration orally in the recommended doses.

Degeneration of dopaminergic neurons in either CNS or ENS causes GI dysmotility. Thus, it is critical to understand the specific role of DA in gut motility and, most importantly, to characterize specifically the enteric neuropathology occurring in neurodegenerative disorders, such as Parkinson's disease, which is associated with the loss of dopaminergic neurons.

5.2 Effect of Dopamine on Gastrointestinal Motility

In the GI tract, the catecholamine DA is produced by gastric parietal cells, gut immune cells, a subpopulation of enteric neurons, and even gut microbiota. DARs are widely distributed in the ENS and enteric SMCs. DA regulates GI contraction activities from esophagus to colon, gastric myoelectricity, antroduodenal coordination, and GI transit via DARs or adrenergic receptors on enteric SMCs or neurons of myenteric plexus. Besides, central DA also can control GI motility.

5.2.1 Esophageal Motility

The function of the esophagus is to transport the swallowed contents into the stomach. The esophagus is divided into three motor regions: the upper esophageal sphincter, the esophageal body, the LES. The esophagus has longitudinal and circular muscle layers. Following swallow, the two sphincters relax and open, and peristaltic contraction sweeps behind the bolus autonomously through the entire length of the esophagus. During peristalsis, the two muscle layers contract in perfect synchrony, while during transient LES relaxation, longitudinal muscle contracts independent of the circular muscle. In some motility disorders of the esophagus, there is discoordination between the two muscle layers (Mittal 2013). Endoscopy, barium radiography, esophageal electrical impedance recordings, and manometry are used as the diagnostic mainstays for esophageal motor disorders. Spastic esophageal motor disorders and ineffective esophageal motility are associated with reflux esophagitis (Smout 2008). Esophageal motor disorders often lead to dysphagia and chest pain.

From 1980, researchers have confirmed that DA plays different roles on the esophageal motility via D_1 receptors (D_1Rs) or D_2Rs. Intravenous (i.v.) infusion of DA (10–40 mg/kg/min) completely inhibits contractile activity of the LES during the digestive and interdigestive states in healthy conscious dogs by means of chronically implanted force transducers (Itoh et al. 1980). From rat study, DA inhibits intraluminal LES pressure and in vitro contraction of LES helical strips via D_2Rs, but stimulates contraction of LES helical strips via D_1Rs (Sigala et al. 1994). Pharmacological characterization with selective D_1R and D_2R agonists and antagonists strongly suggests that D_1Rs are involved in LES helical strips contraction via stimulating cyclic AMP formation, while D_2Rs mediate the relaxation of the sphincter via inhibiting cyclic AMP formation. The selective D1-like agonist fenoldopam (40 mg/kg, i.v.) increases the LES pressure; on the other hand, the D2-like agonist bromocriptine (10 mg/kg, i.v.) induces a decrease of the resting LES pressure (Sigala et al. 1994). Besides, a study from binding experiments suggests the participation of the α_2-adrenergic receptors in the contractile response elicited by DA in the isolated chicken esophagus (Sánchez et al. 1990).

D_2R antagonists are widely applied in both clinical and experimental medicine. D_2R antagonist domperidone (0.5, 1.0, and 2.0 mg/kg) alone has no activity upon normal contractions of the LES, but it antagonizes the effect of DA on both postprandial and interdigestive contractions in a dose-related fashion (Itoh et al. 1980). Another D_2R antagonist metoclopramide (0.25, 0.5, and 1.0 mg/kg) can abolish normal interdigestive contractions of esophagus. DA-induced inhibition is antagonized by metoclopramide during the digestive state, but not the interdigestive state (Itoh et al. 1980). In clinic, metoclopramide causes a small and nonsignificant increase in LES pressure, without effect on esophageal peristalsis during gastroesophageal reflux (GER) in infants.

5.2.2 *Gastric Motility*

The stomach has longitudinal, circular, and oblique muscle layers. The oblique layer helps mix and grind the chyme. The stomach can be divided into two distinct motor regions: a proximal region (the gastric fundus and the oral third of the gastric corpus, which serves as gastric reservoir) and a distal region (the distal two thirds of the corpus, the antrum, and the pylorus). During fasting, the stomach exhibits interdigestive migrating motor (myoelectric) complexes (MMCs). When food passes down the pharynx and the esophagus, the proximal stomach receives and stores food by receptive relaxation whereby the smooth muscle relaxes with limited increases in intragastric pressure. The mechanism of receptive relaxation is mediated by a vago-vagal reflex. When food enters the stomach, the fundus dilates by adaptive relaxation with small increases in intragastric pressure. Adaptive relaxation is an intragastric pressure-induced reflex. Stretch of the gastric wall activates the mechanoreceptors in gastric mucosa, then the capsaicin-sensitive afferent sensory neuron generates impulses. The sensory neuron directly activates the inhibitory efferent neuron or indirectly activates it via interneurons of the myenteric plexus. This causes relaxation of circular muscle. The proximal stomach then pushes the stored contents into the corpus and antrum by its slow, sustained, tonic contractions. By pacemaker potentials and the peristaltic contractions, the distal stomach mixes and triturates the gastric contents (chyme) into particles of about 0.1 mm for passage through the pyloric sphincter (Kumral and Zfass 2018). The pylorus serves as a regulator of gastric emptying into the duodenum. During trituration, the pyloric tone is increased to retain food with antral mixing and to prevent duodenogastric reflux. Gastric emptying occurs only when there is a sufficient pressure difference between the stomach and the duodenum to overcome the resistance to flow across the pylorus. Liquid emptying is fundal with less importance on pyloric opening, due to proximal gastric tonic contractions resulting in a pressure gradient extending from the fundus to the pylorus (Solnes et al. 2018). However, solid emptying occurs in the antrum where phasic contractions grind food into small particles (<2 mm) so they can pass through the pylorus. It is determined to a large extent by the resistance to their passage across the pylorus. Antral and pyloric contractions determine the resistance at the pylorus. Disorders of gastric reservoir function result in symptoms of early satiety and anorexia. Diseases and operations that impair distal gastric contractions may cause slow gastric emptying of solids.

5.2.2.1 Gastric Contractile Activity

From clinical experiments, we know that graded doses of DA decrease gastric tone (Levein et al. 1999a), decrease intragastric pressure, and delay gastric emptying (Valenzuela and Liu 1982; 5 mg/kg/min, Levein et al. 1999a) in healthy or vagotomized subjects. DA interrupts the fed motility pattern and inhibits the high antral waves in humans (Marzio et al. 1990). DA infusion (1 mg/kg/hr., i.v.) causes the disappearance of postprandial gastric motility (Iwanaga et al. 1996). Besides, oral levodopa (L-DOPA, 1000 mg), a DA precursor, also inhibits gastric emptying of a mixed solid–liquid meal in normal subjects (Berkowitz and McCallum 1980).

Evidence from animal studies suggests that DA causes decrease in intragastric pressure (Valenzuela 1976), inhibits interdigestive-state gastric motility in anesthetized rats (Zheng et al. 2014), inhibits gastric contractile activity (Itoh et al. 1980) and antral motility (Bech et al. 1982) in a dose-related manner in conscious dogs and rats (Takeuchi et al. 1988), decreases spontaneous activity of isolated guinea pig stomach (0.25 μg/ml) (Van Nueten et al. 1978), and inhibits the amplitude of gastric contraction initially followed by a secondary prolonged period of increase in contractile activity in fasted ponies (King and Gerring 1988). DA even at low dose (5 mg/kg/min) significantly diminishes gastric emptying, which is mainly caused by a reduction in the contraction force of the antral waves. DA at higher dose (10–15 mg/kg/min) induces retrograde power contractions accompanied by enterogastric reflux (Ehrlein 1988). During stress loading, DA suppresses the gastric motility and increases of gastric acid secretion in a dose-dependent fashion up to 25 mg/kg/min (Nagahata et al. 1990).

The inhibitory effect of DA on gastric motility is mediated by D_2Rs. The D_2R antagonist domperidone alone (0.5, 1.0, and 2.0 mg/kg, i.v.) has no activity upon normal contractions of the stomach, but it antagonizes the effect of DA on in vivo postprandial and interdigestive contractions or in vitro isolated gastric strip contractions in a dose-related fashion. Domperidone completely prevents the DA effects on intragastric pressure, gastric emptying (Valenzuela and Liu 1982), and antral waves (Marzio et al. 1990) in healthy and even vagotomized subjects or dogs (Itoh et al. 1980). Moreover, Dr. Zhu JX's lab has reported that up to 50 μg/kg domperidone can enhance the antral motility in normal rats (Zheng et al. 2014). An in vitro study shows that domperidone (0.16 μg/ml) selectively inhibits the effect of DA on isolated guinea pig stomach (Van Nueten et al. 1978). D_2R antagonist haloperidol (0.16 μg/ml) also selectively inhibits the effect of DA on gastric contraction in guinea pig (Van Nueten et al. 1978). Metoclopramid, a D_2R antagonist, increases the rate of gastric emptying of a liquid meal in infants with GER (Hillemeier et al. 1986) and adult rats (5–10 mg/kg, subcutaneous (s.c.) injection) (Cooper and McRitchie 1985), although 10 mg metoclopramid is not enough to fully reverse DA inhibitory effects on gastric tone in healthy volunteers (Levein et al. 1999a). Besides, intramuscular (i.m.) injection of metoclopramide (10 mg) antagonizes effect of oral L-DOPA (1000 mg) in restoring gastric emptying to normal in healthy subjects (Berkowitz and McCallum 1980). In conscious dogs, inhibition of antral motility by DA is significantly blocked by the peripherally acting dopaminergic blocker domperidone and by cis-flupenthixol, which blocks both peripheral and central DARs. While the effect of DA is not significantly altered by the β_1-adrenoceptor blocker practolol, the α-adrenoceptor blocker phentolamine, or the combined α and β-adrenoceptor blocker labetalol (Bech and Hovendal 1982; Bech et al. 1984). Gastric emptying is decreased by peripherally administering a DAR agonist apomorphine (0.1–0.5 mg/kg s.c.), which is antagonized by pretreatment with D_2R antagonist haloperidol, domperidone, or (−)-sulpiride (Costall et al. 1983). DAR agonist 6,7-ADTN (1–50 mg/kg) also delays gastric emptying of a liquid test meal in conscious rats with chronic gastric fistula. Although there is a report showing that the inhibitory effects of DA on gastric motor activity can be significantly reversed by pretreatment with yohimbine, an inhibitor of α_2-adrenoceptors

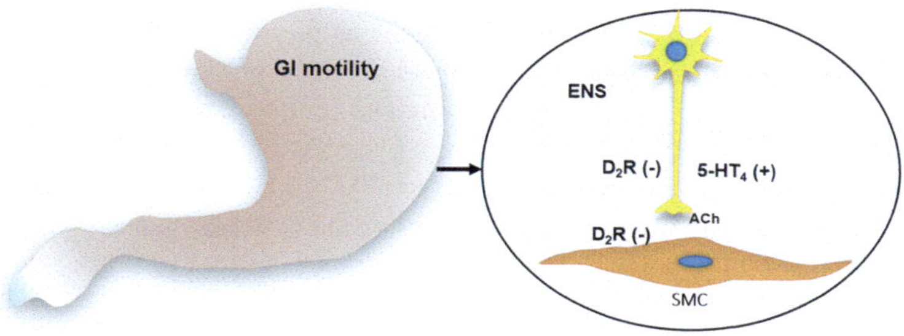

Fig. 5.1 Schematic of the effects of dopamine (DA) on gastric motility. Both D_2 receptors (D_2R) and 5-HT_4 receptors are located on cholinergic neurons in the gastric wall. DA inhibits gastric motility by reducing acetylcholine (ACh) release, which is mediated by D_2R on the cholinergic neurons. However, the 5-HT_4 receptor activation can evoke ACh release to enhance the gastric motility. *ENS* enteric nervous system, *GI* gastrointestinal, *SMC* smooth muscle cell

(5 mg/kg, s.c.) (Takeuchi et al. 1988), DA actions on gastric motility and gastric emptying are mainly mediated by the DARs.

DA exerts their inhibition of gastric motor activity partly via intramural cholinergic neural pathways (Ulvestad and Gerner 1984). Study from in vitro guinea pig antral pouches shows that DA significantly impairs antral pressure responses to cholinesterase inhibitor physostigmine, whereas the ACh-induced response is not significantly reduced. It suggests that DA to some extent acts on intramural cholinergic neural pathways by reducing spontaneously release of ACh (Ulvestad and Gerner 1985). Vagotomy and splanchnicectomy have no effect on the inhibitory effect of DA, which implies that inhibition of gastric motility by DA is independent of extrinsic innervation (Takahashi et al. 1991; Nagahata et al. 1995). In vitro DA inhibits [^{3}H]-ACh release in a dose-dependent manner (Nagahata et al. 1995). Pertussis toxin prevents activation of the inhibitory guanine nucleotide binding protein coupled to adenylate cyclase. Pretreatment with pertussis toxin blocks the inhibitory action of DA on the release of [^{3}H]-ACh evoked by electric field stimulation (Takahashi et al. 1991). The inhibitory effect of DA is antagonized by D_2R antagonist domperidone but not by phentolamine, propranolol, or SCH-23390. Thus, inhibition of gastric motility by DA is mediated by D_2Rs on cholinergic neurons in the gastric wall (Fig. 5.1).

Besides the peripheral effect, intracerebroventricular (i.c.v.) and intrahypothalamic injection of metoclopramide can enhance field stimulation-induced contractions of gastric strips and gastric emptying in the guinea pig (Costall et al. 1987). Gastric emptying is facilitated by i.c.v. administrations of DAR antagonists metoclopramide and clebopride and is inhibited by i.c.v. DAR agonist apomorphine; the response to i.c.v. apomorphine is antagonized by i.c.v. pretreatment with DAR antagonists haloperidol, domperidone, (−)-sulpiride, metoclopramide, and clebopride. Facilitation of emptying by i.c.v. metoclopramide is prevented by peripheral pretreatment with atropine (Buéno et al. 1983). Microinjection of DA into caudate nucleus could also inhibit gastric spike potentials and motility, an effect again being blocked by co-injected DA antagonist haloperidol or atropine (Jing et al. 1995). DA microinjection in the dorsal vagal complex (DVC) in the brain stem decreases gastric tone and phasic contractions in both corpus and antrum in rats, and the effect is abolished by ipsilateral vagotomy and fourth ventricular treatment with the selective D_2R antagonist L741,626 but not by application of the selective D1-like receptor antagonist SCH-23390 (Anselmi et al. 2017a). Hence, DA in the CNS also can inhibit gastric motility via D_2Rs. DA inhibits gastric emptying at least in part, via vagal cholinergic pathway.

5.2.2.2 Gastric Electrical Activity

Phasic contractions of the GI smooth muscle result from cyclical depolarization/ repolarization cycles of the membrane potential, known as electrical slow waves. ICCs are considered as the pacemaker of slow waves and are electrically coupled to GI smooth muscle cells (Sanders 2019). The dominant pacemaker in the stomach is localized in the myenteric pacemaker region of the proximal portion of its body. The myenteric pacemaker generates small membrane potential oscillations, spreading into both circular and longitudinal muscle layers. When the ICC network becomes impaired, slow waves are abolished and their spreading is not measurable. The phasic electrical activity in GI muscles relies on the pacemaker ICC activity and the response of the muscles. The excitability of muscle cells facilitates the entry and spreading of pacemaker activity. Normal slow wave frequency is an intrinsic activity of the ICCs. Any change in slow wave frequency is often considered as an indicator of interventions into the pacemaker activity in intact muscles. The abnormal slow wave frequency as well as the absence of slow wave activity are associated with abnormal GI transit and gastroparesis (Chen et al. 1996). DA induces concentration-dependent inhibition in frequency and propagation of slow waves in isolated gastric muscle of cats, which is significantly inhibited by domperidone and phentolamine but not by propranolol, hexamethonium, and tetrodotoxin (TTX) (Jo et al. 1989). It indicates that DAR plays a role in gastric electrical abnormality. However, domperidone does not prevent gastric dysrhythmia or the symptoms of motion sickness induced with spatial disorientation (Kono et al. 1999).

5.2.3 Duodenal Motility

DA shows a dose-dependent inhibition on the duodenal hypermotility following cysteamine; and at 30 mg/kg, DA almost completely abolishes the development of contractions (Takeuchi et al. 1987). In addition to inhibitory effect, DA stimulates duodenal motility by producing a phase III-like activity of the MMC in human, while sulpiride and domperidone can prevent the effect of DA (Marzio et al. 1986). DA can also stimulate the excitatory α-adrenergic receptors in opossum duodenal muscle

(Anuras 1981) and excite isolated rat duodenum by acting on serotoninergic receptors directly (Miachon et al. 1978).

In the dog, DA acts centrally to modify the food-induced disruption of the MMC pattern and the frequency of the MMC. I.c.v. administration of DA (10 mg/kg) increases the duration of the interval between MMC at the duodenum, while the duration of the postprandial disruption of the MMC pattern is significantly reduced. All of the central effects persist after vagotomy (Fioramonti et al. 1984). Thus, the central effect of DA on the duodenum is vagus-independent.

5.2.4 Antroduodenal Coordination

Antroduodenal coordination contributes to sequential motor activity of the stomach and duodenum. Coordination of antral and duodenal motor activities is strictly regulated and governed by GI electrical activity and neural connectivity through enteric nerves and vagal efferent nerves (Hausken et al. 1992). Abnormal gastroduodenal motor coordination is thought to be part of many motility disorders. DA inhibits antroduodenal coordination. Intra-arterially administered DA (0.25 mg/L) reduces phasic activity of duodenal bulb and blocks antroduodenal coordination on isolated guinea-pig stomach preparations. The D_2R antagonist, domperidone or metoclopramide, increases the phasic amplitude and improves antroduodenal coordination in guinea pig (Schuurkes and Van Nueten 1981) and dog (Schuurkes and Van Nueten 1984b). Domperidone effectively antagonizes the inhibitory effects of DA in a competitive way. Both TTX and atropine, but not hexamethonium, prevent the stimulatory effects of domperidone. These results indicate that DA controls antroduodenal coordination via specific DARs, D_2Rs, located on cholinergic intramural nerves in the gut wall (Schuurkes and Van Nueten 1984a; Van Nueten and Schuurkes 1984).

5.2.5 Small Intestinal Motility

The small intestine produces a variety of patterns of contraction that promote the propulsion of intestinal contents. The small intestine motility is accomplished by both the circular and the longitudinal smooth muscle layers. Intestine motility is divided into tonic and phasic contractions, which are mediated by an increase in intracellular calcium. The spike potentials on top of the slow waves induce phasic muscle contraction. Slow waves and spike potentials, and thus, phasic contraction, can be influenced by neurotransmitters, hormones, and drugs. Ingestion of food initiates peristaltic propulsive activity. The peristaltic reflex consists of two successive phases: descending relaxation of circular muscle distal to the bolus and ascending contraction of circular muscle proximal to the bolus. In-between meals, the MMC prevails, which helps to clear the luminal contents of the intestine. Small intestinal motility can be recorded by manometry and by transit tests (Husebye 1999). Abnormalities in small bowl motility can result in increased transit, decreased transit, or nonpropulsive activity. Patients with small intestinal dysmotility may experience symptoms like bloating, nausea, abdominal pain, constipation, or diarrhea (Kuemmerle 2000).

Infusion of DA (1, 5, 10, or 28–38 mg/kg/min, i.v.) inhibits the motility of ileum in ponies (King and Gerring 1988) or sheep (Cebrat et al. 1989). After DA administration, the small intestine shows a prolonged period of irregular activity (phase II) and a marked increase in the interval between successive phase IIIs in fasted ponies (King and Gerring 1988). D1-like receptor antagonist SCH-23390, but not D2-like receptor antagonist (sulpiride or domperidone), can increase basal tone and amplitude of spontaneous contractions of mouse ileum, and antagonize the responses to DA, which implies that D1-like receptor plays a major role in the DA modulation of mouse ileal contractility (Zizzo et al. 2010). The lower concentration of DA activates the DARs, while medium to high concentrations of DA can activate the adrenoceptors. In guinea pig (Zar et al. 1982) and sheep (Cebrat et al. 1989), the inhibitory action of DA in the ileum is antagonized by phentolamine (5 μM in vitro or 0.6–0.8 mg/kg i.v.), an α_1/α_2-adrenoceptor antagonist, but not D_2R antagonists (metoclopramide 90 μM in vitro, primozide 1 μM in vitro, or domperidone 0.4–0.8 mg/kg i.v.) or β_1/β_2-adrenoceptor antagonist (propranolol 0.5–0.7 mg/kg i. v.). In the myenteric plexus and longitudinal muscle tissues of the guinea pig ileum, there are prejunctional inhibitory α_2, postjunctional inhibitory α_2 and β, and postjunctional excitatory α_1 adrenoceptors (Bauer and Kuriyama 1982). In vitro studies on the longitudinal muscle of the guinea pig ileum show that DA (1–100 mM) inhibits cholinergic transmission by inhibiting neuronal ACh release via prejunctional α_2- and postjunctional β_2-adrenoceptor (Bauer and Kuriyama 1982). In summary, DA plays an inhibitory effect on the ileum via different receptors in different animals.

In 2-day-old mice, DA induces a TTX-insensitive contractile effect in small intestine via D1-like receptor. TTX is a potent non-protein specific blocker of voltage-gated sodium channels. In 90-day-old mice, the relaxation effect of DA is inhibited by both D1-like and D2-like receptor antagonists, and mimicked by D1-like and D2-like receptor agonists. D_1R, D_2R, and D_3R expression shows a significant increase in proximal intestine with age. In mouse small intestine, the response to DA undergoes developmental changes shifting from contraction to relaxation at weaning, due to D2-like receptor recruitment and increased expression of D_1Rs (Zizzo et al. 2016). By acting on D2-like receptors, DA inhibits indomethacin-induced intestinal hypermotility to protect the small intestine against indomethacin-induced damage in rats (Miyazawa et al. 2003).

Besides, some hormones are also able to act centrally to control duodenal or jejunal motility through dopaminergic pathway, such as human pancreatic growth hormone-releasing factor (hpGRF) (Bueno et al. 1985) and neurotensin (NT) (Fargeas et al. 1990). Injected ventricularly 10–15 min after a daily meal, hpGRF (0.1 mg/kg, i.c.v) reduces the duration of the jejunal fed pattern in dogs. This effect of hpGRF is abolished by a previous i.v. injection of D_2R antagonists metoclopramide (1 mg/kg) (Bueno et al. 1985). NT (10 mg, i.c.v) delays the occurrence of postprandial disruption of duodenal motility and blocked it on the jejunum in conscious rats. The central effect of NT on the fed pattern involves dopaminergic neurons as it is mimicked by DA, blocked by D_2R antagonists haloperidol (Fargeas et al. 1990). Injection of amphetamine (3 mg/ml/kg, i.p.), a central nervous stimulant, inhibits gastric emptying and intestinal transit in male rats. Haloperidol completely prevents amphetamine induced inhibition of gastric emptying and intestinal transit in male rats. The selective D1-like receptor antagonist, SCH-23390, also significantly attenuates the amphetamine-induced inhibition of gastric emptying and intestinal transit (Huang and Chien 2012). These findings clarify the influence of CNS stimulant on GI tract via the DA pathway.

5.2.6 *Colonic Motility*

The luminal content in the colon becomes progressively harder and thicker after water being absorbed. Compared to the small intestine, the colon requires stronger contractions with larger amplitude and longer duration to propel and mix semisolid to solid contents. The colonic motor activity can be divided into three basic categories: individual phasic contractions, organized groups of contractions, and special propulsive contractions (Sarna 1991). In practice, measurements of colonic transit are the best characterized and most widely available tests of colonic motor function. Scintigraphy, experimental stable isotope measurement of orocecal transit, and wireless motility can be used to measure small-bowel and colonic transit (Szarka and Camilleri 2012). Colonic high-resolution manometry is recently used in human to characterize colonic propagating and nonpropagating motor patterns (Dinning 2018). The relevant symptoms of colonic dysmotility are diarrhea, constipation, and pain (O'Brien and Phillips 1996).

The effect of DA on colonic motility is controversial. In 1979, Lechin et al. has found that sulpiride, a D_2R antagonist, increases human sigmoidal motility by blocking postsynaptic DARs (Lechin and van der Dus 1979). Measurements performed on the rat and guinea pig distal colon and even dog proximal colon (Bueno et al. 1984) have demonstrated that DA in a dose-dependent manner inhibits the spontaneous motility (Gonzalez and Sarna 2001), and prolongs colonic transit time for the artificial feces in the colon (Lim et al. 2008). DA infused intravenously (29–38 mg/kg/min) inhibits the cecum motility of sheep; however, domperidone (0.2–0.4 mg/kg), a D_2R antagonist without crossing the brain–blood barrier, does not change its movements (Zieba et al. 1991). The amplitude and frequency of spontaneous giant contractions in circular muscle strips of middle colon are significantly increased by the D1-like receptor antagonist SCH-23390 (Gonzalez and Sarna 2001). Dr. Zhu JX's lab has reported that TTX, a voltage-gated sodium channels blocker, increases contraction frequency of isolated distal colon, but fails to affect the inhibition effect of DA on the motility (Fig. 5.2) (Zhang et al. 2012), indicating a nonneural mechanism of DA on colonic motility. In the presence of TTX,

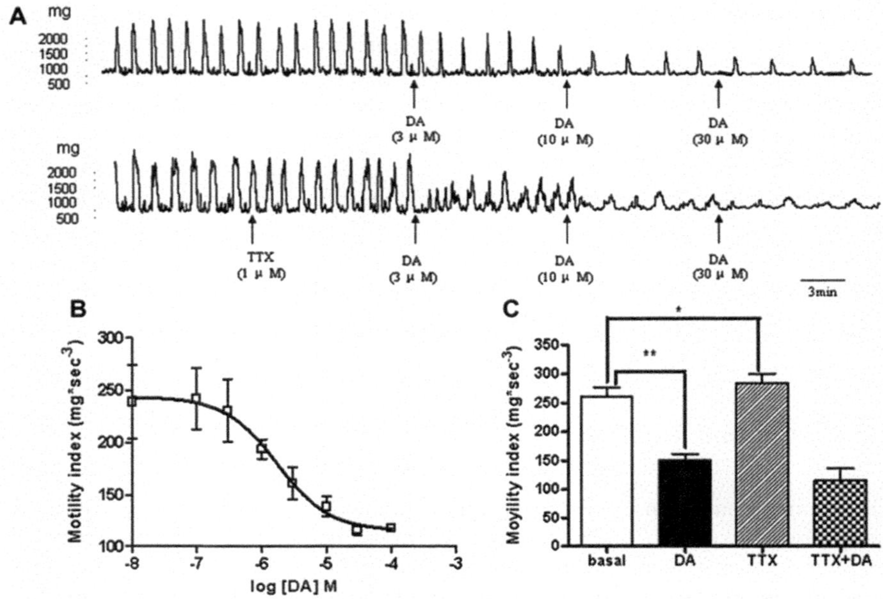

Fig. 5.2 The inhibitory action of dopamine (DA) on the spontaneous motility of rat distal colon with or without tetrodotoxin (TTX). (**a**) Original tracings showing the effects of DA on the colonic motility in the absence or presence of TTX. (**b**) Concentration–response curve of DA-induced inhibition of motility in rat distal colon. (**c**) The effects of DA, TTX, and DA + TTX on colonic contraction. (Zhang X, et al., Transl Res 2012, 159(5), 407–414)

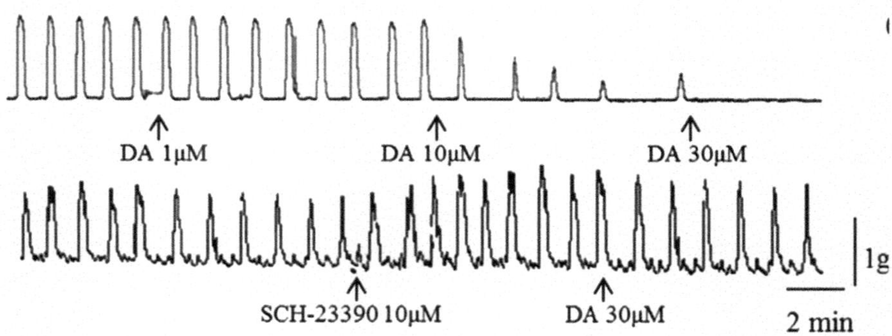

Fig. 5.3 Representative recordings showing the effects of dopamine (DA) on the motility of isolated rat colon in the presence or absence of the D1-like receptor antagonist SCH-23390. (Zhang X, et al., Transl Res 2015, 166(2), 152–162)

SCH-23390, but not sulpiride, antagonizes the inhibition of DA on colonic muscle strips (Fig. 5.3) (Zhang et al. 2015). It suggests that DA inhibits colonic motility via non-neuronal D_1Rs. While Aguilar et al. have reported that DA exhibits a relaxant effect on the isolated rat colon via atypical β-adrenoceptors (Aguilar et al. 2005).

However, in contrast to its known inhibitory effect, DA (5–100 mg/kg) is also reported to stimulate rectosigmoid motility in healthy subjects (Wiley and Owyang 1987). Administration of D_2R antagonist sulpiride (100 mg, i.t.) significantly reduces the postprandial increase in colonic motor activity in patients with irritable colon (Lanfranchi et al. 1983). In the dog, i.v. infusion of DA (0.5 and 1 mg/kg/h) and bromocriptine (10–40 mg/kg/hr), a non-selective D2-like receptor agonist, stimulates the motility of the descending colon and whole colon, respectively. The excitatory effect can be blocked by D_2R antagonists haloperidol and domperidone, but not α- and β-adrenoceptor blocking agents (Bueno et al. 1984). In a brief summary, DA primarily exerts an inhibitory role on colonic motility via nonneural D1-like receptors or an excitatory effect via D2-like receptors.

In addition, i.c.v. injection of DA (100 mg/kg) increases colonic spike burst frequency in rats. Injection of SKF-38393 (5 mg/kg, i.c.v.), a selective D1-like receptor agonist, or quinpirole (5 mg/kg), a selective D2-like receptor agonist, also increases colonic spike burst frequency (Gué et al. 1994). It suggests that DA acts on both D1- and D2-like receptors on the central neurons to increase colonic spike burst. Intrathecal (i.t.) application of DA into the L6-S1 spinal cord, where the lumbosacral defecation center is located, causes propulsive contractions of the colorectum via D2-like receptors, but not D1-like receptors. Inactivation of spinal neurons using TTX blocks the effect of DA. The spinal neurons show an inward current in response to DA in patch clamp recordings from spinal cord slices. The effect of DA is mediated primarily by D2-like receptors expressed on choline acetyltransferase (ChAT) immunoreactive neurons. It suggests that DA activates sacral parasympathetic preganglionic neurons via D2-like receptors and causes propulsive motility of the colorectum in rats (Naitou et al. 2016).

5.2.7 Gastrointestinal Transit

GI transit includes the standardized esophageal transit, solid and liquid gastric emptying, small-bowel transit, colon transit, and whole-gut transit. GI transit process involves an intricate coordination of smooth muscle contractions leading to mixing and forward propulsion. The nervous system plays a critical role in these activities. Many factors alter GI transit. Functional GI disorders (FGID) are often associated with abnormalities of gut transit (e.g., delayed gastric emptying, constipation, diarrhea). The gastric emptying rate and the intestinal transit rate are also important factors affecting the plasma concentration profile of orally administered drugs and the drug absorption. GI transit tests can be performed using several methods, including scintigraphy (gastric emptying, small bowel transit, and colonic transit), breath test (gastric emptying and small bowel transit), wireless motility capsule (gastric emptying, small bowel transit, and colonic transit), tracking with radio-opaque markers (colonic transit), and emerging magnetic resonance imaging (MRI) approaches (Lin et al. 2005; Rao et al. 2011; Kumral and Zfass 2018; Al-Saffar et al. 2019).

Orocaecal transit time is prolonged during continuous infusion of DA (5 mg/kg/min, i.v.) in volunteers (Levein et al. 1999b). In vivo application of metoclopramide, a D_2R antagonist, significantly accelerates GI transit in mice and gastric emptying in rats (Lobbezoo et al. 1985), and DARs agonist apomorphine and morphine produce a dose-dependent decrease in GI transit of charcoal dust in rats (Dhasmana et al. 1989). Intracisternal (i.c.) administration of DA and apomorphine produces dose-dependent decreases in GI transit amenable to blockade by the classical D_2R antagonist haloperidol (i.c.). Administration of DA (i.t.), apomorphine (s.c. or i.t.), or bromocriptine (s.c.) decreases GI transit, which is significantly antagonized by the D2-like receptor antagonist domperidone or alizapride (s.c.) (Dhasmana et al. 1993). Mice lacking D_2Rs and D_2Rs plus D_3Rs have greater stool frequency, water content, and mass than wild-type littermates due to the abnormal motility without D_2Rs (Li et al. 2006). Mice lacking D_3Rs do not have significant change in GI transit time (Li et al. 2006). Hence, the net inhibitory effect of DA on GI motility is physiologically significant. DA can prolong the GI transit time via both peripheral and central D2-like receptors.

5.3 Dopamine and Gastrointestinal Dysmotility

DA is a key element of the "brain–gut axis" and represents a potentially important target for pharmacotherapeutic exploitation. DA not only prolongs gastric emptying time and GI transit time and inhibits GI motility in normal people but also in patients, such as patients with partial gastric resection (Andreica et al. 1991) and mechanically ventilated patients (Dive et al. 2000). Low-dose DA adversely affects gastroduodenal motility in mechanically ventilated critically ill patients. The number of contractions is lower in the antrum, and phase III motor activity is higher in the duodenum during DA infusion (4 mg/kg/min) than with placebo in mechanically ventilated patients during critical illness (Dive et al. 2000). DA inhibits the GI motility mainly via DARs. When given with DA, the specific DAR antagonists metoclopramide and domperidone counteract DA effects. When given alone, metoclopramide and domperidone elicit an increased LES pressure and stimulate gastroduodenal motility and gastric emptying. Thus, metoclopramide and domperidone have consequently proved to be of value in certain cases of gastroparesis and in relieving nausea and vomiting. Besides D2-like receptors, 5-hydroxytryptamine (5-HT) receptors also modulate GI motility. Gastroprokinetic agents that act on several 5-HT receptor subtypes and/or D2-like receptors are used clinically. Prokinetic agents (erythromycin, domperidone, metoclopramide) and antiemetics (phenothiazines, serotonin antagonists, butyrophenones) are the most widely used medications. The data from literature suggest that both domperidone and motilin-agonist erythromycin appear to be the most effective with regard to the acceleration of gastric emptying (Sturm et al. 1999).

5.3.1 DA Agonist, Antagonist, and GI Dysmotility

Antidopaminergic GI prokinetics, such as bromopride, clebopride, domperidone, levosulpiride, and metoclopramide, have been exploited clinically for the management of motor disorders of the upper GI tract, including functional dyspepsia (FD), gastric stasis of various origins, and emesis. The prokinetic effect of these drugs is mediated through the blockade of enteric (neuronal and muscular) inhibitory D2-like receptors. The pharmacological profiles of the marketed compounds differ in terms of their molecular structure, affinity at D2-like receptors, ability to interact with other receptor systems (5-HT_3 and 5-HT_4 receptors for metoclopramide; 5-HT_4 receptors for levosulpiride), and ability to permeate the blood-brain barrier (domperidone does not easily cross the barrier) (Tonini et al. 2004). Among all D2-like receptor antagonists, we would like to elaborate on domperidone, metoclopramide, levosulpiride, and haloperidol. From a therapeutic point of view, metoclopramide and domperidone are used in gastric motility disorders, and haloperidol is used in psychotic alterations (Velasco and Luchsinger 1998).

5.3.1.1 DA Receptor Agonists

DAR agonists mimic the effects of DA on GI motility by eliciting significant inhibition of gastric motility, relaxation of intestine, and reduction in GI transit (Table 5.1). Among the D1-like receptor agonists, we discuss fenoldopam, octopamine, and SKF 38393 in detail. Among the D2-like receptor agonists, we discuss quinpirole, PD128907, R(+)-7-OH-DPAT, and PD 168,077 in details. The selective D_1R agonist fenoldopam reduces GI transit partially via D_1R (Dhasmana et al. 1993). Apomorphine (Fig. 5.4) is a strong D_1R and D_2R agonist, which elicits decreases in rat GI transit (Dhasmana et al. 1993), inhibition of goat ruminal strip motility (Maas et al. 1982), and inhibition of rat gastric phasic contractions by s.c. or i.v. routes (Koga et al. 2003). These effects of apomorphine are mediated by central and/or peripheral D_2Rs but not D_1Rs (Dhasmana et al. 1993; Koga et al. 2003). Octopamine (Fig. 5.4), a biogenic monoamine structurally related to noradrenaline, is highly concentrated in neurons in several invertebrate species (Axelrod and Saavedra 1977). In mammals, octopamine is localized in nerves in peripheral tissues and brain and formed from tyramine through β-hydroxylation in sympathetic nerves (Saavedra and Axelrod 1976). Octopamine acts on intestinal D_1R sites to produce relaxation of rabbit jejunum through an increase of cAMP (Cheng and Hsieh-Chen 1988). S.c., i.p., or i.t. treatment of the mixed D_2R and D_3R agonist quinpirole (Fig. 5.4) or D_2R agonist apomorphine mimics the effects of DA and elicits significant reductions in GI transit (Dhasmana et al. 1989, 1993). A selective D_3R agonist, PD128907 (s.c.) (Fig. 5.4), dose-dependently delays gastric emptying in rats and reduces electrical field stimulation-induced relaxation of pyloric strips. The D_3R antagonist partially reverses the effect of DA on gastric emptying (Kashyap et al. 2009). Other D_3R agonists (e.g., R(+)-7-OH-DPAT [0.03–1 mg/kg, s.c.]) also delay

Table 5.1 The effects of DA receptor agonists on GI motility

Drug	Characteristics	Dose and route	Function	Reference
Fenoldopam	D_1R agonist	5, 10, 20 mg/kg, (s.c.)	↓ Gastrointestinal transit time in rats	Dhasmana (1993)
Octopamine	D_1R agonist		Relaxation of rabbit jejunum	Cheng and Hsieh-Chen (1988)
SKF-38393	Selective D_1R and D_5R agonist	0.01–10 mg/kg (s.c.)	(-) Gastric emptying in rats	Yoshikawa and Yoshida (2010)
Apomorphine	D_2R agonist	0.002–0.1 mg/kg (i. v.)	↓ Gastric phasic contractions, ↑ frequency of small phasic contractions, and ↑ gastric tone in rats	Koga (2003)
		3, 3.5, 5, 10 mg/kg (s.c.); 10, 30, 100 g (i.t.)	↓ Gastrointestinal transit in rats	Dhasmana (1989, 1993)
Quinpirole	Mixed D_2R and D_3R agonist	8 mg/kg (i.p.); 4 and 8 mg/kg (s.c.); 200 mg (i.t.)	↓ Gastrointestinal transit in rats	Dhasmana (1993)
PD128907	D_3R agonist	0.01–1 mg/kg (s.c.)	↓ Gastric emptying, ↓ electrical field stimulation-induced relaxation of pyloric strips in rats	Kashyap (2009)
R(+)-7-OH-DPAT	D_3R agonist	0.03–1 mg/kg (s.c.)	↓ Gastric emptying in rats	Yoshikawa and Yoshida (2010)
PD 128907	D_3R agonist	0.3 mg/kg (s.c.)	↓ Gastric emptying, but administration into the lateral cerebral ventricle does not affect gastric emptying in rats	Yoshikawa and Yoshida (2010)
PD 168007	D_4R agonist	0.01–10 mg/kg (s.c.)	(-) Gastric emptying in rats	Yoshikawa and Yoshida (2010)

Note: *D_1R* dopamine D_1 receptor, *D_2R* dopamine D_2 receptor, *D_3R* dopamine D_3 receptor, *D_4R* dopamine D_4 receptor, *D_5R* dopamine D_5 receptor, *i.p.* intraperitoneal injection, *i.t.* intrathecal injection, *i.v.* intravenous injection, *s.c.* subcutaneous injection. (-), no change; ↓, decrease; ↑, increase

gastric emptying in rats. Both the selective D_1R and D_5R agonist SKF-38393 (Fig. 5.4) and the selective D_4R agonist PD 168,077 (Fig. 5.4) fail to delay gastric emptying in rats (Yoshikawa and Yoshida 2010). The selective D_3R antagonist (+)-S 14297 (10 mg/kg, s.c.) partially inhibits the S(+)-PD 128,907-induced delay in gastric emptying. Although administration of S(+)-PD 128,907 (1–100 μg/kg) into

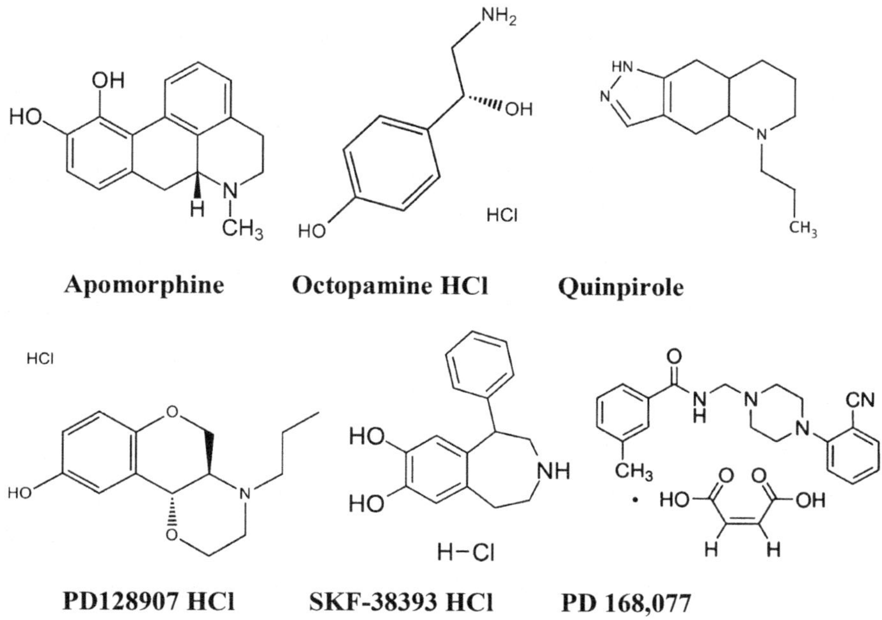

Fig. 5.4 Chemical structures of dopamine receptor (DAR) agonists

the fourth cerebral ventricle partially and dose-dependently delays gastric emptying in rats, its administration into the lateral cerebral ventricle does not affect gastric emptying (Yoshikawa and Yoshida 2010).

5.3.1.2 D_2 Receptor Antagonists

DA antagonists such as domperidone appear to have maximal prokinetic effect in the proximal GI tract and are effective for such conditions as gastroparesis and GER (Valenzuela and Dooley 1984), but they appear to have little physiologic effect in normal colon or in colonic motility disorders (Table 5.2). These compounds endowed with DA antagonism have the disadvantage of causing neuroendocrine side effects and/or extrapyramidal dyskinetic reactions (seen especially after metoclopramide) (Scarpignato 1997). Metoclopramide, a combined cholinergic agonist and DA antagonist, is used exclusively for proximal motility dysfunction (Longo and Vernava 1993). Metoclopramide is a central and peripheral acting DA antagonist. It also stimulates motility in the upper GI tract and increases LES pressure. In the pediatric population, it is used extensively as an antiemetic and in the treatment of GER disease. Domperidone is a pure DA antagonist. It accelerates gastric emptying but has less effect on bowel motility. It is a prokinetic agent used as a second-line treatment option for gastroparesis in those unable to tolerate metoclopramide. Dopaminergic blocking agents haloperidol and sulpiride strongly

Table 5.2 The effects of D_2 receptor antagonists on GI motility

Drug	Characteristics	Dose and route	Function	Reference
Domperidone	D_2-receptor antagonist, not cross the BBB	10 mg (p.o.)	No effect on gastric emptying in healthy adults	Markey and Shafat (2012)
		0.4 mg/kg (p.o.)	Decreases the amplitude of rectosigmoid contractions induced by DA but has no effect on the frequency; attenuates postprandial gastrocolic response in healthy adults	Wiley and Owyang (1987)
		20–120 mg for 3–48 months, or 0.9 mg/kg for 6–8 weeks (p.o.)	Increases gastric emptying in patients with delayed gastric emptying; reverses gastric electrical abnormalities; and improves dyspeptic symptoms in children with diabetic gastroparesis	Davis (1988), Soykan et al. (1997a, b), Lidor (2014), Franzese (2002)
		20 mg (i.v.) or 10 mg, t.i.d. (p.o.)	EGG returns to normal in functional dyspepsia patients, but gastric basal motility is not improved in IBS or functional dyspepsia patients	Lanfranchi (1985), Zhao (2012)
		0.056 mg/kg (p.o.)	Increases antrum motility in pentobarbital anesthetized mongrel dogs; blocks emesis induced by L-DOPA in Parkinson beagle dogs; restores the decreased intestinal motility induced by L-DOPA in Parkinson mice	Shuto (1980a, b)
		5.0 mg/kg (p.o.)	Increases gastric emptying, but causes no change in contractile activities of circular and	Nieto (2013)

(continued)

Table 5.2 (continued)

Drug	Characteristics	Dose and route	Function	Reference
			longitudinal muscle strips from the intestine or colon in healthy horses	
		0.06 mg/kg (i.v.)	Prevents vomiting produced by ionizing radiation, but does not alter the accompanying delay of gastric emptying in Parkinson dogs	Dubois (1984)
		0.3–1 M (in vitro)	Increases gastric contractions; decreases contractile frequency and disrupts antroduodenal coordination in guinea pigs	Schuurkes (1985)
Metoclopramide	D_2-receptor antagonist and 5-HT_4 receptor agonist	p.o.	Increases lower esophageal sphincter pressure and rate of gastric emptying in gastroparesis and gastroesophageal reflux disease patients	Saltzman and McCallum (1983), DiPalma (1990)
		10 mg (p.o.)	Increases gastric emptying in healthy male volunteers; increases antral emptying of a standardized semisolid meal; no effect on fasting antral emptying in patients with functional dyspepsia	Dalvi (1990), Dumitraşcu (1998)
		10 mg (i.v.)	Increases gastric emptying in critically ill patients or coronary artery bypass graft surgery patients with early enteral feeding; increases gastric motility and	Jooste (1999), MacLaren (2001), Sustić (2005)

(continued)

Table 5.2 (continued)

Drug	Characteristics	Dose and route	Function	Reference
			tolerance to intragastric enteral nutrition in critically ill patients	
		i.m.	Decreases gastric transit time but does not change small bowel transit time of capsule endoscopy	Zhang (2011)
		1.35 mg/kg (p.o./i.m.)	Causes strong but short enhancement of antrum motility in mongrel dogs; stimulates gastric emptying in rats and Charles Foster rats	Shuto (1980b), Megens (1991), Tamhane (1997)
		p.o.	Decreases intraruminal pressure peaks of rumen dorsal sac in Holstein calves; returns transit to normal and restores coordination of gastric and small intestinal activity in postoperative ileus in ponies	Guard (1988), Gerring and Hunt (1986)
		i.v.	Ameliorates gastric emptying caused by endotoxin in horses	Doherty (1999)
		1 μg/ml (in vitro)	Increases gastric muscle strip contraction in adult rats, but not in newborn fundic muscle	Kasirer (2014)
Levosulpiride	D_2 receptor antagonist and 5-HT_4 receptor agonist	25 mg, t.i.d. (p.o.) for 6 months	Improves the rate of gastric emptying and dyspeptic symptoms in patients with gastroparesis	Melga (1997), Stanghellini (2004), Serra (2010)
Haloperidol	D_2 receptor antagonist	p.o.	Has analgesic and antiemetic effects in patients with gastroparesis	Ramirez (2017)
		0.5 mg/kg for 3 weeks (i.p.)	Reduces the activity of complement and	Cheremnykh (2019)

(continued)

Table 5.2 (continued)

Drug	Characteristics	Dose and route	Function	Reference
			induces anti-inflammatory transformation of peritoneal macrophages in rats	
Itopride	D_2 receptor antagonist and acetylcholinesterase inhibitor	100, 200 mg for 7 days, t.i.d. (p.o.)	Has anti-emetic effect in patients with functional dyspepsia; decreases total postprandial gastric volume but has no effect on gastric emptying and gastric motor sensory function in healthy adults and patients with functional dyspepsia	Choung (2007)
		10–100 mg/kg (p.o.)	Inhibits apomorphine (0.1 mg/kg, s.c.)-induced vomiting in dogs	Iwanaga (1996)
HSR-803	Anti-D_2 receptor activity with a strong anticholinesterase activity	1–100 mM (in vitro)	Increases the amplitude of longitudinal muscle contractions and the frequency of peristalsis in isolated segments of guinea pig ileum	Iwanaga (1993)
BRL 24924	D_2 receptor antagonist	2.2 mg (p.o.)	Increases mean lower esophageal sphincter pressure but has no effect on either frequency or duration of provoked post-prandial gastroesophageal reflux in healthy adults	Robertson (1988)
		0.1–1.0 mg/kg (i.v.)	Stimulates GI motor activity during the phase I period of MMC in conscious dogs	Yoshida (1991)

Note: *BBB* blood-brain barrier, *DA* dopamine, *EGG* electrogastrogram, *IBS* irritable bowel syndrome, *i.m.* intramuscular injection, *i.v.* intravenous injection, *MMC* migrating motor complexes, *p.o.* per os, *s.c.* subcutaneous injection, *t.i.d.* three times a day

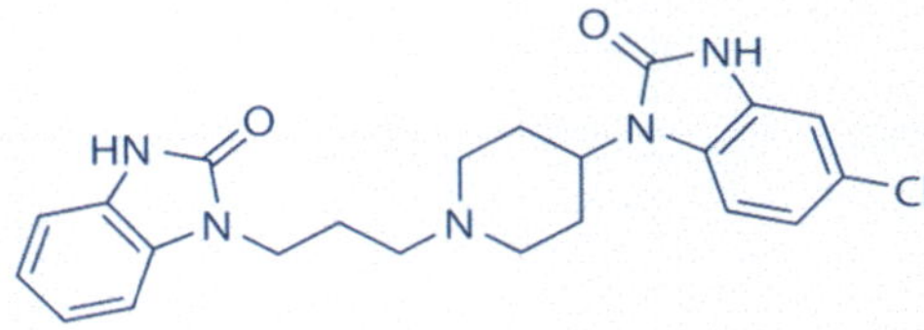

Fig. 5.5 Domperidone

inhibit distal colon motility in most nonpsychotic subjects. Only sulpiride displays distal colon motility inhibition, and such inhibition is observed in only 10% of schizophrenic subjects (Lechin et al. 1980).

Domperidone
Domperidone (KW-5338) (Fig. 5.5), a peripheral D_2R antagonist, is considered to be an agent to cross the blood–brain barrier with difficulty. It regulates the motility of gastric and small intestinal smooth muscle and has been shown to have some effects on the motor function of the esophagus. It also has antiemetic activity as a result of blockade of DARs in the CTZ. A major side effect of domperidone is prolactinemia, allowing it to be used off-label for the purpose of inducing lactation. Despite that domperidone is not currently approved by the U.S. Food and Drug Administration (FDA) due to its risks of sudden cardiac death and ventricular arrhythmia, it is still being widely used in China, Canada, and Europe for both gastroparesis and inducing lactation. Domperidone treatment enhances the quality of life in 88% patients with gastroparesis. Chronic domperidone treatment in patients with gastroparesis significantly reduces GI symptoms and hospitalizations, enhances quality of life, and accelerates gastric emptying of a solid meal to a normal rate (Phan et al. 2014).

In clinical trials, domperidone acts as both an antiemetic and an upper GI tract prokinetic agent. Domperidone successfully treats gastroparesis on a long-term outcome basis and has an excellent safety profile (Soykan et al. 1997a, b). Domperidone shows no CNS side effects. It appears that domperidone, at therapeutic dosages, does not pass the blood–brain barrier, but it can stimulate the pyloric muscle (Stellamor and Hruby 1983). Domperidone provides better relief of symptoms of anorexia, nausea, vomiting, abdominal pain, early satiety, bloating, and distension in patients with diabetic gastropathy. Domperidone also provides short-term relief of symptoms in patients with dyspepsia or GER, prevents nausea and vomiting associated with emetogenic chemotherapy, and minimizes GI and emetic adverse effects of antiparkinsonian drugs (Barone 1999).

In healthy adults, domperidone has no effect on gastric emptying. Acute administration of domperidone does not change gastric emptying or appetite sensations following a high-fat meal (Wiley and Owyang 1987; Markey and Shafat 2012). In dyspeptic patients without antroduodenal pathology, patients with gastroparesis, lung transplant recipients, and even children with insulin-dependent diabetes mellitus, domperidone increases the number of antral contractions, reduces the number of isolated duodenal contractions, improves gastroparesis symptom scores, and accelerates gastric emptying (Eyre-Brook et al. 1984; Davis et al. 1988; Soykan et al. 1997a; Franzese et al. 2002; Lidor et al. 2014). It is worth to mention that no

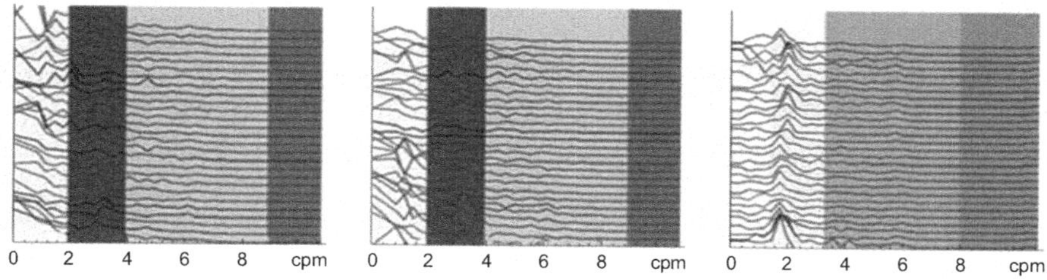

Fig. 5.6 Dynamic impedance spectra of gastric motility in patients with functional dyspepsia before treatment (left), and 1 week after (middle), and 3 week after (right) treatment with domperidone. (Zhao S, et al., World J Gastroenterol 2012, 18(25), 3282–3287)

adverse effects are observed in any patients treated with domperidone. In FD patients (Fig. 5.6), after 1-week or 3-week administration of domperidone (10 mg, tid), the electrogastrogram (EGG) and impedance of gastric motility of the patients all return to normal (Zhao et al. 2012). In patients with dyspeptic symptoms, idiopathic gastroparesis, or diabetic gastroparesis, domperidone is superior to cisapride in reversing gastric emptying delay and gastric electrical abnormalities, as well as in improving dyspeptic symptoms and diabetic metabolic control (Franzese et al. 2002). It appears that domperidone has no effect on sigmoid or rectosigmoid motor activity (Lanfranchi et al. 1985; Wiley and Owyang 1987). In irritable bowel syndrome (IBS) patients, domperidone does not induce any significant change in sigmoid basal motor activity, but prevents the increase in motor activity produced by the infusion of DA and attenuates the postprandial gastrocolic response (Lanfranchi et al. 1985; Wiley and Owyang 1987). Additionally, the use of domperidone can increase the completion rate of small bowel capsule endoscopy (Koulaouzidis et al. 2015).

In animal studies, domperidone enhances the amplitude of antrum contractions with moderately reduced contractile frequency, augments the gastric body motility with increased tonus, enhances antroduodenal coordination in a dose-dependent manner in guinea pig (Schuurkes and Van Nueten 1982; Schuurkes et al. 1985) and anesthetized mongrel dogs (Shuto et al. 1980b), increases gastric emptying in healthy horses (Nieto et al. 2013), and decreases both DA-induced gastric relaxation in mongrel dogs (Shuto et al. 1980b) and DA-induced contractile activity of midjejunal longitudinal muscle strips in horses (Nieto et al. 2013). Oral administration of domperidone (5.0 mg/kg) in healthy horses does not alter contractile activities of circular and longitudinal muscle of the small intestine or colon (Nieto et al. 2013). The domperidone-induced augmentation of the gastric motility is mainly mediated by enteric neurons, because its effect is diminished after TTX treatment, but only becomes weaker and shorter after vagotomy and splanchnicotomy (Shuto et al. 1980b). The above studies suggest that domperidone activates gastric motility mainly by stimulating intramural neurons.

In Parkinson's disease, domperidone shows a strong anti-emetic action against L-DOPA-induced emesis in beagle dogs and restores L-DOPA-induced depression of

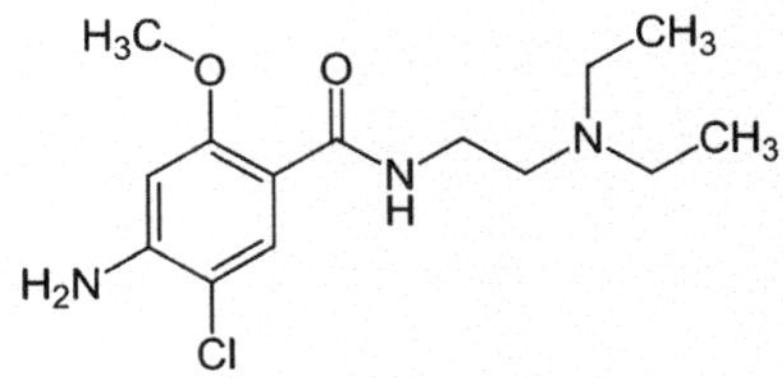

Fig. 5.7 Metoclopramide

intestinal motility in mice without reduction in therapeutic effects of L-DOPA (Shuto et al. 1980a). Domperidone prevents vomiting produced by ionizing radiation but does not alter the accompanying delay of gastric emptying in dogs (Dubois et al. 1984). D_2Rs and 5-HT_2 receptors located in the emetic central pattern generator but not peripheral 5-HT_2 or DARs mediate vomiting correlated with GI motor in dog (Lang and Marvig 1989).

Metoclopramide

Metoclopramide (4-amino-5-chloro-N-2-methoxybenzamide) (Fig. 5.7) is a compound of benzamides. It is principally a D_2R antagonist but also acts as an agonist on serotonin 5-HT_4 receptors and causes weak inhibition of 5-HT_3 receptors (Fig. 5.8). Metoclopramide has wide applications in both clinical and experimental medicines. Metoclopramide has unique properties of increasing LES pressure and gastric emptying rate. These GI motility actions of metoclopramide are important in the treatment of diabetic gastroparesis, severe GER, and postoperative ileus (Saltzman and McCallum 1983; DiPalma 1990). In the United States, metoclopramide remains the first-line prokinetic therapy, because it is the only FDA-approved medication for gastroparesis. Metoclopramide also shows a profile of a centrally active D_2R antagonist. It is being used increasingly in the management of nausea and vomiting, and at high doses can significantly relieve cytotoxic agent-induced emesis (Desmond and Watson 1986).

In healthy male volunteers, the half-time of gastric emptying is not significantly altered (McClelland and Sutton 1986) or reduced by metoclopramide (Dalvi et al. 1990). Metoclopramide facilitates human gastric antrum circular muscle contractions without affecting submaximal contractions to carbachol (Fig. 5.9, Broad et al. 2014). These effects are not altered in the absence and presence of the nitric oxide (NO) synthase inhibitor nitro-L-arginine methyl ester (L-NAME). In clinic, metoclopramide improves gastric emptying (Jooste et al. 1999), enhances gastric motility, improves tolerance to intragastric enteral nutrition in critically ill patients, and decreases gastric transit time of capsule endoscopy (MacLaren et al. 2001; Zhang et al. 2011) or in coronary artery bypass graft surgery patients with early enteral feeding (Sustić et al. 2005). In patients with FD, antral emptying of a standardized semisolid meal is accelerated by metoclopramide, while fasting antral emptying is not changed by metoclopramide and the postprandial antral area has a trend to decrease (Dumitraşcu et al. 1998). It suggests that the effect of metoclopramide is less likely due to the enhancement of antral motility. Furthermore, combination therapy with metoclopramide and neostigmine decreases gastric

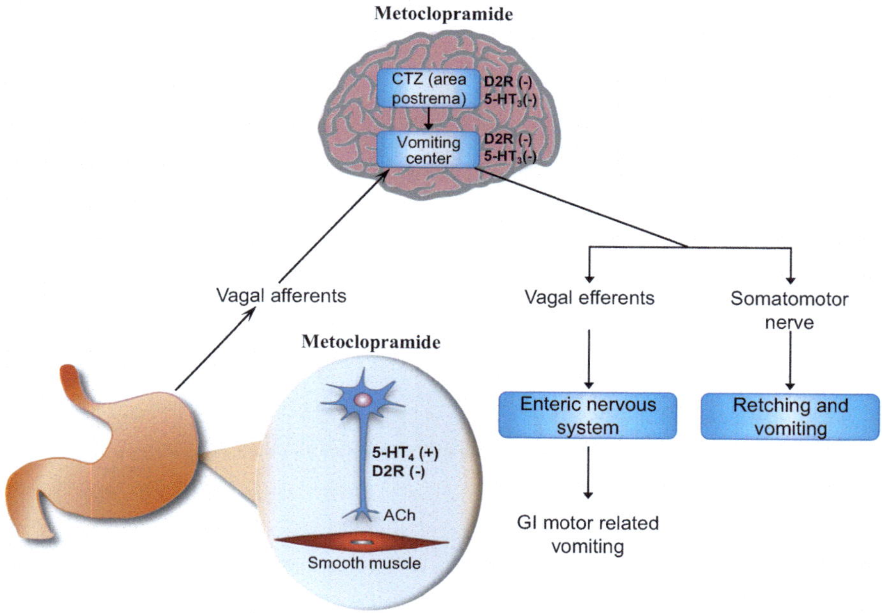

Fig. 5.8 Scheme of the mechanism of metoclopramide action on vomiting and gastrointestinal (GI) smooth muscle contraction. Receptors of the chemoreceptor trigger zone (CTZ) that have been identified to induce emesis include D_2 receptor (D_2R) and 5-HT_3 receptor. Metoclopramide antagonizes the activity of D_2R in the CTZ in the central nervous system to prevent nausea and vomiting. At higher doses, metoclopramide can also act as a 5-HT_3 antagonist to play the antiemetic effect. The gastroprokinetic activity of metoclopramide is mediated by inhibiting D_2R and stimulating 5-HT_4 receptor in the enteric nervous system, which may also contribute to the antiemetic effect. *ACh* acetylcholine

residual volumes in critically ill patients with a higher efficacy than monotherapies (Baradari et al. 2017).

Measurements performed in dogs, rats, and calves have demonstrated that metoclopramide shows strong but short enhancement of antrum motility (Shuto et al. 1980b), a stimulatory effect on gastric emptying (Megens et al. 1991; Tamhane et al. 1997), and an inhibitory effect on intraruminal pressure peaks of rumen dorsal sac (Guard et al. 1988). Metoclopramide restores gastric and small intestinal motility in postoperative ileus and returns GI transit to normal in ponies (Gerring and Hunt 1986). Metoclopramide significantly ameliorates endotoxin-induced delay in liquid gastric emptying in horses (Doherty et al. 1999). There is scant D_2R expression in neonates, but increased expression in adult rat gastric tissue. Metoclopramide increases electrical field stimulation-induced muscle strip contraction by promoting ACh release from the cholinergic nerves in adult rats, but no significant effect is observed in newborn rat fundic muscle (Kasirer et al. 2014). In addition to D_2R antagonizing effect, metoclopramide has high affinity at the 5-HT_3 receptor as it

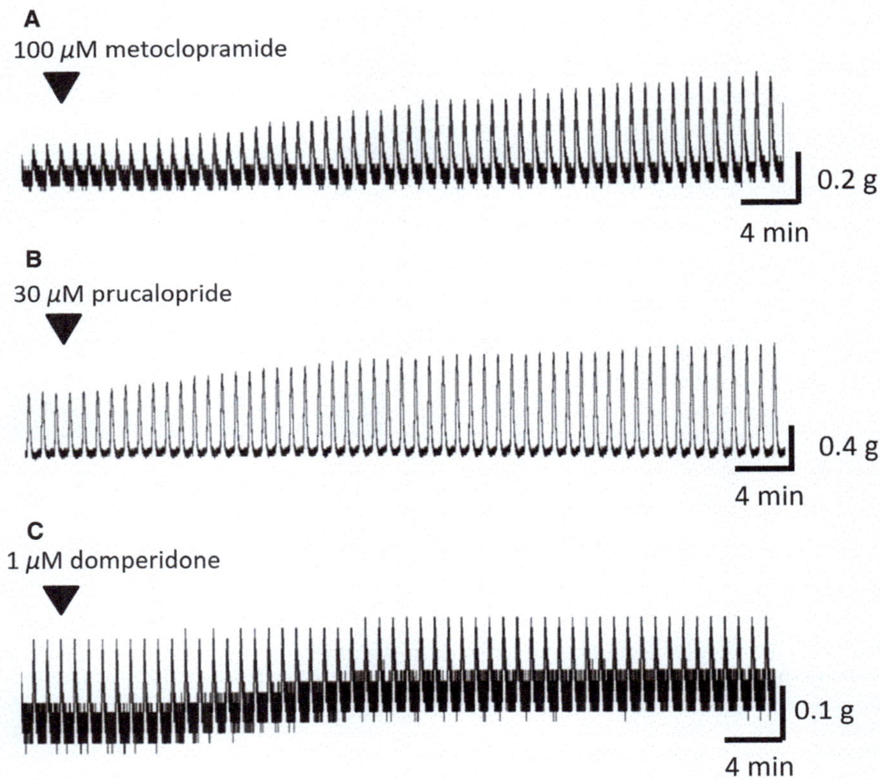

Fig. 5.9 Effect of metoclopramide, prucalopride, and domperidone on human gastric antrum. Example traces showing the effects of metoclopramide (**a**), prucalopride (**b**), and domperidone (**c**) on contractions evoked by electrical field stimulation in circular muscle strips from human gastric antrum. (Broad J, et al., Neurogastroenterol Motil 2014, 26(6), 851–861)

displaces the 5-HT_3 antagonist [^{3}H]GR65630 from cortical membranes (Ki = 232 nM) and blocks the 5-HT-induced Bezold–Jarisch reflex. Metoclopramide activates a 5-HT_4 receptor in the ileum to increase cholinergic contractions (Linnik et al. 1991).

Levosulpiride

Levosulpiride (Fig. 5.10) is the l-enantiomer of sulpiride, which has replaced sulpiride in several European Union countries. Levosulpiride is a D_2R antagonist and a serotonin 5-HT_4 receptor agonist with cholinergic effect. Clinical trials have shown that levosulpiride is more effective in reducing the symptoms of dyspepsia, while its effect is similar or superior to that of other DA antagonists. After 6 months of levosulpiride therapy, the rate of gastric emptying and dyspeptic symptoms are improved in insulin-dependent diabetes mellitus subjects (Melga et al. 1997). At a dosage of 25 mg three times daily, levosulpiride accelerates gastric and gallbladder

Fig. 5.10 Levosulpiride

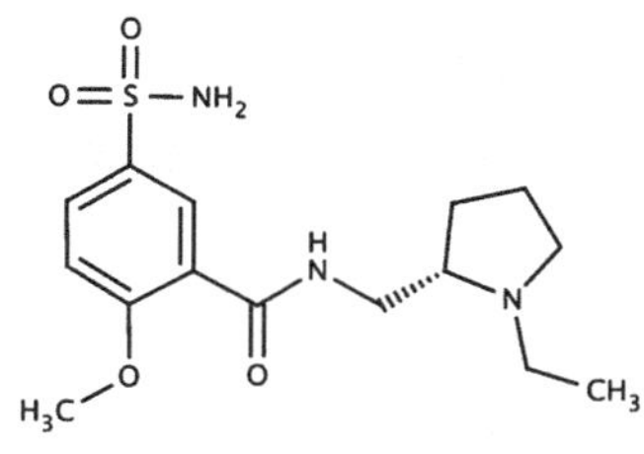

Fig. 5.11 Itopride hydrochloride

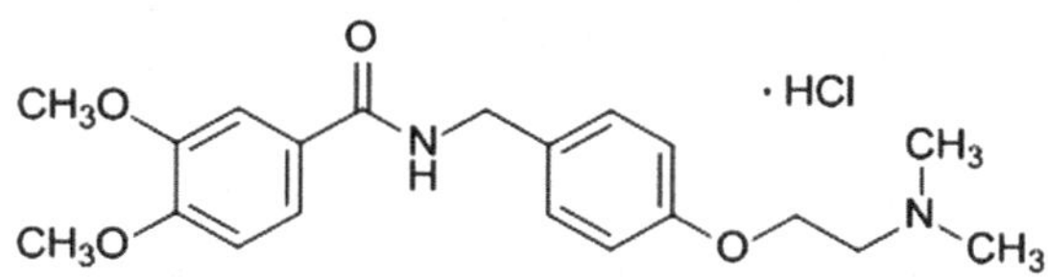

emptying. The safety profile of levosulpiride is good, and the frequency of adverse events is similar to that of other D_2R antagonists. Therefore, this drug is a useful therapeutic option in the management of patients with FD, as well as in those with delayed gastric emptying (Stanghellini et al. 2004; Serra 2010).

Itopride

Itopride (Fig. 5.11), a D_2R antagonist and acetylcholinesterase inhibitor, is an orally active gastroprokinetic with moderate anti-emetic action and no extrapyramidal side effects (Giurcan and Voiosu 2010). Itopride hydrochloride is also called HSR-803. HSR-803 stimulates endogenous ACh release by antagonizing the D_2Rs on the postsynaptic cholinergic neurons, and the anticholinesterase activity of HSR-803 may cause the released ACh to accumulate at cholinergic receptor sites (Iwanaga et al. 1990). In healthy subjects, HSR-803 reduces total postprandial gastric volume (Choung et al. 2007). HSR-803 may be useful in the management of FD (Mizuta et al. 2006). HSR-803 significantly alleviates satiation, bloating, and nausea by accelerating gastric emptying in patients with FD (Iwanaga et al. 1996; Giurcan and Voiosu 2010). HSR-803 at doses of 10–100 mg/kg, p.o. inhibits apomorphine (0.1 mg/kg, s.c.)-induced vomiting in dogs (Iwanaga et al. 1996). In isolated segments of guinea pig ileum (Iwanaga et al. 1993), HSR-803 concentration-dependently increases the amplitude of longitudinal muscle contractions and the frequency of peristalsis, mainly due to prevention of ACh hydrolysis.

Renzapride

Renzapride (BRL 24924) is a GI prokinetic agent with properties similar to metoclopramide but with increased potency and devoid of side effects associated with blockade of DARs in the CNS. Injection of BRL 24924 (0.1–1.0 mg/kg, i.v.) stimulates GI motor activity during the phase I period (Yoshida et al. 1991) and increases mean LES pressure, but it has no effect on either the frequency or the duration of provoked post-prandial GER (Robertson et al. 1988).

Haloperidol

Haloperidol acts as a selective D_2R antagonist. Haloperidol has been shown to have antipsychotic and antiemetic properties (Ramirez et al. 2017). From a therapeutic point of view, haloperidol is used in psychotic alterations (Velasco and Luchsinger 1998). Besides, haloperidol can also reduce the activity of complement and induce the anti-inflammatory transformation of peritoneal macrophages in rats (Cheremnykh et al. 2019).

5.3.1.3 Other Non-dopaminergic Prokinetics

Erythromycin is a macrolide antibiotic that have gained widespread use. Animal and human studies have demonstrated a prokinetic effect of erythromycin on the GI tract at typically a quarter or less of the antibiotic dose. Significantly, erythromycin has no serious adverse effects as a prokinetic agent. Erythromycin is a motilin agonist that exerts its prokinetic effect by stimulating propagative contractile activity in preterm infants with or at risk of feeding intolerance or in children with GI dysmotility (Ng and Shah 2008; Curry et al. 2001). Erythromycin effectively increases gastric emptying in a dose–response manner in healthy subjects (Boivin et al. 2003) and gastroparetic rats (Qin et al. 2003). D_3R and neuropeptide Y Y5 receptor may also be involved in prokinetic action of erythromycin (Qin et al. 2003).

Prucalopride, mosapride, cisapride, and tegaserod, 5-HT_4 receptor agonists, are potent prokinetic agents. Prucalopride has colonic prokinetic activity. It is recently approved by the FDA for the treatment of chronic idiopathic constipation (Vijayvargiya and Camilleri 2019). Prucalopride is also able to enhance the gastric emptying rate (Carbone et al. 2019). Mosapride, cisapride, and tegaserod can decrease esophageal reflux, increase gastric emptying, and promote intestinal peristalsis (Komura et al. 2017). Cisapride can speed emptying of fasting gastric contents, particularly when emptying is slowed by DA (Müller-Lissner et al. 1986). Cisapride can temporarily improve the symptoms of lower GI dysmoility in Parkinson's disease (Jost 1997). However, both cisapride and tegaserod have serious adverse effects. Hitherto, cisapride has been withdrawn, while tegaserod has been placed under restricted use.

5.3.2 Gastrointestinal Dysmotility and Altered Gut Microbiome in Parkinson's Disease Patients

Parkinson's disease (PD) is a common neurodegenerative disorder, characterized by dopaminergic neuron degeneration in the substantia nigra (SN). In 1817, James Parkinson was clearly aware that GI dysfunction was part of the clinical spectrum of PD (Pfeiffer 2018). Most of PD patients experience GI motility disorders, manifesting mainly as dysphagia, gastroparesis, small intestinal dysfunction, delayed colon transit, constipation, and defecatory dysfunction (Jost 1997; Jost and Eckardt 2003; Pfeiffer 2018, 2011). GI dysfunction is now often considered to be a cardinal PD symptom that can dominate the clinical outlook for some patients (Natale et al. 2010). Gastroparesis and constipation are very frequent in the preclinical stage of PD and have also been identified as premotor markers of early PD (Abbott et al. 2001; Ueki and Otsuka 2004).

5.3.2.1 Gastroparesis in Parkinson's Disease Patients

Gastroparesis is a disorder characterized by a delay in gastric emptying of a meal in the absence of a mechanical gastric outlet obstruction. Diagnosis of gastroparesis is based on the presence of symptoms such as nausea, vomiting, and postprandial abdominal fullness and on an objectively determined delay in gastric emptying. Gastric emptying can be assessed by scintigraphy and stable isotope breath tests in clinic. Prevalence of gastroparesis in PD ranges from 70% to 100% (Heetun and Quigley 2012), although not all affected individuals have symptoms. Gastroparesis can be present in both early and advanced PD, with delayed gastric emptying of solid meals (Goetze et al. 2006). The most likely causes of these disorders are central neuronal degeneration and degeneration of the myenteric plexus (Braak et al. 2006). Impaired antral contractions or failure of the pyloric sphincter to relax may be responsible. It is also reported that in early motor stage PD off dopaminergic medication, there is no gastric emptying slowing after a high caloric test meal. Epprecht et al. speculate that clinically relevant delayed gastric emptying in early PD is related to the iatrogenic effect of L-DOPA treatment (Epprecht et al. 2015). Symptomatic treatment of PD is based on the DA precursor L-DOPA. Gastroparesis can also be responsible for inconsistent responses to L-DOPA, including both delayed response and complete dose failure, since L-DOPA is only absorbed in the small intestine, where transit time is approximately 3 h and the plasma elimination half-life is short. Therefore, gastric emptying is a major determining factor for the onset of symptom relief (Nyholm and Lennernäs 2008). Other processes, such as competition with protein for absorption in the small intestine, *Helicobacter pylori* infection (Niehues and Hensel 2009; Lyte 2010; Pierantozzi et al. 2001; Rahne et al. 2013; Dobbs et al. 2010, 2013; Hashim et al. 2014; Narożańska et al. 2014; Çamcı and Oğuz 2016; Mridula et al. 2017; Liu et al. 2017; Dardiotis et al. 2018), and small intestinal bacterial overgrowth may also produce erratic responses to L-DOPA (Fasano et al. 2013). Although gastric emptying is commonly delayed in PD patients with long-term L-DOPA therapy, delayed gastric emptying does not differ between PD patients with or without motor fluctuations (Tanaka et al. 2009). Thus, treatment of gastroparesis in PD patient is very important. Management of gastroparesis consists of dietary and lifestyle measures and/or pharmacological interventions (prokinetics, antiemetics, intrapyloric botulinum toxin injection) or other interventions that focus on adequate nutrient intake through a nasoduodenal tube, percutaneous gastrostomy, or jejunostomy (Haans and Masclee 2007). Domperidone therapy significantly reduces upper GI symptoms and accelerates gastric emptying of a solid meal in PD patients, except improvement in dysphagia and constipation. Patients' global assessment of PD remains stable or is improved during domperidone therapy (Soykan et al. 1997a, b). Domperidone is the drug of choice for motility disorders of the upper GI tract, while administration of cisapride (5-HT_4 agonist) leads to a marked temporary improvement in symptoms in lower GI disorders (Jost 1997). Another D_2R antagonist metoclopramide cannot be used in PD patients because it crosses the blood–brain barrier and blocks DARs within the CNS.

5.3.2.2 Intestinal Bacterial Overgrowth in Parkinson's Disease Patients

The small intestinal motility has received very little research attention in PD. However, small intestinal bacterial overgrowth (SIBO) has been reported to be highly prevalent in several studies, with the presumption that impaired small intestinal motility may be responsible for it (Davies et al. 1996; Gabrielli et al. 2011; Fasano et al. 2013; Tan et al. 2014; Niu et al. 2016; Vizcarra et al. 2018). Glucose and/or lactulose breath test has been used as a tool for the diagnosis of bacterial overgrowth. All participants underwent the glucose and/or lactulose breath test to assess SIBO. About 55% of the 182 Chinese patients with PD (Tan et al. 2014), 54.5% of the 33 (Fasano et al. 2013), and 54.17% of the 48 (Gabrielli et al. 2011) Italian patients with PD were SIBO positive. SIBO is characterized by increased bacterial density in the small intestine caused by proximal migration of colonic bacteria (Gasbarrini et al. 2007). It may be characterized by flatulence and a sense of bloating (Gabrielli et al. 2011). Motor fluctuations present higher in PD patients with SIBO than in patients without SIBO (Fasano et al. 2013; Tan et al. 2014). Patients with SIBO have longer wearing-off time (sudden wearing off of medication effects) daily and more episodes of delayed-on (delayed onset of symptom relief) and no-on (complete dose failure). The eradication of SIBO results in improvement in motor fluctuations without affecting the pharmacokinetics of L-DOPA (Fasano et al. 2013). There is reduced absorption of non-mediated uptake (mannitol) in PD patients while lactulose absorption is similar to control ones (Davies et al. 1996). If sufficiently severe, SIBO may cause malabsorption, which could conceivably explain the weight loss so frequently evident in PD.

5.3.2.3 Constipation in Parkinson's Disease Patients

Constipation is a frequent and disabling non-motor feature of PD, and it can precede motor symptoms by as much as 20 years. Constipation can produce discomfort and affect quality of life. Constipation has been documented in PD by numerous investigators, with longer transit times than in control individuals (Jost and Schimrigk 1991; Edwards et al. 1994; Ashraf et al. 1997; Li et al. 2010; Guo et al. 2013a, b; Knudsen et al. 2017; Qin et al. 2019). Qin et al. reported that 53 of 108 Chinese patients at the early and middle stages of PD suffer from constipation (49.07%) (Qin et al. 2019). Slowed colon transit is evident in approximately

25% ~ 80% of individuals with PD (Guo et al. 2013a; Stocchi and Torti 2017; De Pablo-Fernández et al. 2019; Frazzitta et al. 2019), and the slowed transit presumably is responsible for less frequent bowel movements. Constipation is more common in late-onset PD patients, and late-onset PD patients are likely to present with more and severe symptom than early onset PD patients (Guo et al. 2013a, b). A decreased number of neurons per ganglion and atrophic/pycnotic neurons are present in the submucosal and/or myenteric plexus within the colon (Lebouvier et al. 2010; Ohlsson and Englund 2019). The causes of constipation in patients with PD may be due to the presence of Lewy bodies (Kupsky et al. 1987; Wakabayashi et al. 1988, 1990; Pouclet et al. 2012), α-synuclein deposition (Shannon et al. 2012; Hilton et al. 2014; Chahine et al. 2018), and the loss of dopaminergic neurons within myenteric plexus in colonic tissue (Singaram et al. 1995; Lebouvier et al. 2008), or Lewy body-related α-synucleinopathy in spinal cord (Nardone et al. 2019). For the treatment of slow transit constipation in PD, lubiprostone and polyethylene glycol are first-line compounds recommended. Secondary, pelvic floor dyssynergia in PD can be treated by L-DOPA or apomorphine injections and nonpharmacological interventions (such as biofeedback therapy or functional magnetic stimulation) (Rossi et al. 2015).

5.3.2.4 Altered Gut Microbiome in Parkinson's Disease Patients

The gut microbiome in GI system and PD has been the focus of research in recent years. Several groups have reported differences in microbiota composition (-β-diversity) between PD patients and controls, but the results are heterogeneous (Scheperjans et al. 2015; Cassani et al. 2015; Unger et al. 2016; Hill-Burns et al. 2017; Petrov et al. 2017; Li et al. 2017; Mertsalmi et al. 2017; Minato et al. 2017; Heintz-Buschart et al. 2018; Lin et al. 2018, 2019; Qian et al. 2018; van Kessel et al. 2019). Alterations in the gut microbiome play an important role in the pathogenesis of PD, and progression of PD is associated with gut dysbiosis (Minato et al. 2017). Fecal microbial composition is determined by sequencing the V3-V4 region of the 16S ribosomal RNA gene extracted from the stool. In the PD patients, the levels of Akkermansia (Keshavarzian et al. 2015; Hill-Burns et al. 2017), Lactobacillus (Hasegawa et al. 2015; Hill-Burns et al. 2017; Petrov et al. 2017; Lin et al. 2019), Bifidobacterium (Unger et al. 2016; Hill-Burns et al. 2017; Petrov et al. 2017; Lin et al. 2018), and Enterobacteriaceae (Scheperjans et al. 2015; Unger et al. 2016) are elevated in the fecal samples of PD patients, but the levels of Lachnospiraceae (Keshavarzian et al. 2015; Unger et al. 2016; Hill-Burns et al. 2017; Lin et al. 2018), Prevotellaceae (Scheperjans et al. 2015; Petrov et al. 2017), Faecalibacterium (Unger et al. 2016; Petrov et al. 2017; Li et al. 2017), Ruminococcus (Petrov et al. 2017; Li et al. 2017), genera Blautia (Petrov et al. 2017; Li et al. 2017), Bacteroidetes (Unger et al. 2016; Petrov et al. 2017), and Prevotella (Mertsalmi et al. 2017; Petrov et al. 2017; Lin et al. 2019) are reduced. Enterococcaceae are reported significantly reduced in one study (Unger et al. 2016), nevertheless opposite results are reported in another study (Li et al. 2017). Genera Clostridium IV, Aquabacterium, Holdemania, Sphingomonas, Clostridium XVIII, Butyricicoccus, and Anaerotruncus are enriched in the feces of Chinese PD patients after adjusting for age, gender, body mass index, and constipation (Qian et al. 2018). In southern China, the abundance of Lachnospiraceae is reduced by 42.9% in patients with PD (Lin et al. 2018). PD patients with IBS-like symptoms have a lower fecal abundance of Prevotella bacteria than those without IBS-like symptoms (Mertsalmi et al. 2017). A 2-year follow-up study shows that the deteriorated PD patient group demonstrates accelerated lowering of Bifidobacterium, *B. fragilis*, and *Clostridium leptum* than the stable group (Minato et al. 2017). Fecal short chain fatty acids (SCFA) (one main metabolic product of notably Lachnospiraceae) concentrations are significantly reduced in PD patients compared to controls. The reduction in SCFA might, theoretically, induce alterations in the ENS and contribute to the GI dysmotility in PD (Unger et al. 2016). Besides, the higher relative abundance of bacterial tyrosine decarboxylases at the site of L-DOPA absorption, proximal small intestine, restricts levels of L-DOPA in the plasma in the treatment of PD (van Kessel et al. 2019). Altered microbiota is correlated with elevated plasma concentrations of IFNγ and TNFα in PD patients (Lin et al. 2019), suggesting a contribution of aberrant immune responses to inflammatory processes in PD. The microbiota–gut–brain axis may be implicated in the GI dysfunction of PD patients.

5.3.3 Parkinson's Disease Models with Gastrointestinal Dysmotility

PD is a multi-system, complex disorder with an uncertain etiology that affects the central and peripheral nervous systems. GI dysfunction is one of the most common non-motor symptoms in PD patients, which has a greater impact on their life quality. The factors that trigger the pathophysiology of PD are unknown. In addition to genetic predisposition and aging, environmental factors, such as pesticides, also play an important role. The majority of GI symptoms in PD are related to slow motility in the GI tract. Delayed and uncoordinated emptying of stomach contents, slowed colon transit, and defecatory dysfunction are common in early PD and increase in frequency as the disease advances (Greene 2011). An accurate understanding of symptoms and pathology of PD patients is required to provide the basis for validation of any proposed PD animal models. Considering diverse PD animal models with different drug, dosage, route of administration, or differential genome, it is difficult to evaluate all the symptoms and investigate mechanisms of PD in one distinct animal model. This section briefly summarizes the data reported from several common PD model systems referable to the GI tract.

5.3.3.1 6-OHDA Rat Model

Idiopathic PD is a late-onset, chronic, and progressive motor dysfunction attributable to loss of nigrostriatal dopaminergic neurons. Microinjection of 6-hydroxy-dopamine (6-OHDA) into the substantia nigra pars compacta (SNpc) or the medial forebrain bundle (MFB) to destroy the midbrain dopaminergic system has been used to induce parkinsonism (Table 5.3). Dr. Zhu JX's lab has reported that the rats microinjected with 6-OHDA (or lipopolysaccharide (Zheng et al. 2013)) bilaterally into the SNpc display selective loss of midbrain dopaminergic neurons, decreased ChAT, and enhanced tyrosine hydroxylase (TH) expression in the DMV (Fig. 5.12). The model shows delayed gastric emptying, impaired gastric motility (Fig. 5.13), decreased daily fecal production, decreased fecal water content, and decreased amplitude of colonic spontaneous contraction (Fig. 5.14) (Tian et al. 2008; Zheng et al. 2011, 2014; Zhang et al. 2015; Li et al. 2015; Garrido-Gil et al. 2018). Besides, several studies have also reported GI dysmotility in rat models with unilateral microinjection of 6-OHDA into SNpc (Toti and Travagli 2014; Anselmi et al. 2017b) or MFB (Blandini et al. 2009; Colucci et al. 2012; Vegezzi et al. 2014; Levandis et al. 2015; Pellegrini et al. 2016, 2017; Karasawa et al. 2014).

Decreased ACh content, increased expression of TH, dopamine transporter (DAT), and D_2Rs, β_1 receptors, and enhanced DA content are found in the stomach of the 6-OHDA rats (Tian et al. 2008; Zheng et al. 2011, 2014). The duodenum shows increased TH and DAT expression. Decreased ACh content, catechol-*O*-methyltransferase (COMT), D_1R, D_2R, angiotensin II type 2 receptor (AT2), and 5-HT_4 receptor expression, but elevated DA content, IL-1β content, nicotinamide adenine dinucleotide phosphate (NADPH) activity, oxidative stress p47phox, β_3, and angiotensin II type 1 receptor (AT1) expression are found in the colon (Table 5.3). Dr. Zhu JX's lab has further studied the underlying mechanism of GI dysmotility in 6-OHDA rats. They have found that in the stomach, upregulated D_2Rs and β_1 receptors in the muscularis externa of the gastric corpus are responsible for the hypomotility of gastric muscle, and administration of the D_2R antagonist domperidone can relieve gastric dysmotility in 6-OHDA rats (Zheng et al. 2014; Song et al. 2014). Subdiaphragmatic vagotomy prior to 6-OHDA treatment significantly reduces the increase in TH and D_2R expression in the stomach, accompanied by the improvement of gastric motility in 6-OHDA rats (Zheng et al. 2014). Pellegrini et al. have found that central dopaminergic neurodegeneration triggered by 6-OHDA injection into the MFB might lead to an inflammatory abdominal condition through an impairment of the DMV-vagus nerve anti-inflammatory pathway. Peritoneal macrophages polarize toward a pro-inflammatory phenotype. The bowel inflammation is associated with increased oxidative stress, increase in pro-inflammatory cytokine levels, activation of enteric glia and inflammatory cells, and enhancement of colonic excitatory tachykininergic motility (Fig. 5.15) (Pellegrini et al. 2016). In the MFB lesion model, electrically evoked contractions in colonic strips are significantly reduced in animals after 8 weeks from 6-OHDA injection (Fig. 5.16). Treatment with L-DOPA ameliorates its colonic motility

Table 5.3 6-OHDA PD rat model

Dose/time	Characteristics				Reference
	Nigra-striatum	DVC	Muscle layer in GI tract	Function	
4μg injected unilaterally into two sides of the SNpc (AP: −5.0/−5.6, ML: 2.4/1.6, DV: −7.6/−7.8)	Loss of dopaminergic neurons (96%) in SNpc (4 weeks)	ChAT-IR↓ in DMV or NAmb; NOS-IR↓ in cNTS	ChAT-IR↓, NOS-IR↑ in stomach	Gastric dysmotility; tyramine and TRH injected into the DVC attenuate gastric dysmotility	Toti and Travagli (2014), Anselmi et al. (2017b)
16.5 μg injected unilaterally into the MFB ((1) AP: −4.0, ML: −0.8, DV: −8.0, tooth bar at +3.4 (9 μg/3 μl); or (2) AP: −4.4, ML: −1.2, DV: −7.8, tooth bar at −2.4 (7.5 μg/3 μl))	Loss of dopaminergic striatal terminals (98%) and dopaminergic nigral neurons (95%) (4/8 weeks)		VIP-IR↑, NOS-IR↓, D2R mRNA↓ in ileum GFAP-IR↑, SP-IR↑, NK1↑, VIP-IR↑, DA↑, NOS-IR↓, ChAT-IR↓, D2R mRNA↓, D2R-IR↓, MDA↑, TNF↑, IL-1β↑, eosinophils↑, mast cells↑ in colon; Peritoneal macrophages polarized towards a pro-inflammatory phenotype	Delayed gastric emptying (6 weeks); reduction in daily fecal output (4 weeks); constipation (8 weeks); decreased colonic transit; decreased intraluminal pressure and frequency of peristalsis in the distal colon; response of colonic motility to dopaminergic stimulation is subverted; increased electrically and SP-evoked tachykininergic phasic contractions in longitudinal strips of distal colon (8 weeks)	Blandini et al. (2009), Colucci et al. (2012), Vegezzi et al. (2014), Levandis et al. (2015), Pellegrini et al. (2016, 2017)
12 μg injected unilaterally into the MFB (AP: −4.4, ML: −1.6, DV: −8.0)				Water intake↓; fecal output↓; delayed gastric emptying; HM01 (ghrelin agonist) alleviates reduced fecal output and water content and L-DOPA delays gastric emptying	Karasawa et al. (2014)
8 μg injected bilaterally	Loss of dopaminergic		ACh↓, TH↑, DAT↑, DA↑,	Delayed gastric emptying;	Tian et al. (2008),

(continued)

Table 5.3 (continued)

	Characteristics				
Dose/time	Nigra-striatum	DVC	Muscle layer in GI tract	Function	Reference
into the SNpc (AP: −5.5/5.6, ML: ±1.9/2.0, DV: −7.2/7.5)	neurons (70%) in SNpc (4 weeks)	ChAT↓, TH↑ in DMV	D2R↑, β1↑ in stomach; TH↑, DAT↑ in duodenum; ACh↓, COMT↓, D1R↓, D2R↓, AT2↓, 5-HT4↓, DA↑, IL-1β↑, NADPH activity↑, oxidative stress p47phox↑, β3↑, AT1↑ in colon	impaired gastric motility; decreased daily fecal production and fecal water content; decreased amplitude of colonic spontaneous contraction; D2R antagonist domperidone and subdiaphragmatic vagotomy relieved gastric dysmotility; upregulation of β1-AR is responsible for enhanced NE reactivity in gastric motility	Zheng et al. (2011, 2014), Song et al. (2014), Zhang et al. (2015), Li et al. (2015), Garrido-Gil et al. (2018)

ACh acetylcholine, *AP* anteroposterior, *AT1* angiotensin II type 1 receptor, *AT2* angiotensin II type 2 receptor, *β1* β1-adrenoceptor, *β3* β3-adrenoceptor, *ChAT* choline acetyltransferase, *cNTS* the subnucleus centralis of the nucleus tractus solitarius, *COMT* catechol-*O*-methyltransferase, *DA* dopamine, *DAT* dopamine transporter, *DMV*the dorsal motor nucleus of the vagus, *DV* dorsoventral, *DVC* the dorsal vagal complex, *D_1R* dopamine D_1 receptor, *D_2R* dopamine D_2 receptor, *GFAP* glial fibrillary acidic protein, *IL-1β* interleukin-1 beta, *IR* immunoreactivity, *L-DOPA* levodopa, *MDA* malondialdehyde, *MFB* the medial forebrain bundle, *ML* medio-lateral, *NADPH* nicotinamide adenine dinucleotide phosphate, *NAmb* the nucleus ambiguus, *NE* norepinephrine, *NK1* neurokin 1, *NOS* nitric oxide synthase, *PD* Parkinson's disease, *RC* rostro-caudal, *SNpc* substantia nigra pars compacta, *SP* substance P, *TH* tyrosine hydroxylase, *TNF* tumor necrosis factor, *TRH* thyrotropin-releasing hormone, *VIP* vasoactive intestinal peptide, *6-OHDA* 6-hydroxy-dopamine

through a normalization of myenteric cholinergic neurotransmission, along with an improvement of colonic inflammation (Pellegrini et al. 2017). Systemic administration of 6-OHDA to rhesus monkeys significantly reduces catecholaminergic expression in the colonic myenteric plexus without inducing constipation (Shultz et al. 2016). These findings support the concept that ENS catecholaminergic loss is not responsible for constipation in PD. Central dopaminergic neurodegeneration may lead to GI disturbances (Zheng et al. 2014) and the increase in the GI inflammatory response through direct SN-DMV pathway (Anselmi et al. 2017b) or indirect SN-lateral hypothalamus (LH)-DMV pathway (Wang et al. 2014; Yang et al. 2019).

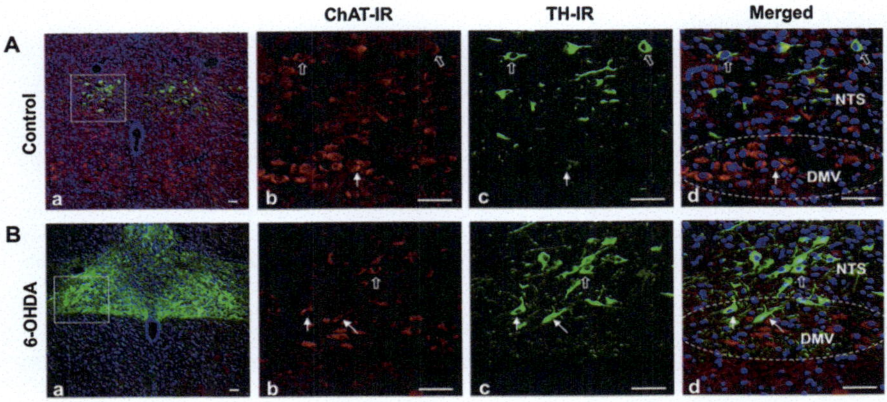

Fig. 5.12 Distribution of tyrosine hydroxylase (TH) and choline acetyltransferase (ChAT) immunoreactive neurons in the dorsal motor nucleus of the vagus (DMV) of control and 6-OHDA rat. TH (TH-IR, green) and ChAT immunoreactivity (ChAT-IR, red) in the DMV of a control (**A**) and 6-OHDA rat (**B**). (Zheng LF, et al., Brain Res 2011, 1420, 59–67)

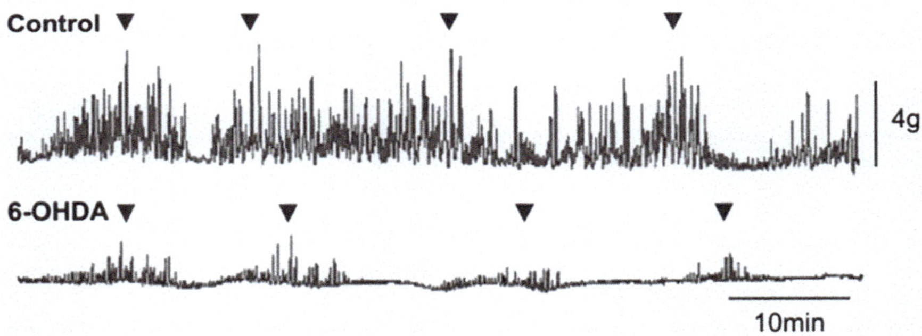

Fig. 5.13 Gastric migrating motor complex (MMC) in control and 6-OHDA rats. The black triangles show the phase III-like activity of gastric MMC. (Zheng LF, et al., Brain Res 2011, 1420, 59–67)

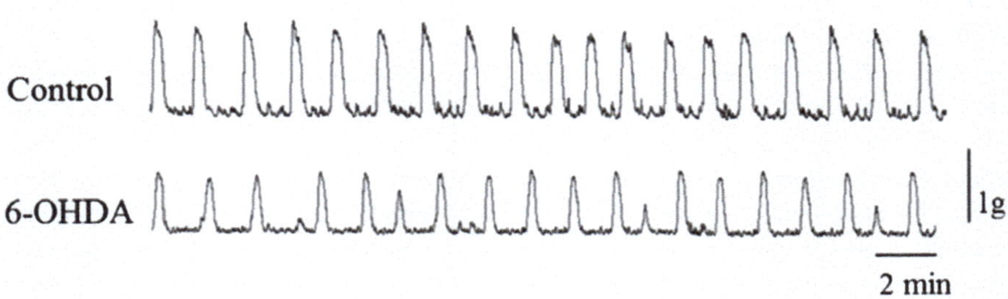

Fig. 5.14 Representative spontaneous contraction tracings of the isolated colonic smooth muscle strips of control and 6-OHDA rats. (Zhang X, et al., Transl Res 2015, 166(2), 152–162)

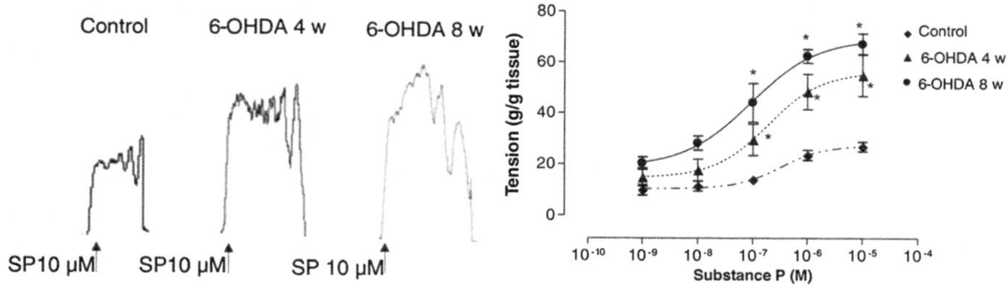

Fig. 5.15 Effect of substance P (SP) on longitudinal muscle strip motility in the distal colon of control or 6-OHDA rats. Exogenous SP (10μM)-induced colonic excitatory motility is enhanced in 6-OHDA rats (4 weeks or 8 weeks after 6-OHDA injection into the medial forebrain bundle) in the presence of tetrodotoxin, as shown in the representative tracings (left) and dose–response curve (right). (Pellegrini C, et al., J Neuroinflammation 2016, 13(1), 146)

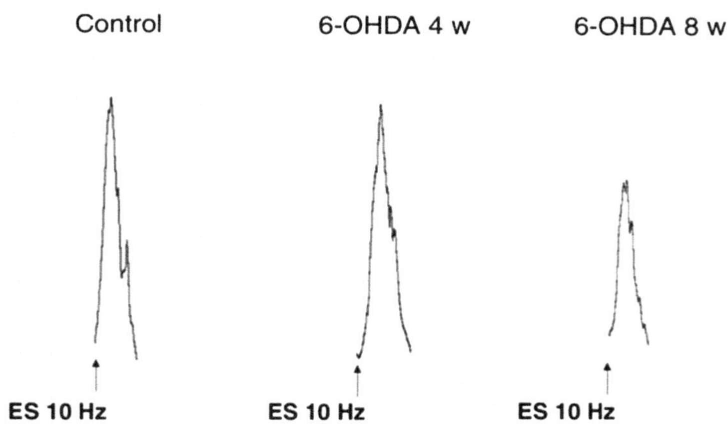

Fig. 5.16 Representative tracings of colonic contractile responses elicited by electrical stimulation (ES) of longitudinal muscle strips in the distal colon of control animals or 6-OHDA rats. ES-evoked contractions in colonic strips are reduced in animals 8 weeks after 6-OHDA injection into the medial forebrain bundle. ES: 0.5 ms, 30 mA, 10 s, 10 Hz. (Pellegrini C, et al., J Neuroinflammation 2016, 13(1), 146)

5.3.3.2 MPTP Models

The prototypical parkinsonian neurotoxin, MPTP (1-methyl 4-phenyl 1,2,3,6-tetrahydropyridine), is a selective dopaminergic neuron toxin in the ENS. Acute administration of MPTP causes decrease of DA level in jejunal myenteric plexus, disruption of intestinal MMC (Eaker et al. 1987), and inhibition (of over 60%) of GI motility (Haskel and Hanani 1994). The action of MPTP on GI motility can be fully blocked by administration of β-adrenergic blocker propranolol and dopaminergic blocker haloperidol together (Haskel and Hanani 1994). Long-time MPTP-lesioned mice show the loss of dopaminergic neurons both in the SN and in the ENS (Anderson et al. 2007; Tian et al. 2008; Natale et al. 2010; Lai et al. 2018; Choi et al. 2018a; Torres et al. 2018; Jarras et al. 2020). MPTP induces macrophage activation, apoptosis, and nuclear factor kappa b (NF-κB) activation in the GI tract (Natale et al. 2010; Côté et al. 2011, 2015a, b; Poirier et al. 2016; Huh et al. 2020; Sun et al. 2018; Lai et al. 2018; Choi et al. 2018a; Torres et al. 2018; Jarras et al. 2020). There are no differences in the density of gut cholinergic or nitric oxide neurons between control and MPTP groups, and the contents of gut NE and 5-HT are not affected in the MPTP group (Natale et al. 2010). In the animals treated with high dose of MPTP (60 mg/kg), dopaminergic damage transiently causes colonic hypermotility 2–3 days after treatment, but the stool frequency is similar to controls 8–10 days after MPTP (Anderson et al. 2007). However, most of the studies about MPTP mice with moderate dosage of MPTP (4.75, 8, or 20 mg/kg) demonstrate that the mice present delayed gastric emptying, intestinal dysmotility, decreased stool frequency, reduced fecal water content, impaired intestinal barrier, and gut microbial dysbiosis (Natale et al. 2010; Côté et al. 2011, 2015a, b; Poirier et al. 2016; Lai et al. 2018; Jarras et al. 2020; Sampath et al. 2019). Although the GI dysmotility in MPTP models is debated, the MPTP mice really display GI pathology (Table 5.4). Thus, MPTP model is suitable to investigate the pathology and disorder in PD derived from peripheral lesion.

5.3.3.3 Paraquat Model

Paraquat is a widely used herbicide responsible for large numbers of deaths worldwide. Paraquat can alter cellular and organ function in a variety of ways. It is a pro-oxidant that generates the superoxide free radical and has been used to induce oxidative stress in a wide range of tissue/cell types to mimic diseases. The extensive degeneration of dopaminergic neurons in the SNpc and the presence of cytoplasmic aggregates of misfolded α-synuclein are pathophysiologic hallmarks of PD (Table 5.5). The upper GI tract, and the stomach in particular, is influenced heavily by vagal inputs originating from brainstem nuclei located in the DVC (i.e., nucleus tractus solitarius, DMV, and area postrema). Motor commands to the upper GI tract arise from the spontaneously active, preganglionic cholinergic parasympathetic motoneurons in the DMV (Travagli et al. 1991). These neurons integrate signals from higher centers (Browning and Travagli 2014). The efferent vagal fibers project to postganglionic myenteric neurons of the ENS that ultimately control the motility response of the GI tract (Travagli and Anselmi 2016). Anselmi et al. identified a nigro-vagal monosynaptic pathway in rats that controls gastric tone and motility. Stimulation of the SNpc increases gastric tone and motility via activation of D_1Rs in the DVC (Anselmi et al. 2017a). Rats are given i.p. injections of 10 mg/kg of paraquat once weekly for 3 weeks to induce features of Parkinsonism. The model rats display reduced gastric tone and motility, accompanied by decreased TH-immunoreactivity (IR) and increased α-synuclein aggregates in the SNpc, and

Table 5.4 MPTP PD models

Animal	Dose/time	Characteristics			Reference
		Nigra-striatum	Muscle layer in the GI tract	Function and gut microbiota	
Mouse	i.p. injection of 30 mg/kg/day MPTP for 5 days	DA↓, 5-HT↓, DOPAC↓, HVA↓, 5-HIAA↓, TLR-4↑, TBK1↑, NF-κB↑, TNF-α↑ in striatum; TH-IR↓, GFAP↑, Iba1↑ in SNpc	TH↓ in the stomach, duodenum, and colon; ZO-1↓, occludin↓, CD11b↑, macrophage activation, iNOS↑, COX-2↑, TNF-α↑, IL-1β↑, p-ERK↑, apoptosis (bax↑, bcl-2↓, cytochrome C↑, caspase-3 cleavage levels↑, HO-1↓), TLR4↑,TBK1↑, NF-κB↑(colon)	Phylum Firmicutes↓, order Clostridiales↓; phylum Proteobacteria↑, order Turicibacterales↑, Enterobacteriales↑ (fecal samples)	Tian et al. (2008), Huh et al. (2020), Sun et al. (2018)
	i.p. injection of 30 mg/kg/day MPTP for 7 days		Nrf2↓, oxidant detoxifying systems↓, nNOS↓, dihydrofolate reductase (DHFR) protein↓, biopterin levels↓ in stomach and colon	Delayed gastric emptying; reduced intestinal transit time; nitrergic relaxation↓ in gastric tissue and colon tissue	Sampath et al. (2019)
	i.p. injection of 10 mg/kg MPTP for 5 days, 2-day break, then followed by a 20 mg/kg MPTP dosage for 5 days	TH↓ in striatum		No difference in microbiome α-diversity	Torres et al. (2018)
	i.p. 3/10 injections of 25 mg/kg MPTP and 100 mg/kg probenecid at an interval of 3.5 days between consecutive doses	TH↓ in SNpc and striatum	TH-IR↓, ChAT-IR↓, iNOS-IR↑, α-Syn-IR↑, Iba1↑, TH-IR↓, VIP-IR↓, M1 macrophages↑, IL-1β↑, IL-6↑ in ileum	Intestinal dysmotility	Choi et al. (2018a)
	i.p. injection of 18 mg/kg MPTP twice per week for 5 weeks	DA↓ in striatum, TH↓, α-Syn↓, Iba1↑, IL-17↑, IL-1β↑, TNF-α↑ in SN	DA↓, TH↑, α-Syn↑, iNOS↑, IL-17↑, IL-1β↑, TNF-α↑ in ileum	1-h stool frequency↓, fecal water content↓ at 34 days, impaired intestinal barrier, gut microbial dysbiosis (Lachnospiraceae↓,	Lai et al. (2018)

(continued)

Table 5.4 (continued)

Animal	Dose/time	Characteristics			Reference
		Nigra-striatum	Muscle layer in the GI tract	Function and gut microbiota	
				Prevotellaceae↑, Erysipelotrichaceae↑, Clostridiales↓, Erysipelotrichales↑, Proteobacteria↓)	
	6.5–8 mg/kg MPTP given in four i.p. injections 2 h apart on a single day		No difference in total neuronal density, TH^+ neuron↓, dopaminergic neuron↓, MHCII-expressing pro-inflammatory macrophages↑ in distal ileum		Jarras et al. (2020)
	60 mg/kg MPTP given in four i.p. injections 2 h apart on a single day	After 10 days, decreases in TH-IR in the SN and striatum of 57% and 52%, respectively	40% reduction of TH-IR neurons (distal ileum)	Relaxation defect; a transient increase in colon motility; no changes in gastric emptying or small intestine transit	Anderson et al. (2007)
	i.p. injection of 4.75, 8, or 20 mg/kg MPTP (twice or 3 times, 2 h apart) on a single day	TH-IR↓ in striatum and SN	TH-IR↓, DAT-IR↓, DA↓, α-Syn↑ in duodenum; TH-IR↓, VIP-IR↓, M1 macrophages↑, IL-1β↑, IL-6↑ in ileum	Daily stool weight↓, constipation	Natale et al. (2010), Côté et al. (2011, 2015a, b), Poirier et al. (2016)
Monkey	i.v. injection of 0.2 mg/kg MPTP hydrochloride daily until they displayed parkinsonian symptoms		Hu-IR neurons/ganglia↑, nNOS -IR↑, TH-IR↓ in colon	No constipation	Chaumette et al. (2009)

α-Syn alpha-synuclein, *ChAT* choline acetyltransferase, *COX-2* cyclooxygenase-2, *DA* dopamine, *DAT* dopamine transporter, *DOPAC* 3,4-dihydroxyphenylacetic acid, *GFAP* glial fibrillary acidic protein, *HVA* homovanillic acid, *Iba1* ionized calcium binding adaptor molecule 1, *IL-1β* interleukin-1 beta, *IL-6* interleukin-6, *IL-17* interleukin-17, *iNOS* inducible nitric oxide synthase, *i.p.* intraperitoneal, *IR* immunoreactivity, *i.v.* intravenous injection, *MPTP* 1-methyl 4-phenyl 1,2,3,6-tetrahydropyridine, *NF-κB* nuclear factor kappa b, *nNOS* neuronal nitric oxide synthase, *PD* Parkinson's disease, *SN* substantia nigra, *SNpc* substantia nigra pars compacta, *TBK1* TANK-binding kinase 1, *TH* tyrosine hydroxylase, *TLR-4* toll-like receptor 4, *TNF-α* tumor necrosis factor α, *VIP* vasoactive intestinal peptide, *5-HIAA* 5-hydroxyindoleaceticacid, *5-HT* 5-hydroxytryptamine

Table 5.5 Paraquat PD models

	Characteristics				
Dose/time	Nigra-striatum	DVC	Myenteric plexus/ muscle layer	Function	Reference
i.p. injection of 10 mg/kg paraquat once a week for 3 consecutive weeks	TH-IR↓, α-Syn aggregates in the SNpc	ChAT-IR↓, α-Syn aggregates in the DMV	No change of PGP9.5-IR and ChAT-IR in gastric myenteric ganglia	Reduced gastric tone and motility	Anselmi et al. (2017a, b)

α-Syn alpha-synuclein, *ChAT* choline acetyltransferase, *DMV* the dorsal motor nucleus of the vagus, *DVC* the dorsal vagal complex, *i.p.* intraperitoneal, *IR* immunoreactivity, *PD* Parkinson's disease, *PGP9.5* protein gene product 9.5, *SNpc* substantia nigra pars compacta

decreased ChAT-IR and increased α-synuclein aggregates in the DMV (Anselmi et al. 2017b).

In the paraquat-induced model of parkinsonism, the above-mentioned nigro-vagal pathway is compromised during the early stages of motor deficit development (Anselmi et al. 2017b). It suggests the presence of tonically active dopaminergic inputs originated from the SNpc that modulate vagal efferent outflow, which might be involved in the prodromal gastric dysmotility observed in patients with early-stage PD.

5.3.3.4 Rotenone Models

An environmental toxicant rotenone, a mitochondrial complex I inhibitor, has been widely used to establish PD models to replicate the GI dysfunction seen clinically (Table 5.6). Chronic models of rats or mice are induced by i.p. (Drolet et al. 2009; Johnson et al. 2018), s.c. (Bu et al. 2019), intrastriatal administration (Perez-Pardo et al. 2018), or oral gavage (Dodiya et al. 2020; Perez-Pardo et al. 2018) of 2–10 mg/kg rotenone (dissolved in DMSO/sunflower oil, daily or 5 days/week) for 4, 6 weeks, or even 6 months. Both oral and intrastriatal administration of rotenone can induce similar PD-like GI dysfunction (Perez-Pardo et al. 2018). Rotenone-induced PD rats demonstrate weight loss, decrease of the dopaminergic neurons in SN (Fig. 5.17), loss of enteric myenteric plexuses neurons (Fig. 5.18), GI inflammation (Morais et al. 2018; Perez-Pardo et al. 2018), delayed gastric emptying (Johnson et al. 2018), impairment of GI movement (Figs. 5.18 and 5.19), and a decrease in fecal pellet output (Fig. 5.20) (Drolet et al. 2009; Greene et al. 2009; Perez-Pardo et al. 2018; Morais et al. 2018; Bu et al. 2019). Rotenone disrupts the colonic expression of the tight junction protein ZO-1 and increases oxidative stress (N-tyrosine) (Dodiya et al. 2020). Increased enteric glial cell glial fibrillary acidic protein (GFAP) expression, α-synuclein accumulation (Fig. 5.21) (Dodiya et al. 2020; Perez-Pardo et al. 2018), but decreased expression of ChAT, neuronal nitric oxide synthase (nNOS), and TH are existed in the myenteric plexus of colon (Fig. 5.22) (Bu et al. 2019; Dodiya et al. 2020) from rotenone models. Rotenone-treated rats also have a higher α-synuclein

Table 5.6 Rotenone PD models

Animal models	Dose/time	Characteristics			Reference
		Nigra-striatum	Myenteric plexus/muscle layer	Function	
Rotenone Lewis rats	i.p. injections of 2.0 mg/kg rotenone (dissolved in DMSO), 5 days/week for 6 months		α-Syn↑, loss of 25% neurons in the myenteric plexuses (small intestine)	GI transit time↑	Drolet et al. (2009)
Rotenone SD rats	i.p. injections of 2.75–3 mg/kg rotenone, 5 days/week for 4 weeks			Delayed gastric emptying; diarrhea; wet tissue weight of the stomach↑, small intestine↑, and colon↑; a trend for an overall increase in bacterial levels in the small intestine and colon	Johnson et al. (2018), Greene et al. (2009)
	Subcutaneous injection of rotenone (2.5 mg/kg/day, dissolved in sunflower oil) for 6 weeks	TH-IR↓ in SN	ChAT↓, TH↓, nNOS↓ in the colon	Constipation, body weight↓	Bu et al. (2019)
Rotenone mice	Oral gavage of 10 mg/kg rotenone once a day for 4 weeks; intrastriatal administration of 5.4μg of rotenone dissolved in 2μl DMSO	Dopaminergic cell loss in SN	Inflammation, and α-Syn accumulation in the colon	Delayed intestinal transit; reduced colon length; dysfunctional barrier integrity in both the ileum and colon (oral rotenone)	Perez-Pardo et al. (2018)
	Oral gavage of a freshly prepared rotenone solution at a dose of 10 mg/kg/day suspended in vehicle (4% carboxymethylcellulose and 1.25% chloroform) for 6 weeks	Dystrophic/phagocytic microglia↑, FJ-C (cell death marker)↑, TH-IR↓, LPS reactivity↑ in SN; DA↓, no significant changes of DOPAC and HVA in striatum	TH-IR↓, GFAP↑, α-Syn↑ in colon	Intestinal permeability↑, tight junction barrier↓ in colon; relative abundance of Akkermansia (gram-negative) in fecal microbiota↑	Dodiya et al. (2020)

	Daily oral administration of rotenone (30 mg/kg, dissolved in 4% chloroform and 0.5% carboxymethyl cellulose sodium salt) for 4 weeks	TH-IR↓ in SN	α-Syn↓, but lack of ENS neurodegeneration in jejunum	Decrease in stool frequency, but normal gastric emptying and total transit time	Tasselli et al. (2013)

α-Syn alpha-synuclein,*ChAT* choline acetyltransferase, *DA* dopamine, *DMSO* dissolved in dimethyl sulfoxide, *DOPAC* dopamine carboxylase, *ENS* enteric nervous system, *FJ-C* fluorojade-C, *GFAP* glial fibrillary acidic protein, *HVA* homovanillic acid, *i.p.* intraperitoneal, *LPS* lipopolysaccharides, *nNOS* neuronal nitric oxide synthase, *SD* Sprague-Dawley, *SN* substantia nigra, *TH* tyrosine hydroxylase

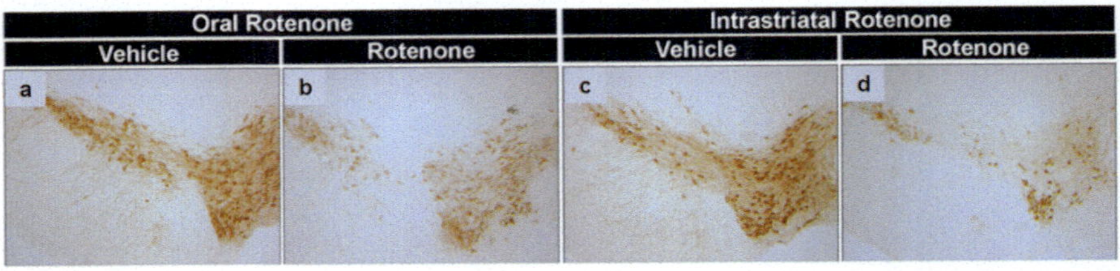

Fig. 5.17 Both oral and intrastriatal administration of rotenone decreased the number of tyrosine hydroxylase immunoreactive (TH-IR) cells in the substantia nigra (SN). (**a**) The SN from oral gavage of vehicle rat; (**b**) the SN from oral gavage of rotenone; (**c**) the SN from intrastriatal injection of vehicle; (**d**) the SN from intrastriatal injection of rotenone. (Perez-Pardo P, et al., Nutr Neurosci 2018, 21(6), 391–402)

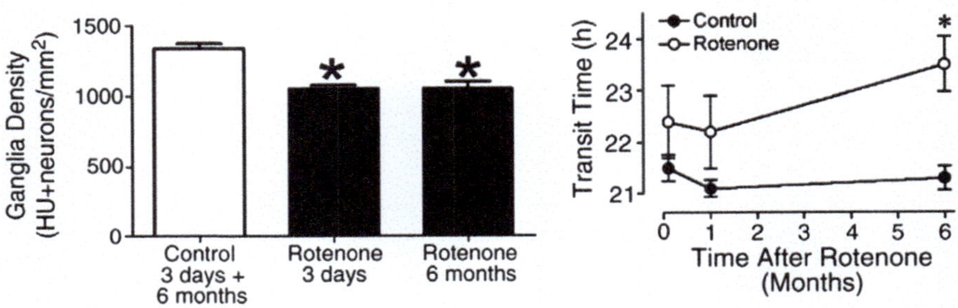

Fig. 5.18 Enteric neuron density in the small intestine and time-course of gastrointestinal transit times in control and rotenone-treated rats. Left: Density of myenteric neurons is reduced in rats with intraperitoneal injection of rotenone 3 days or 6 months later. Right: Gastrointestinal transit times are much longer in rotenone-treated rats than control rats. (Drolet RE, et al., Neurobiol Dis 2009, 36 (1), 96–102)

protein level and moderate but permanent loss of neurons in the small intestine myenteric plexuses (Drolet et al. 2009). Wet tissue weight of the stomach, small intestine, and colon is increased in rotenone rats. Microbiota analysis reveals an overall increase of bacterial levels in the small intestine and colon of rotenone-treated rats (Johnson et al. 2018), especially an abundance of gram-negative Akkermansia (Dodiya et al. 2020). Histological analysis demonstrates colonic mucosal thickening and goblet cell hyperplasia in rotenone rats (Johnson et al. 2018), which may be an adaptive response to the changes in GI microbiota. Chronic oral treatment with rotenone in mice can only reduce fecal pellet output and decrease α-synuclein expression in jejunal myenteric plexus (Tasselli et al. 2013). In summary, chronic rotenone-treated rodent may be a good model for investigating the mechanisms of GI symptoms in PD.

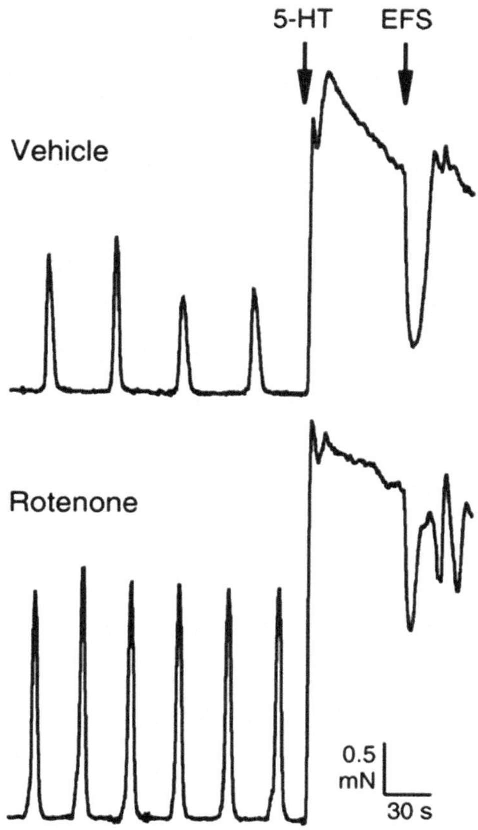

Fig. 5.19 Representative force recording of longitudinal colon muscle contraction from a vehicle-treated rat and rotenone (intraperitoneal injection) model of Parkinson's disease. The relaxation in proximal colon induced by electrical field stimulation (EFS) is weakened in rotenone-treated rats. (Greene JG, et al., Exp Neurol 2009, 218(1), 154–161)

5.3.3.5 α-Synuclein Mouse Model

PD is characterized by the accumulation of α-synuclein inclusions, which are the major component of Lewy bodies. Robust transgenic mouse models have been generated which overexpress the wild type/mutant human α-synuclein (Table 5.7). A transgenic mouse model of PD, in which mutant (A53T) human α-synuclein is expressed under control of the prion promoter (AS mice), develops age-dependent motor deficits. These mice show that GI expression of human α-synuclein is limited to efferent vagal fibers projecting from the DMV to the ENS, with an age-related disruption in these fibers. The mice develop impaired gastric emptying, colon motility, and age-related decline in stool frequency (Kuo et al. 2010; Noorian et al. 2012; Farrell et al. 2014; Wang et al. 2018; Rota et al. 2019). The severity of impaired colon motility as well as anxiety-like and depressive-like behaviors is correlated with the genomic DNA levels of A53T mutant α-synuclein (Wang et al. 2018). The study provides direct evidence that the genomic DNA levels of mutant α-synuclein correlate with non-motor symptoms in an A53T mouse model. Twelve-

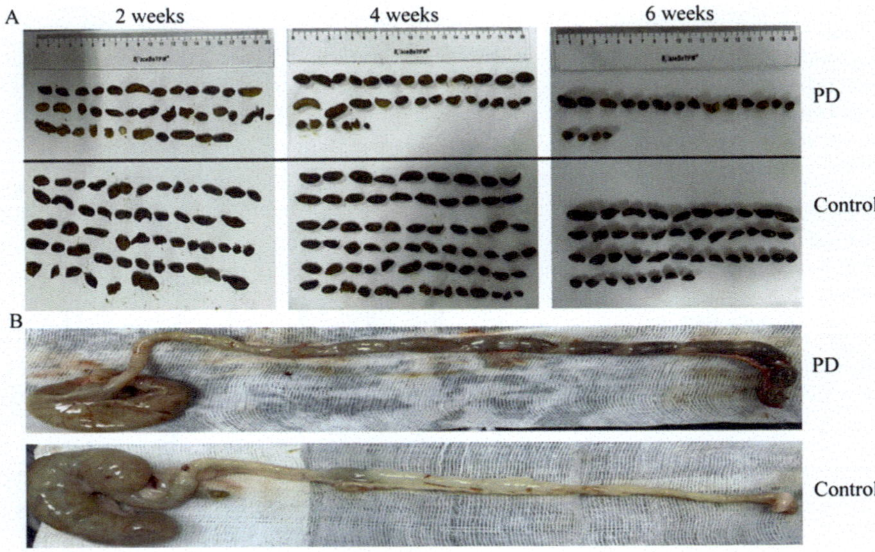

Fig. 5.20 Constipation in subcutaneous injection of rotenone-induced Parkinson's disease (PD) model. (**a**) The fecal production during 12 h period in PD and control rats. (**b**) Residual colonic content 12 h after a meal in PD and control rats. (Bu J, et al., Life Sci 2019, 231, 116581)

month-old mice overexpressing human wild-type α-synuclein under the Thy1 promoter (Thy1-aSyn) display colonic motor dysfunction. Eight- to ten-month-old Thy1-aSyn mice display increased α-synuclein in the myenteric plexus with abundant varicose terminals surrounding pChAT-immunoreactive neurons (Wang et al. 2008, 2012). A mouse model expressing human wild-type α-synuclein fused with the green fluorescent protein (GFP) also has an age-dependent progressive expression and accumulation of α-synuclein-GFP and phosphorylation of α-synuclein at serine 129 in the enteric neurons of submucosal and myenteric plexuses (Fig. 5.23) in the colon, and α-synuclein is distinctly co-localized with calretinin (Zhong et al. 2017; Chen et al. 2018).

In non-transgenic mice, intrastriatal injection of human preformed fibril α-synuclein (PFFs) induces Lewy body-like pathology (pathologic α-synuclein inclusions) in various brain regions (SNpc, striatum and hippocampus) and the ENS (small intestine) at 5 months post injection, accompanied with microglial activation, astrogliosis, and increased infiltration of B, $CD4^+$ T, $CD8^+$ T, and natural killer cells in the brain (Earls et al. 2019). However, they did not observe the GI motor dysfunction. Inoculation of PFFs into the mouse duodenal and pyloric muscularis layer causes the development of phosphorylated α-synuclein (p-α-Syn)-positive Lewy body-like aggregates first in the DMV, then in the locus coeruleus, and much later in basolateral amygdala, dorsal raphe nucleus, and SNpc (Kim et al. 2019). Uemura et al. have found that p-α-Syn only presents in the DMV and the ambiguous nucleus, but not in the SNpc after inoculation of PFFs into mouse gastric

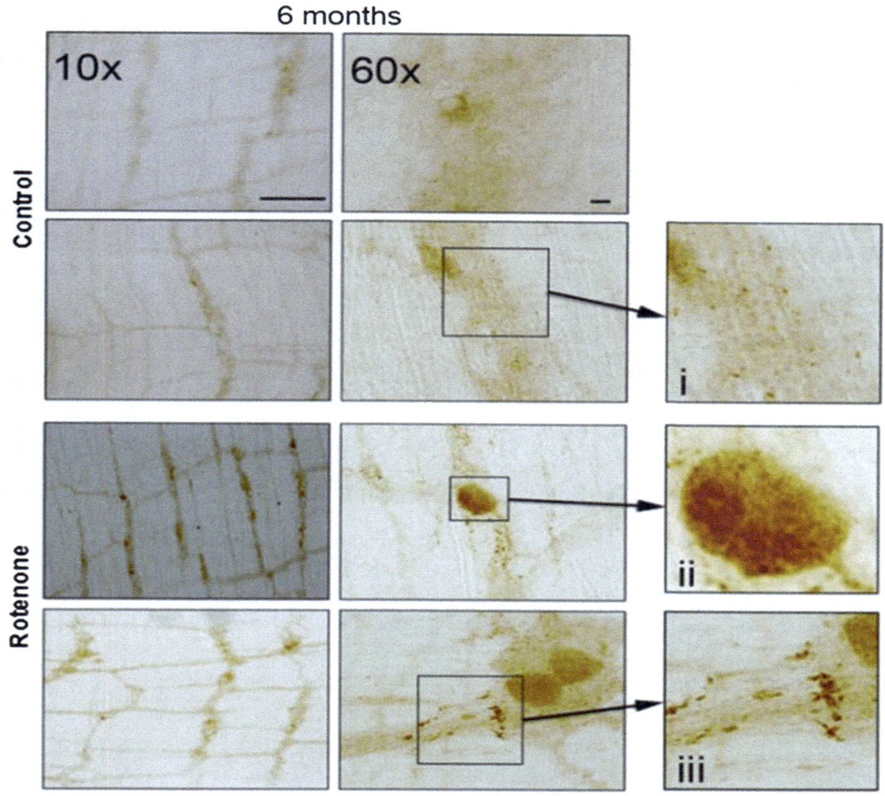

Fig. 5.21 α-Synuclein immunoreactivity in the small intestine myenteric plexus of control rats and rotenone-treated (intraperitoneal injection) rats. (Drolet RE, et al., Neurobiol Dis 2009, 36(1), 96–102)

wall in wild-type mice (Uemura et al. 2018) or even in bacterial artificial chromosome (BAC) transgenic mice harboring the human *α-Syn* gene with an A53T mutation (Uemura et al. 2020), which does not support caudo-rostral spread in accordance with Braak's hypothesis. Further, Challis et al. have found that inoculation of PFFs into the duodenum in aged mice, but not younger mice, results in progression of α-Syn histopathology to the midbrain (Challis et al. 2020). However, all of the studies have also observed p-α-Syn-positive aggregation in the myenteric plexus of the upper duodenum and/or stomach (Uemura et al. 2018; Kim et al. 2019; Challis et al. 2020). Gut-to-brain spread of α-synucleinopathy is retrogradely induced via the vagus nerve (Uemura et al. 2018; Kim et al. 2019; Challis et al.

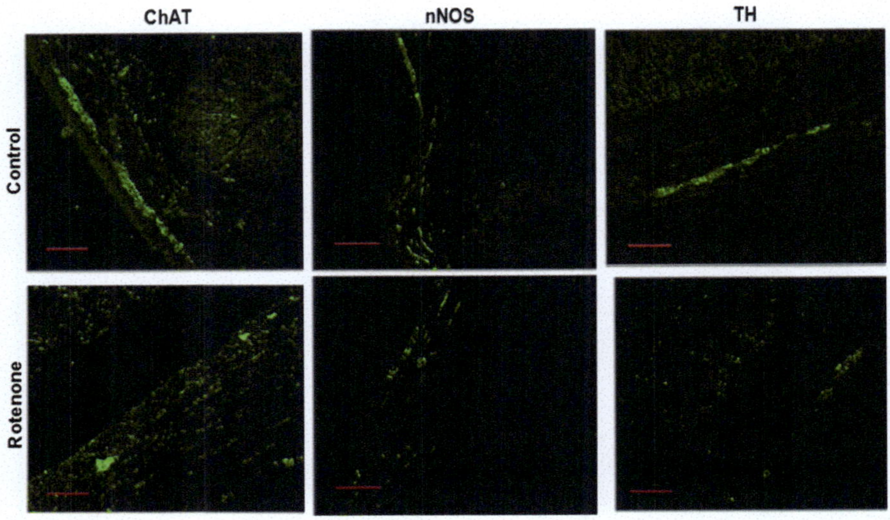

Fig. 5.22 Representative micrographs showing immunofluorescent staining of choline acetyltransferase (ChAT), neuronal nitric oxide synthase (nNOS), and tyrosine hydroxylase (TH) in the colon of control and rotenone-treated rats at 6 weeks. (Bu J, et al., Life Sci 2019, 231, 116581)

2020). Taken together, the α-Syn PFF injection into the mouse GI tract can induce α-Syn pathology resembling that of GI tract and the DMV at very early PD, while further progression of PD pathology from the DMV to the SNpc is needed to be confirmed in both aged and younger mice.

In summary, the α-synuclein mouse models possess the potential for studying GI dysfunction in age-related PD with pathologic α-synuclein inclusions in the ENS.

5.3.3.6 VMAT2-Deficient Mice

The type 2 vesicular monoamine transporter (VMAT2), a neuronal H^+-ATPase-coupled antiporter, regulates the storage of monoamines transmitters into synaptic vesicles. It has a neuroprotection role against endogenous and exogenous toxicants. Increasing evidence suggests that impairment of VMAT2 contributes to the pathogenesis of PD. The mice with a 95% genetic reduction in VMAT2 expression (VMAT2-deficient, VMAT2 LO, Table 5.8) display progressive loss of striatal DA, L-DOPA-responsive motor deficits, α-synuclein accumulation, and nigral dopaminergic cell loss. VMAT2-deficient mice also demonstrate progressive deficits in delayed gastric emptying, olfactory discrimination, altered sleep latency, anxiety-like behavior, and age-dependent depressive behavior (Taylor et al. 2009).

Table 5.7 Alpha-synuclein PD mouse models

Treatment	Characteristics				Reference
	Substantia Nigra	DVC	Myenteric plexus	Function	
Transgenic mice expressing human A53T αS (or A30P) under the control of the mouse prion protein (PrP) promoter	α-Syn (−), DAT↓	α-Syn (−)	Cholinergic transmission↓, α-Syn↑ in stomach, small intestine (aging mice) and colon	Constipation, the whole gut transit time↑, reduced number of stools with longer pellets excreted, wet and dry stool weight↓, delayed GI transit, food intake (−), colonic contractile activity↓, decreased electrically evoked colonic motor response but not in ileum	Wang et al. (2008, 2012, 2018), Kuo et al. (2010), Farrell et al. (2014), Rota et al. (2019), Noorian et al. (2012)
Transgenic mice expressing human full-length wild-type α-Syn gene was inserted into the pAcGFP1-C1 vector at the initiation codon of the mouse α-Syn gene			In colon: α-Syn↑, p-α-Syn accumulation↑ with age; α-Syn highly co-localized with calretinin, to a lesser extent with VIP, but not with nNOS; VIP↑, calretinin↑, nNOS (−)		Zhong et al. (2017), Chen et al. (2018)
Unilateral injection of human PFF α-Syn (5 μg in 1 μl) into the right striatum (AP, +0.3 mm; ML, +2.3 mm; DV, −3.5 mm)	α-Syn↑, TH-IR (−), microglial activation and astrogliosis (Iba-1↑, MHC-II↑, GFAP↑), TNF-α↑, IL-1β↑, CD80↑, CD19+↑, CD4+↑, CD8		α-Syn↑ in small intestine		Earls et al. (2019)

(continued)

Table 5.7 (continued)

Treatment	Characteristics				Reference
	Substantia Nigra	DVC	Myenteric plexus	Function	
	+↑, NK cell↑, CD11b+ CD45high↑				
Intramuscular α-Syn PFF injection into the mouse duodenum (12.5 μg) or/ and pyloric stomach (6.25 μg)	pSer129-α-Syn↑	pSer129-α-Syn↑	P-α-Syn-positive aggregates↑ in the myenteric plexus of upper duodenum and pyloric stomach	The whole gut transit time↑, pellet weight↓, water weight per pellet↓	Kim et al. (2019), Challis et al. (2020)
Intramuscular α-Syn PFF injection into the mouse gastric wall (12.5 μg)	pSer129-α-Syn (−)	p-α-Syn aggregates present in the DMV at 45 days post inoculation	P-α-Syn-positive aggregates↑ in the myenteric plexus of stomach		Uemura et al. (2018, 2020)

α-Syn α-synuclein, *DAT* dopamine transporter, *DMV* the dorsal motor nucleus of the vagus, *GFAP* glial fibrillary acidic protein, *Iba1* ionized calcium binding adaptor molecule 1, *IL-1β* interleukin-1β, *nNOS* neuronal nitric oxide synthase, *p-α-syn* phosphorylated α-synuclein, *PFF* preformed fibril, *TNF-α* tumor necrosis factor α, *VIP* vasoactive intestinal peptide

5.3.3.7 Gut Bacterium

Given PD patients display alterations in gut microbiota composition (Schwiertz and Schäfer 2016), investigators have explored mechanism between gut microbial dysbiosis and pathogenesis of PD using animal models. Sequencing of 16S rRNA

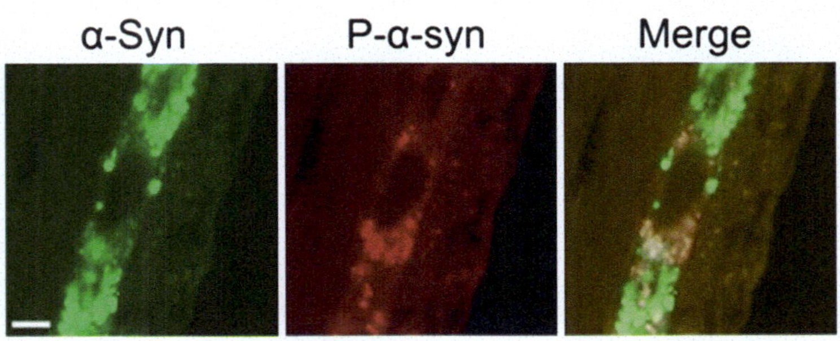

Fig. 5.23 Partial phosphorylation-α-synuclein (red) and pan α-synuclein expression (green) in the myenteric plexus of mouse colon. (Chen QQ, et al., Transl Neurodegener 2018, 7, 13)

Table 5.8 Other PD animal models

Animal models	Dose/time	Characteristics				Reference
		Nigra-striatum	DVC	Muscle layer	Function	
VMAT2-deficient mice	The VMAT2 locus (SLC18A2) cloned from the 129/Sv genomic library and a 2.2 kb PvuII fragment from the third intron of the VMAT2 gene, and cloned into the blunt-ended NotI site of the construct	Monoamine↓, DOPAC/DA↑, 5-HIAA/5-HT↑			Delayed gastric emptying; increased stool frequency	Taylor et al. (2009)
Gut bacterium Proteus mirabilis	Proteus mirabilis (p.o.) for 5 day	TH-IR↓, Iba1↑, α-Syn-IR↑ in striatum and SN		α-Syn↑ in distal colon	Colonic barrier disruption and inflammation (LPS↑ in feces and serum, occludin↓, TNF-α↑, TLR4↑)	Choi et al. (2018b)
LPS rats	Bilateral injection of 2 μl LPS (2 g/l) into the SN (AP, −5.6; ML, ±2.0; DV,−7.5 (mm))	TH-IR↓, Iba1↑	TH-IR↑, ChAT-IR↓		Gastric motility↓, maximal contraction amplitude of MMC↓	Zheng et al. (2013)

AP anteroposterior, *α-Syn* alpha-synuclein, *ChAT* choline acetyltransferase, *DA* dopamine, *DOPAC* 3,4-dihydroxyphenylacetic acid, *DV* dorsoventral, *DVC* the dorsal vagal complex, *Iba1* ionized calcium binding adaptor molecule 1, *IR* immunoreactivity, *LPS* lipopolysaccharide, *ML* mediolateral, *MMC* migrating motor complexes, *p.o.* per os, *SN* substantia nigra, *TH* tyrosine hydroxylase, *TLR-4* toll-like receptor 4, *TNF-α* tumor necrosis factor α, *VMAT2* the type 2 vesicular monoamine transporter, *5-HIAA* 5-hydroxyindoleaceticacid, *5-HT* 5-hydroxytryptamine

reveals that phylum Firmicutes and order Clostridiales decrease, while phylum Proteobacteria, order Turicibacterales and Enterobacteriales increase in fecal samples of PD mice, along with increased fecal short-chain fatty acids (SCFAs). Fecal microbiota transplantation (FMT) reduces gut microbial dysbiosis, decreases fecal SCFAs, alleviates physical impairment, increases striatal DA and 5-HT content, suppresses neuroinflammation (FMT alleviates microglial and astrocyte activation in the SN), and reduces TLR4/TNF-α signaling in gut and brain (Sun et al. 2018). Oral

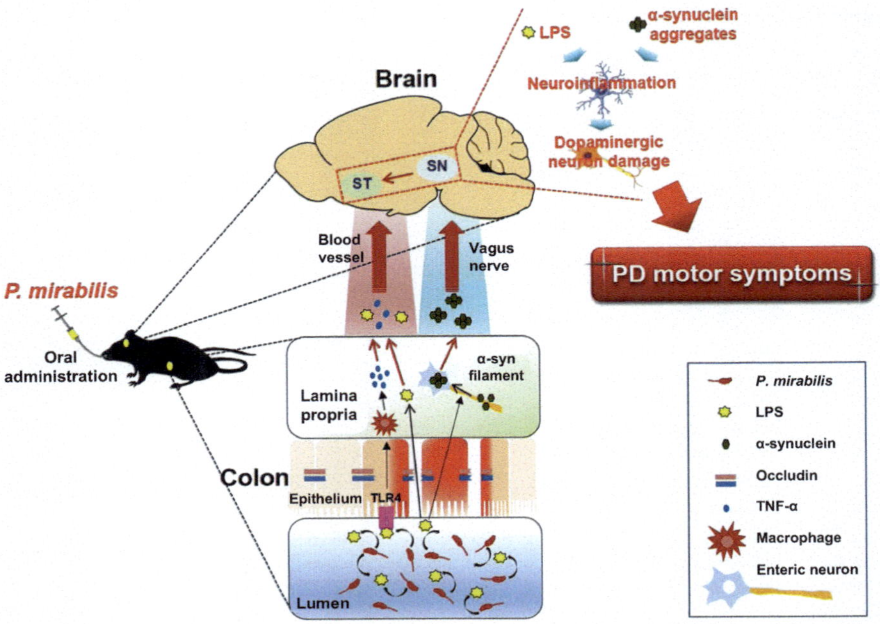

Fig. 5.24 The microbiota–gut–brain axis in Parkinson's disease (PD). Schematic summary of *P. mirabilis*, a specific gut bacterium, -induced PD pathogenesis in the colon and in the brain of mouse. (Choi JG, et al., Sci Rep 2018b, 8(1), 1275)

administration of *Proteus mirabilis* damages dopaminergic neurons and motor functions in mice (Table 5.8). Administration of *P. mirabilis* isolated from PD mice significantly induces motor deficits, selectively causes dopaminergic neuronal damage and inflammation in SN and striatum, and stimulates α-synuclein aggregation in the brain as well as in the colon. It is reported that lipopolysaccharides, a virulence factor of *P. mirabilis*, may be associated with these pathological changes via gut leakage and inflammatory actions. These results suggest a role of *P. mirabilis* on PD pathogenesis in the brain (Fig. 5.24) (Choi et al. 2018b).

In summary, the results from PD-related animal models suggest the existence of possible bidirectional communication between the gut and the brain for the genesis of PD-like phenotype and pathology (Fig. 5.25) (Perez-Pardo et al. 2018).

5.4 Perspective

The effect of DA and its receptor antagonists on normal and abnormal GI motility is an area of intense research. DA system lesion is associated with PD, accompanied with a series of GI disorders including gastroparesis and constipation. However, the

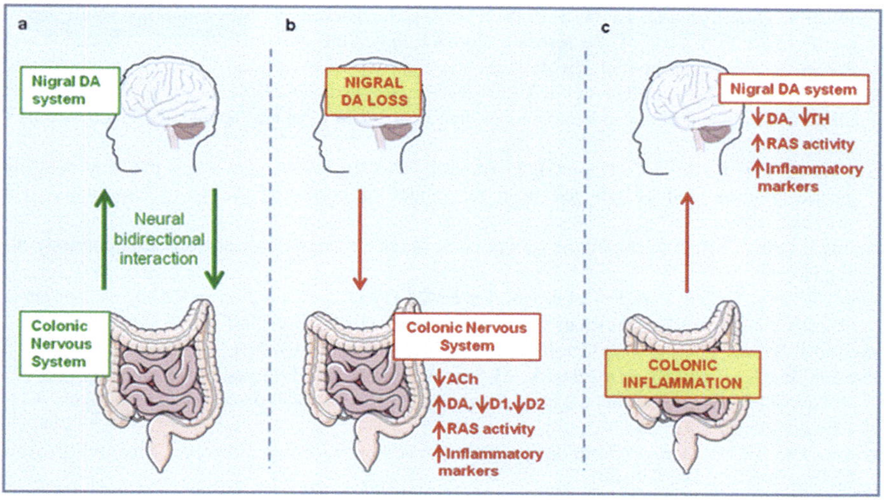

Fig. 5.25 The possible mechanisms for PD. Schematic model summarizing the bidirectional neural crosstalk between nigral and colonic lesions. Central dopaminergic lesions or gut inflammatory lesions induce alterations in this mutual interaction that lead to gastrointestinal or nigral functional dysregulation, respectively. *ACh* acetylcholine, *D1* dopamine type 1 receptors, *D2* dopamine type 2 receptors, *DA* dopamine, *RAS* renin-angiotensin system, *TH* tyrosine hydroxylase. (Garrido-Gil P, et al., Mol Neurobiol 2018, 55(9), 7297–7316)

underlying mechanisms are not fully understood. Although a number of studies are investigating the alterations of the ENS and the DMV for GI dysmotility in PD patients or animal models, the research focus has partially been shifted to gut microbiota. The interaction between altered central DA system and GI motility disorders is progressively being unrevealed. Pharmacological agents for the treatment of GI disorders in PD, including D_2R antagonist domperidone, metoclopramide, and 5-HT_4 receptor agonists, are under evaluation.

References

Abbott RD, Petrovitch H, White LR, Masaki KH, Tanner CM, Curb JD, Grandinetti A, Blanchette PL, Popper JS, Ross GW (2001) Frequency of bowel movements and the future risk of Parkinson's disease. Neurology 57(3):456–462

Aguilar MJ, Estañ L, Martínez-Mir I, Martínez-Abad M, Rubio E, Morales-Olivas FJ (2005) Effects of dopamine in isolated rat colon strips. Can J Physiol Pharmacol 83(6):447–452

Al-Saffar A, Takemi S, Saaed HK, Sakata I, Sakai T (2019) Utility of animal gastrointestinal motility and transit models in functional gastrointestinal disorders. Best Pract Res Clin Gastroenterol 40-41:101633

Anderson G, Noorian AR, Taylor G, Anitha M, Bernhard D, Srinivasan S, Greene JG (2007) Loss of enteric dopaminergic neurons and associated changes in colon motility in an MPTP mouse model of Parkinson's disease. Exp Neurol 207(1):4–12

Andreica V, Dumitraşcu D, Tamaş S, Cotul S (1991) The determination by computed scintigraphy of the evacuation time of the resected stomach under the administration of salbutamol and dopamine. Med Interna 43(1–2):124–128

Anselmi L, Toti L, Bove C, Travagli RA (2017a) Vagally mediated effects of brain stem dopamine on gastric tone and phasic contractions of the rat. Am J Physiol Gastrointest Liver Physiol 313 (5):G434–G441

Anselmi L, Toti L, Bove C, Hampton J, Travagli RA (2017b) A nigro-vagal pathway controls gastric motility and is affected in a rat model of parkinsonism. Gastroenterology 153 (6):1581–1593

Anuras S (1981) Effect of dopamine on opossum duodenal smooth muscle. Gastroenterology 80 (1):51–54

Ashraf W, Pfeiffer RF, Park F, Lof J, Quigley EMM (1997) Constipation in Parkinson's disease: objective assessment and response to psyllium. Mov Disord 12(6):946–951

Axelrod J, Saavedra JM (1977) Octopamine. Nature 265(5594):501–504

Baradari AG, Khajavi MR, Firouzian A, Alipour A, Daneshpour H, Panahi Y, Sahebkar A (2017) Effects of combined prokinetic administration on gastric emptying in critically ill patients. Arab J Gastroenterol 18(1):30–34

Barone JA (1999) Domperidone: a peripherally acting dopamine2-receptor antagonist. Ann Pharmacother 33(4):429–440

Bauer V, Kuriyama H (1982) Homogeneous and non-homogeneous distribution of inhibitory and excitatory adrenoceptors in the longitudinal muscle of the guinea-pig ileum. Br J Pharmacol 76 (4):603–611

Bech K, Hovendal CP (1982) Effect of dopamine on bethanechol-stimulated gastric antral motility in dogs with gastric fistula. Scand J Gastroenterol 17(7):945–951

Bech K, Hovendal CP, Andersen D (1982) Effect of dopamine on pentagastrin-stimulated gastric antral motility in dogs with gastric fistula. Scand J Gastroenterol 17(1):103–107

Bech K, Hovendal CP, Gottrup F, Andersen D (1984) Dopaminergic and beta-adrenergic effects on gastric antral motility. Scand J Gastroenterol Suppl 89:65–70

Berkowitz DM, McCallum RW (1980) Interaction of levodopa and metoclopramide on gastric emptying. Clin Pharmacol Ther 27(3):414–420

Blandini F, Balestra B, Levandis G, Cervio M, Greco R, Tassorelli C, Colucci M, Faniglione M, Bazzini E, Nappi G, Clavenzani P, Vigneri S, De Giorgio R, Tonini M (2009) Functional and neurochemical changes of the gastrointestinal tract in a rodent model of Parkinson's disease. Neurosci Lett 467(3):203–207

Boivin MA, Carey MC, Levy H (2003) Erythromycin accelerates gastric emptying in a dose-response manner in healthy subjects. Pharmacotherapy 23(1):5–8

Braak H, de Vos RA, Bohl JR, Del Tredici K (2006) Gastric alpha-synuclein immunoreactive inclusions in Meissner's and Auerbach's plexuses in cases staged for Parkinson's disease-related brain pathology. Neurosci Lett 396:67–72

Broad J, Góralczyk A, Mannur K, Dukes GE, Sanger GJ (2014) Drugs acting at 5-HT4, D2, motilin, and ghrelin receptors differ markedly in how they affect neuromuscular functions in human isolated stomach. Neurogastroenterol Motil 26(6):851–861

Browning KN, Travagli RA (2014) Central nervous system control of gastrointestinal motility and secretion and modulation of gastrointestinal functions. Compr Physiol 4:1339–1368

Bu J, Qiao X, He Y, Liu J (2019) Colonic electrical stimulation improves colonic transit in rotenone-induced Parkinson's disease model through affecting enteric neurons. Life Sci 231:116581

Buéno L, Sorraing JM, Fioramonti J (1983) Influence of dopamine on rumino-reticular motility and rumination in sheep. J Vet Pharmacol Ther 6(2):93–98

Bueno L, Fargeas MJ, Fioramonti J, Honde C (1984) Effects of dopamine and bromocriptine on colonic motility in dog. Br J Pharmacol 82(1):35–42

Bueno L, Fioramonti J, Primi MP (1985) Central effects of growth hormone-releasing factor (GRF) on intestinal motility in dogs: involvement of dopaminergic receptors. Peptides 6(3):403–407

Çamcı G, Oğuz S (2016) Association between Parkinson's disease and Helicobacter pylori. J Clin Neurol 12(2):147–150

Carbone F, Van den Houte K, Clevers E, Andrews CN, Papathanasopoulos A, Holvoet L, Van Oudenhove L, Caenepeel P, Arts J, Vanuytsel T, Tack J (2019) Prucalopride in gastroparesis: a randomized placebo-controlled crossover study. Am J Gastroenterol 114(8):1265–1274

Cassani E, Barichella M, Cancello R, Cavanna F, Iorio L, Cereda E, Bolliri C, Zampella Maria P, Bianchi F, Cestaro B, Pezzoli G (2015) Increased urinary indoxyl sulfate (indican): new insights into gut dysbiosis in Parkinson's disease. Parkinsonism Relat Disord 21(4):389–393

Cebrat E, Wegrzyn T, Zieba D, Leroch Z (1989) Effect of dopamine on small-intestinal motility in sheep. Pol Arch Weter 29(3–4):71–78

Chahine LM, Beach TG, Seedorff N, Caspell-Garcia C, Coffey CS, Brumm M, Adler CH, Serrano GE, Linder C, Mosovsky S, Foroud T, Riss H, Ecklund D, Seibyl J, Jennings D, Arnedo V, Riley L, Dave KD, Mollenhauer B (2018) Feasibility and safety of multicenter tissue and biofluid sampling for α-synuclein in Parkinson's disease: the systemic synuclein sampling study (S4). J Parkinsons Dis 8(4):517–527

Challis C, Hori A, Sampson TR, Yoo BB, Challis RC, Hamilton AM, Mazmanian SK, Volpicelli-Daley LA, Gradinaru V (2020) Gut-seeded α-synuclein fibrils promote gut dysfunction and brain pathology specifically in aged mice. Nat Neurosci 23(3):327–336

Chaumette T, Lebouvier T, Aubert P, Lardeux B, Qin C, Li Q, Accary D, Bézard E, Bruley d, Varannes S, Derkinderen P, Neunlist M (2009) Neurochemical plasticity in the enteric nervous system of a primate animal model of experimental parkinsonism. Neurogastroenterol Motil 21 (2):215–222

Chen JD, Lin Z, Pan J, McCallum RW (1996) Abnormal gastric myoelectrical activity and delayed gastric emptying in patients with symptoms suggestive of gastroparesis. Dig Dis Sci 41 (8):1538–1545

Chen QQ, Haikal C, Li W, Li MT, Wang ZY, Li JY (2018) Age-dependent alpha-synuclein accumulation and aggregation in the colon of a transgenic mouse model of Parkinson's disease. Transl Neurodegener 7:13

Cheng JT, Hsieh-Chen SC (1988) Octopamine relaxes rabbit jejunal smooth muscle by selective activation of dopamine D1 receptors. Naunyn Schmiedeberg's Arch Pharmacol 338(4):373–378

Cheremnykh EG, Ivanov PA, Sokolov OY, Prokhorova TA, Tereshkina EB, Baymeeva NV, Miroshnichenko II, Kost NV (2019) Haloperidol reduces the activity of complement and induces the anti-inflammatory transformation of peritoneal macrophages in rats. J Neuroimmune Pharmacol 14(3):369–374

Choi JG, Huh E, Ju IG, Kim N, Yun J, Oh MS (2018a) 1-Methyl-4-phenyl-1,2,3,6 tetrahydropyridine/probenecid impairs intestinal motility and olfaction in the early stages of Parkinson's disease in mice. J Neurol Sci 392:77–82

Choi JG, Kim N, Ju IG, Eo H, Lim SM, Jang SE, Kim DH, Oh MS (2018b) Oral administration of Proteus mirabilis damages dopaminergic neurons and motor functions in mice. Sci Rep 8 (1):1275

Choung RS, Talley NJ, Peterson J, Camilleri M, Burton D, Harmsen WS, Zinsmeister AR (2007) A double-blind, randomized, placebo-controlled trial of itopride (100 and 200 mg three times daily) on gastric motor and sensory function in healthy volunteers. Neurogastroenterol Motil 19 (3):180–187

Colucci M, Cervio M, Faniglione M, De Angelis S, Pajoro M, Levandis G, Tassorelli C, Blandini F, Feletti F, De Giorgio R, Dellabianca A, Tonini S, Tonini M (2012) Intestinal dysmotility and enteric neurochemical changes in a Parkinson's disease rat model. Auton Neurosci 169 (2):77–86

Cooper SM, McRitchie B (1985) Role of dopamine and alpha-adrenoreceptors in the control of gastric emptying in the rat: possible involvement in the mechanism of action of metoclopramide. J Auton Pharmacol 5(4):325–331

Costall B, Gunning SJ, Naylor RJ, Simpson KH (1983) A central site of action for benzamide facilitation of gastric emptying. Eur J Pharmacol 91(2–3):197–205

Costall B, Domeney AM, Gunning SJ, Kelly ME, Naylor RJ, Nohria V, Owera-Atepo JB, Simpson KM, Tan CC, Tattersall D (1987) The action of dazopride to enhance gastric emptying and block emesis. Neuropharmacology 26(7A):669–677

Côté M, Drouin-Ouellet J, Cicchetti F, Soulet D (2011) The critical role of the MyD88-dependent pathway in non-CNS MPTP-mediated toxicity. Brain Behav Immun 25(6):1143–1152

Côté M, Bourque M, Poirier AA, Aubé B, Morissette M, Di Paolo T, Soulet D (2015a) GPER1-mediated immunomodulation and neuroprotection in the myenteric plexus of a mouse model of Parkinson's disease. Neurobiol Dis 82:99–113

Côté M, Poirier AA, Aubé B, Jobin C, Lacroix S, Soulet D (2015b) Partial depletion of the proinflammatory monocyte population is neuroprotective in the myenteric plexus but not in the basal ganglia in a MPTP mouse model of Parkinson's disease. Brain Behav Immun 46:154–167

Curry JI, Lander TD, Stringer MD (2001) Review article: erythromycin as a prokinetic agent in infants and children. Aliment Pharmacol Ther 15(5):595–603

Dalvi SS, Nadkarni PM, Gupta KC (1990) Effect of Asparagus racemosus (Shatavari) on gastric emptying time in normal healthy volunteers. J Postgrad Med 36(2):91–94

Dardiotis E, Tsouris Z, Mentis AA, Siokas V, Michalopoulou A, Sokratous M, Dastamani M, Bogdanos DP, Deretzi G, Kountouras J (2018) H. pylori and Parkinson's disease: meta-analyses including clinical severity. Clin Neurol Neurosurg 175:16–24

Davies KN, King D, Billington D, Barrett JA (1996) Intestinal permeability and orocaecal transit time in elderly patients with Parkinson's disease. Postgrad Med J 72(845):164–167

Davis RH, Clench MH, Mathias JR (1988) Effects of domperidone in patients with chronic unexplained upper gastrointestinal symptoms: a double-blind, placebo-controlled study. Dig Dis Sci 33(12):1505–1511

De Pablo-Fernández E, Passananti V, Zárate-López N, Emmanuel A, Warner T (2019) Colonic transit, high-resolution anorectal manometry and MRI defecography study of constipation in Parkinson's disease. Parkinsonism Relat Disord 66:195–201

Demol P, Ruoff HJ, Weihrauch TR (1989) Rational pharmacotherapy of gastrointestinal motility disorders. Eur J Pediatr 148(6):489–495

Desmond PV, Watson KJ (1986) Metoclopramide--a review. Med J Aust 144(7):366–369

Dhasmana KM, Banerjee AK, Zhu YN, Erdmann W, Parmar SS, Salzman SK (1989) Role of dopamine receptors in gastrointestinal motility. Res Commun Chem Pathol Pharmacol 64 (3):485–488

Dhasmana KM, Villalón CM, Zhu YN, Parmar SS (1993) The role of dopamine (D2), alpha and beta-adrenoceptor receptors in the decrease in gastrointestinal transit induced by dopamine and dopamine-related drugs in the rat. Pharmacol Res 27(4):335–347

Dinning PG (2018) A new understanding of the physiology and pathophysiology of colonic motility? Neurogastroenterol Motil 30(11):e13395

DiPalma JR (1990) Metoclopramide: a dopamine receptor antagonist. Am Fam Physician 41 (3):919–924

Dive A, Foret F, Jamart J, Bulpa P, Installé E (2000) Effect of dopamine on gastrointestinal motility during critical illness. Intensive Care Med 26(7):901–907

Dobbs SM, Dobbs RJ, Weller C, Charlett A, Bjarnason IT, Lawson AJ, Letley D, Harbin L, Price AB, Ibrahim MA, Oxlade NL, Bowthorpe J, Leckstroem D, Smee C, Plant JM, Peterson DW (2010) Differential effect of Helicobacter pylori eradication on time-trends in Brady/hypokinesia and rigidity in idiopathic parkinsonism. Helicobacter 15:279–294

Dobbs SM, Charlett A, Dobbs RJ, Weller C, Iguodala O, Smee C, Lawson AJ, Taylor D, Bjarnason I (2013) Antimicrobial surveillance in idiopathic parkinsonism: indication-specific improvement in hypokinesia following Helicobacter pylori eradication and non-specific effect of antimicrobials for other indications in worsening rigidity. Helicobacter 18:187–196

Dodiya HB, Forsyth CB, Voigt RM, Engen PA, Patel J, Shaikh M, Green SJ, Naqib A, Roy A, Kordower JH, Pahan K, Shannon KM, Keshavarzian A (2020) Chronic stress-induced gut dysfunction exacerbates Parkinson's disease phenotype and pathology in a rotenone-induced mouse model of Parkinson's disease. Neurobiol Dis 135:104352

Doherty TJ, Andrews FM, Abraha TW, Osborne D, Frazier DL (1999) Metoclopramide ameliorates the effects of endotoxin on gastric emptying of acetaminophen in horses. Can J Vet Res 63 (1):37–40

Drolet RE, Cannon JR, Montero L, Greenamyre JT (2009) Chronic rotenone exposure reproduces Parkinson's disease gastrointestinal neuropathology. Neurobiol Dis 36(1):96–102

Dubois A, Jacobus JP, Grissom MP, Eng RR, Conklin JJ (1984) Altered gastric emptying and prevention of radiation-induced vomiting in dogs. Gastroenterology 86(3):444–448

Dumitraşcu DL, Ungureanu O, Verzea D, Pascu O (1998) The effect of metoclopramide on antral emptying of a semisolid meal in patients with functional dyspepsia. A randomized placebo controlled sonographic study. Rom J Intern Med 36(1–2):97–104

Eaker EY, Bixler GB, Dunn AJ, Moreshead WV, Mathias JR (1987) Chronic alterations in jejunal myoelectric activity in rats due to MPTP. Am J Phys 253(6 Pt 1):G809–G815

Earls RH, Menees KB, Chung J, Barber J, Gutekunst CA, Hazim MG, Lee JK (2019) Intrastriatal injection of preformed alpha-synuclein fibrils alters central and peripheral immune cell profiles in non-transgenic mice. J Neuroinflammation 16(1):250

Edwards LL, Quigley EMM, Harned RK, Hofman R, Pfeiffer RF (1994) Characterization of swallowing and defecation in Parkinson's disease. Am J Gastroenterol 89(1):15–25

Ehrlein HJ (1988) Dopamine delays gastric emptying and induces retrograde power contractions with enterogastric reflux. Z Gastroenterol 26(3):160–165

Epprecht L, Schreglmann SR, Goetze O, Woitalla D, Baumann CR, Waldvogel D (2015) Unchanged gastric emptying and visceral perception in early Parkinson's disease after a high caloric test meal. J Neurol 262(8):1946–1953

Eyre-Brook IA, Smallwood R, Johnson AG (1984) Human antroduodenal motility, pyloric closure and domperidone. Scand J Gastroenterol Suppl 92:4–7

Fargeas MJ, Fioramonti J, Bueno L (1990) Involvement of dopamine in the central effect of neurotensin on intestinal motility in rats. Peptides 11(6):1169–1173

Farrell KF, Krishnamachari S, Villanueva E, Lou H, Alerte TN, Peet E, Drolet RE, Perez RG (2014) Non-motor parkinsonian pathology in aging A53T α-synuclein mice is associated with progressive synucleinopathy and altered enzymatic function. J Neurochem 128(4):536–546

Fasano A, Bove F, Gabrielli M, Petracca M, Zocco MA, Ragazzoni E, Barbaro F, Piano C, Fortuna S, Tortora A, Di Giacopo R, Campanale M, Gigante G, Lauritano EC, Navarra P, Marconi S, Gasbarrini A, Bentivoglio AR (2013) The role of small intestinal bacterial overgrowth in Parkinson's disease. Mov Disord 28(9):1241–1249

Fioramonti J, Fargeas MJ, Honde C, Bueno L (1984) Effects of central and peripheral administration of dopamine on pattern of intestinal motility in dogs. Dig Dis Sci 29(11):1023–1027

Franzese A, Borrelli O, Corrado G, Rea P, Di Nardo G, Grandinetti AL, Dito L, Cucchiara S (2002) Domperidone is more effective than cisapride in children with diabetic gastroparesis. Aliment Pharmacol Ther 16(5):951–957

Frazzitta G, Ferrazzoli D, Folini A, Palamara G, Maestri R (2019) Severe constipation in Parkinson's disease and in parkinsonisms: prevalence and affecting factors. Front Neurol 10:621

Gabrielli M, Bonazzi P, Scarpellini E, Bendia E, Lauritano EC, Fasano A, Ceravolo MG, Capecci M, Rita Bentivoglio A, Provinciali L, Tonali PA, Gasbarrini A (2011) Prevalence of small intestinal bacterial overgrowth in Parkinson's disease. Mov Disord 26(5):889–892

Garrido-Gil P, Rodriguez-Perez AI, Dominguez-Meijide A, Guerra MJ, Labandeira-Garcia JL (2018) Bidirectional neural interaction between central dopaminergic and gut lesions in Parkinson's disease models. Mol Neurobiol 55(9):7297–7316

Gasbarrini A, Lauritano EC, Gabrielli M, Scarpellini E, Lupascu A, Ojetti V, Gasbarrini G (2007) Small intestinal bacterial overgrowth: diagnosis and treatment. Dig Dis 25(3):237–240

Gerring EE, Hunt JM (1986) Pathophysiology of equine postoperative ileus: effect of adrenergic blockade, parasympathetic stimulation and metoclopramide in an experimental model. Equine Vet J 18(4):249–255

Giurcan R, Voiosu TA (2010) Functional dyspepsia: a pragmatic approach. Rom J Intern Med 48 (1):9–15

Goetze O, Nikodem AB, Wiezcorek J, Banasch M, Przuntek H, Mueller T, Schmidt WE, Woitalla D (2006) Predictors of gastric emptying in Parkinson's disease. Neurogastroenterol Motil 18 (5):369–375

Gonzalez A, Sarna SK (2001) Neural regulation of in vitro giant contractions in the rat colon. Am J Physiol Gastrointest Liver Physiol 281(1):G275–G282

Greene JG (2011) Animal models of gastrointestinal problems in Parkinson's disease. J Parkinsons Dis 1(2):137–149

Greene JG, Noorian AR, Srinivasan S (2009) Delayed gastric emptying and enteric nervous system dysfunction in the rotenone model of Parkinson's disease. Exp Neurol 218(1):154–161

Guard C, Schwark W, Kelton D, Dobson A (1988) Effects of metoclopramide, clenbuterol and butorphanol on ruminoreticular motility of calves. Cornell Vet 78(1):89–98

Gué M, Gleïzes-Escala C, Del Rio-Lacheze C, Junien JL, Buéno L (1994) Reversal of CRF- and dopamine-induced stimulation of colonic motility by CCK and igmesine (JO 1784) in the rat. Br J Pharmacol 111(3):930–934

Guo X, Song W, Chen K, Chen X, Zheng Z, Cao B, Huang R, Zhao B, Wu Y, Shang HF (2013a) Disease duration-related differences in non-motor symptoms: a study of 616 Chinese Parkinson's disease patients. J Neurol Sci 330(1–2):32–37

Guo X, Song W, Chen K, Chen X, Zheng Z, Cao B, Huang R, Zhao B, Wu Y, Shang HF (2013b) Gender and onset age-related features of non-motor symptoms of patients with Parkinson's disease--a study from Southwest China. Parkinsonism Relat Disord 19(11):961–965

Haans JJ, Masclee AA (2007) Review article: the diagnosis and management of gastroparesis. Aliment Pharmacol Ther 26(Suppl 2):37–46

Hasegawa S, Goto S, Tsuji H, Okuno T, Asahara T, Nomoto K, Shibata A, Fujisawa Y, Minato T, Okamoto A, Ohno K, Hirayama M (2015) Intestinal dysbiosis and lowered serum lipopolysaccharide-binding protein in Parkinson's disease. PLoS One 10(11):e0142164

Hashim H, Azmin S, Razlan H, Yahya NW, Tan HJ, Manaf MR, Ibrahim NM (2014) Eradication of Helicobacter pylori infection improves levodopa action, clinical symptoms and quality of life in patients with Parkinson's disease. PLoS One 9:e112330

Haskel Y, Hanani M (1994) Inhibition of gastrointestinal motility by MPTP via adrenergic and dopaminergic mechanisms. Dig Dis Sci 39(11):2364–2367

Hausken T, Odegaard S, Matre K, Berstad A (1992) Antroduodenal motility and movements of luminal contents studied by duplex sonography. Gastroenterology 102(5):1583–1590

Heetun ZS, Quigley EM (2012) Gastroparesis and Parkinson's disease: a systematic review. Parkinsonism Relat Disord 18(5):433–440

Heintz-Buschart A, Pandey U, Wicke T, Sixel-Döring F, Janzen A, Sittig-Wiegand E, Trenkwalder C, Oertel WH, Mollenhauer B, Wilmes P (2018) The nasal and gut microbiome in Parkinson's disease and idiopathic rapid eye movement sleep behavior disorder. Mov Disord 33(1):88–98

Hill-Burns EM, Debelius JW, Morton JT, Wissemann WT, Lewis MR, Wallen ZD, Peddada SD, Factor SA, Molho E, Zabetian CP, Knight R, Payami H (2017) Parkinson's disease and Parkinson's disease medications have distinct signatures of the gut microbiome. Mov Disord 32(5):739–749

Hillemeier C, McCallum R, Oertel R, Gryboski J (1986) Effect of bethanechol and metoclopramide on upper gastrointestinal motility in the kitten. J Pediatr Gastroenterol Nutr 5(1):134–137

Hilton D, Stephens M, Kirk L, Edwards P, Potter R, Zajicek J, Broughton E, Hagan H, Carroll C (2014) Accumulation of α-synuclein in the bowel of patients in the pre-clinical phase of Parkinson's disease. Acta Neuropathol 127(2):235–241

Huang WJ, Chien EJ (2012) The roles of dopamine receptor and adrenoreceptor on the inhibition of gastric emptying and intestinal transit by amphetamine in male rats. Chin J Physiol 55 (4):259–266

Huh E, Choi JG, Noh D, Yoo HS, Ryu J, Kim NJ, Kim H, Oh MS (2020) Ginger and 6-shogaol protect intestinal tight junction and enteric dopaminergic neurons against 1-methyl-4-phenyl 1,2,3,6-tetrahydropyridine in mice. Nutr Neurosci 23(6):455–464

Husebye E (1999) The patterns of small bowel motility: physiology and implications in organic disease and functional disorders. Neurogastroenterol Motil 11(3):141–161

Itoh Z, Aizawa I, Nakamura T (1980) Effect of dopamine and its antagonists on contractile activity of the lower esophageal sphincter and the stomach (author's transl). Nihon Heikatsukin Gakkai Zasshi 16(2):99–107

Iwanaga Y, Miyashita N, Morikawa K, Mizumoto A, Kondo Y, Itoh Z (1990) A novel water-soluble dopamine-2 antagonist with anticholinesterase activity in gastrointestinalmotor activity. Comparison with domperidone and neostigmine. Gastroenterology 99(2):401–408

Iwanaga Y, Suzuki N, Kato K, Kimura T, Morikawa K, Kato H, Ito Y, Gomi Y (1993) Stimulatory effects of HSR-803 on ileal motor activity. Jpn J Pharmacol 62(4):395–401

Iwanaga Y, Miyashita N, Saito T, Morikawa K, Itoh Z (1996) Gastroprokinetic effect of a new benzamide derivative itopride and its action mechanisms in conscious dogs. Jpn J Pharmacol 71 (2):129–137

Jarras H, Bourque M, Poirier AA, Morissette M, Coulombe K, Di Paolo T, Soulet D (2020) Neuroprotection and immunomodulation of progesterone in the gut of a mouse model of Parkinson's disease. J Neuroendocrinol 32(1):e12782

Jing H, Lin KW, Mei MH (1995) Participation of dopamine on the muscarinergic inhibitory effect of substance P on gastric myoelectric activity and motility. Sheng Li Xue Bao 47(3):245–252

Jo YH, Sim SS, Choi H, Kim MS (1989) Effect of dopamine on electrical activity of isolated stomach muscle in cats. Dig Dis Sci 34(4):548–552

Johnson ME, Stringer A, Bobrovskaya L (2018) Rotenone induces gastrointestinal pathology and microbiota alterations in a rat model of Parkinson's disease. Neurotoxicology 65:174–185

Jooste CA, Mustoe J, Collee G (1999) Metoclopramide improves gastric motility in critically ill patients. Intensive Care Med 25(5):464–468

Jost WH (1997) Gastrointestinal motility problems in patients with Parkinson's disease. Effects of antiparkinsonian treatment and guidelines for management. Drugs Aging 10(4):249–258

Jost WH, Eckardt VF (2003) Constipation in idiopathic Parkinson's disease. Scand J Gastroenterol 38(7):681–686

Jost WH, Schimrigk K (1991) Constipation in Parkinson's disease. Klin Wochenschr 69 (20):906–909

Karasawa H, Pietra C, Giuliano C, Garcia-Rubio S, Xu X, Yakabi S, Taché Y, Wang L (2014) New ghrelin agonist, HM01 alleviates constipation and L-dopa-delayed gastric emptying in 6-hydroxydopamine rat model of Parkinson's disease. Neurogastroenterol Motil 26 (12):1771–1782

Kashyap P, Micci MA, Pasricha S, Pasricha PJ (2009) The D2/D3 agonist PD128907 (R-(+)-trans-3,4a,10b-tetrahydro-4-propyl-2H,5H-[1]benzopyrano[4,3-b]-1,4-oxazin-9-ol) inhibits stimulated pyloric relaxation and spontaneous gastric emptying. Dig Dis Sci 54(1):57–62

Kasirer MY, Welsh C, Pan J, Shifrin Y, Belik J (2014) Metoclopramide does not increase gastric muscle contractility in newborn rats. Am J Physiol Gastrointest Liver Physiol 306(5):G439–G444

Keshavarzian A, Green SJ, Engen PA, Voigt RM, Naqib A, Forsyth CB, Mutlu E, Shannon KM (2015) Colonic bacterial composition in Parkinson's disease. Mov Disord 30(10):1351–1360

Kim S, Kwon SH, Kam TI, Panicker N, Karuppagounder SS, Lee S, Lee JH, Kim WR, Kook M, Foss CA, Shen C, Lee H, Kulkarni S, Pasricha PJ, Lee G, Pomper MG, Dawson VL, Dawson TM, Ko HS (2019) Transneuronal propagation of pathologic α-synuclein from the gut to the brain models Parkinson's disease. Neuron 103(4):627–641.e7

King JN, Gerring EL (1988) Biphasic disruption of fasting equine gut motility by dopamine--a preliminary study. J Vet Pharmacol Ther 11(4):354–361

Knudsen K, Federova TD, Bekker AC, Iversen P, Østergaard K, Krogh K, Borghammer P (2017) Objective colonic dysfunction is far more prevalent than subjective constipation in Parkinson's disease: a colon transit and volume study. J Parkinsons Dis 7(2):359–367

Koga T, Kobashi M, Mizutani M, Tsukamoto G, Matsuo R (2003) Area postrema mediates gastric motor response induced by apomorphine in rats. Brain Res 960(1–2):122–131

Komura M, Kanamori Y, Tanaka Y, Kodaka T, Sugiyama M, Terawaki K, Suzuki K, Iwanaka T (2017) Mosapride for gastroesophageal reflux disease in neurologically impaired patients. Pediatr Int 59(3):347–351

Kono T, Tokumaru O, Mizumoto C, Tatsuno J, Chen JD (1999) Impaired gastric slow waves induced by spatial disorientation and effect of domperidone. Am J Gastroenterol 94 (5):1224–1229

Koulaouzidis A, Dimitriadis S, Douglas S, Plevris JN (2015) The use of domperidone increases the completion rate of small bowel capsule endoscopy: does this come at the expense of diagnostic yield? J Clin Gastroenterol 49(5):395–400

Kuemmerle JF (2000) Motility disorders of the small intestine: new insights into old problems. J Clin Gastroenterol 31(4):276–281

Kumral D, Zfass AM (2018) Gut movements: a review of the physiology of gastrointestinal transit. Dig Dis Sci 63(10):2500–2506

Kuo YM, Li Z, Jiao Y, Gaborit N, Pani AK, Orrison BM, Bruneau BG, Giasson BI, Smeyne RJ, Gershon MD, Nussbaum RL (2010) Extensive enteric nervous system abnormalities in mice transgenic for artificial chromosomes containing Parkinson disease-associated alpha-synuclein gene mutations precede central nervous system changes. Hum Mol Genet 19(9):1633–1650

Kupsky WJ, Grimes MM, Sweeting J, Bertsch R, Cote LJ (1987) Parkinson's disease and megacolon: concentric hyaline inclusions (Lewy bodies) in enteric ganglion cells. Neurology 37(7):1253–1255

Lai F, Jiang R, Xie W, Liu X, Tang Y, Xiao H, Gao J, Jia Y, Bai Q (2018) Intestinal pathology and gut microbiota alterations in a methyl-4-phenyl-1,2,3,6-tetrahydropyridine (MPTP) mouse model of Parkinson's disease. Neurochem Res 43(10):1986–1999

Lanfranchi GA, Bazzocchi G, Marzio L, Campieri M, Brignola C (1983) Inhibition of postprandial colonic motility by sulpiride in patients with irritable colon. Eur J Clin Pharmacol 24 (6):769–772

Lanfranchi GA, Bazzocchi G, Fois F, Brignola C, Campieri M, Menni B (1985) Effect of domperidone and dopamine on colonic motor activity in patients with the irritable bowel syndrome. Eur J Clin Pharmacol 29(3):307–310

Lang IM, Marvig J (1989) Functional localization of specific receptors mediating gastrointestinal motor correlates of vomiting. Am J Phys 256(1 Pt 1):G92–G99

Lebouvier T, Chaumette T, Damier P, Coron E, Touchefeu Y, Vrignaud S, Naveilhan P, Galmiche J-P, Bruley des Varannes S, Derkinderen P, Neunlist M (2008) Pathological lesions in colonic biopsies during Parkinson's disease. Gut 57(12):1741–1743

Lebouvier T, Neunlist M, Bruley des Varannes S, Coron E, Drouard A, N'Guyen JM, Chaumette T, Tasselli M, Paillusson S, Flamand M, Galmiche JP, Damier P, Derkinderen P (2010) Colonic biopsies to assess the neuropathology of Parkinson's disease and its relationship with symptoms. PLoS One 5(9):e12728

Lechin F, van der Dus B (1979) Dopamine and distal colon motility. Dig Dis Sci 24(1):86–87

Lechin F, Gômez F, van der Dijs B, Lechin E (1980) Distal colon motility in schizophrenic patients. J Clin Pharmacol 20(7):459–464

Levandis G, Balestra B, Siani F, Rizzo V, Ghezzi C, Ambrosi G, Cerri S, Bonizzi A, Vicini R, Vairetti M, Ferrigno A, Pastoris O, Blandini F (2015) Response of colonic motility to dopaminergic stimulation is subverted in rats with nigrostriatal lesion: relevance to gastrointestinal dysfunctions in Parkinson's disease. Neurogastroenterol Motil 27(12):1783–1795

Levein NG, Thörn SE, Lindberg G, Wattwill M (1999a) Dopamine reduces gastric tone in a dose-related manner. Acta Anaesthesiol Scand 43(7):722–725
Levein NG, Thörn SE, Wattwil M (1999b) Dopamine delays gastric emptying and prolongs orocaecal transit time in volunteers. Eur J Anaesthesiol 16(4):246–250
Li ZS, Schmauss C, Cuenca A, Ratcliffe E, Gershon MD (2006) Physiological modulation of intestinal motility by enteric dopaminergic neurons and the D2 receptor: analysis of dopamine receptor expression, location, development, and function in wild-type and knock-out mice. J Neurosci 26(10):2798–2807
Li H, Zhang M, Chen L, Zhang J, Pei Z, Hu A, Wang Q (2010) Nonmotor symptoms are independently associated with impaired health-related quality of life in Chinese patients with Parkinson's disease. Mov Disord 25(16):2740–2746
Li LS, Liu CZ, Xu JD, Zheng LF, Feng XY, Zhang Y, Zhu JX (2015) Effect of entacapone on colon motility and ion transport in a rat model of Parkinson's disease. World J Gastroenterol 21 (12):3509–3518
Li W, Wu X, Hu X, Wang T, Liang S, Duan Y, Jin F, Qin B (2017) Structural changes of gut microbiota in Parkinson's disease and its correlation with clinical features. Sci China Life Sci 60 (11):1223–1233
Lidor AO, Ensor CR, Sheer AJ, Orens JB, Clarke JO, McDyer JF (2014) Domperidone for delayed gastric emptying in lung transplant recipients with and without gastroesophageal reflux. Prog Transplant 24(1):27–32
Lim HC, Kim YG, Lim JH, Kim HS, Park H (2008) Effect of itopride hydrochloride on the ileal and colonic motility in guinea pig in vitro. Yonsei Med J 49(3):472–478
Lin HC, Prather C, Fisher RS, Meyer JH, Summers RW, Pimentel M, McCallum RW, Akkermans LM, Loening-Baucke V (2005) Measurement of gastrointestinal transit. Dig Dis Sci 50 (6):989–1004
Lin A, Zheng W, He Y, Tang W, Wei X, He R, Huang W, Su Y, Huang Y, Zhou H, Xie H (2018) Gut microbiota in patients with Parkinson's disease in Southern China. Parkinsonism Relat Disord 53:82–88
Lin CH, Chen CC, Chiang HL, Liou JM, Chang CM, Lu TP, Chuang EY, Tai YC, Cheng C, Lin HY, Wu MS (2019) Altered gut microbiota and inflammatory cytokine responses in patients with Parkinson's disease. J Neuroinflammation 16(1):129
Linnik MD, Butler BT, Gaddis RR, Ahmed NK (1991) Analysis of serotonergic mechanisms underlying benzamide-induced gastroprokinesis. J Pharmacol Exp Ther 259(2):501–507
Liu H, Su W, Li S, Du W, Ma X, Jin Y, Li K, Chen H (2017) Eradication of Helicobacter pylori infection might improve clinical status of patients with Parkinson's disease, especially on bradykinesia. Clin Neurol Neurosurg 160:101–104
Lobbezoo MW, Janszen FH, Tulp MT, Zwagemakers JM (1985) Differential effects of metoclopramide and zetidoline on gastrointestinal motility. Eur J Pharmacol 108(2):105–112
Longo WE, Vernava AM 3rd (1993) Prokinetic agents for lower gastrointestinal motility disorders. Dis Colon Rectum 36(7):696–708
Lyte M (2010) Microbial endocrinology as a basis for improved L-DOPA bioavailability in Parkinson's patients treated for Helicobacter pylori. Med Hypotheses 74:895–897
Maas CL, van Duin CT, van Miert AS (1982) Modification by domperidone of dopamine- and apomorphine-induced inhibition of extrinsic ruminal contractions in goats. J Vet Pharmacol Ther 5(3):191–194
MacLaren R, Patrick WD, Hall RI, Rocker GM, Whelan GJ, Lima JJ (2001) Comparison of cisapride and metoclopramide for facilitating gastric emptying and improving tolerance to intragastric enteral nutrition in critically III, mechanically ventilated adults. Clin Ther 23 (11):1855–1866
Markey O, Shafat A (2012) Does domperidone, a D2-antagonist alter gastric emptying rates and appetite sensations in healthy adults with high-fat meal? A block-randomised, single-blind placebo-controlled study. Ir J Med Sci 181(2):215–219

Marzio L, Neri M, Di Giammarco AM, Cuccurullo F, Lanfranchi GA (1986) Dopamine-induced migrating myoelectrical complex-like activity in human duodenum. Dig Dis Sci 31(4):349–354

Marzio L, Neri M, Pieramico O, Delle Donne M, Peeters TL, Cuccurullo F (1990) Dopamine interrupts gastrointestinal fed motility pattern in humans. Effect on motilin and somatostatin blood levels. Dig Dis Sci 35(3):327–332

McClelland GR, Sutton JA (1986) A comparison of the gastric and central nervous system effects of two substituted benzamides in normal volunteers. Br J Clin Pharmacol 21(5):503–509

Megens AA, Awouters FH, Niemegeers CJ (1991) General pharmacology of the four gastrointestinal motility stimulants bethanechol, metoclopramide, trimebutine and cisapride. Arzneimittelforschung 41(6):631–634

Melga P, Mansi C, Ciuchi E, Giusti R, Sciaba L, Prando R (1997) Chronic administration of levosulpiride and glycemic control in IDDM patients with gastroparesis. Diabetes Care 20 (1):55–58

Mertsalmi TH, Aho VTE, Pereira PAB, Paulin L, Pekkonen E, Auvinen P, Scheperjans F (2017) More than constipation - bowel symptoms in Parkinson's disease and their connection to gut microbiota. Eur J Neurol 24(11):1375–1383

Miachon S, Peyrin L, Cier JF, Legheand J (1978) Research on the mechanism of the motor action of dopamine on in vitro isolated rat duodenum. C R Seances Soc Biol Fil 172(2):278–282

Minato T, Maeda T, Fujisawa Y, Tsuji H, Nomoto K, Ohno K, Hirayama M (2017) Progression of Parkinson's disease is associated with gut dysbiosis: two-year follow-up study. PLoS One 12 (11):e0187307

Mittal RK (2013) Longitudinal muscle of the esophagus: its role in esophageal health and disease. Curr Opin Gastroenterol 29(4):421–430

Miyazawa T, Matsumoto M, Kato S, Takeuchi K (2003) Dopamine-induced protection against indomethacin-evoked intestinal lesions in rats--role of anti-intestinal motility mediated by D2 receptors. Med Sci Monit 9(2):BR71–BR77

Mizuta Y, Shikuwa S, Isomoto H, Mishima R, Akazawa Y, Masuda J, Omagari K, Takeshima F, Kohno S (2006) Recent insights into digestive motility in functional dyspepsia. J Gastroenterol 41(11):1025–1040

Morais LH, Hara DB, Bicca MA, Poli A, Takahashi RN (2018) Early signs of colonic inflammation, intestinal dysfunction, and olfactory impairments in the rotenone-induced mouse model of Parkinson's disease. Behav Pharmacol 29(2 and 3-Spec Issue):199–210

Mridula KR, Borgohain R, Chandrasekhar Reddy V, Bandaru VC, Suryaprabha T (2017) Association of Helicobacter pylori with Parkinson's disease. J Clin Neurol 13(2):181–186

Müller-Lissner SA, Fraas C, Härtl A (1986) Cisapride offsets dopamine-induced slowing of fasting gastric emptying. Dig Dis Sci 31(8):807–810

Nagahata Y, Urakawa T, Moritomo H, Ichihara T, Itoh A, Takeda K, Kuroda H, Idei H, Tomonaga K, Saitoh Y (1990) Effects of dopamine on stress ulcer. Nihon Shokakibyo Gakkai Zasshi 87(6):1376–1382

Nagahata Y, Azumi Y, Kawakita N, Wada T, Saitoh Y (1995) Inhibitory effect of dopamine on gastric motility in rats. Scand J Gastroenterol 30(9):880–885

Naitou K, Nakamori H, Shiina T, Ikeda A, Nozue Y, Sano Y, Yokoyama T, Yamamoto Y, Yamada A, Akimoto N, Furue H, Shimizu Y (2016) Stimulation of dopamine D2-like receptors in the lumbosacral defaecation centre causes propulsive colorectal contractions in rats. J Physiol 594(15):4339–4350

Nardone R, Höller Y, Brigo F, Versace V, Sebastianelli L, Florea C, Schwenker K, Golaszewski S, Saltuari L, Trinka E (2019) Spinal cord involvement in Lewy body-related α-synucleinopathies. J Spinal Cord Med:1–14

Narożańska E, Białecka M, Adamiak-Giera U, Gawrońska-Szklarz B, Sołtan W, Schinwelski M, Robowski P, Madaliński MH, Sławek J (2014) Pharmacokinetics of levodopa in patients with Parkinson disease and motor fluctuations depending on the presence of Helicobacter pylori infection. Clin Neuropharmacol 37(4):96–99

Natale G, Kastsiushenka O, Fulceri F, Ruggieri S, Paparelli A, Fornai F (2010) MPTP-induced parkinsonism extends to a subclass of TH-positive neurons in the gut. Brain Res 1355:195–206

Ng E, Shah VS (2008) Erythromycin for the prevention and treatment of feeding intolerance in preterm infants. Cochrane Database Syst Rev 3:CD001815

Niehues M, Hensel A (2009) In-vitro interaction of L-dopa with bacterial adhesins of helicobacter pylori: an explanation for clinicial differences in bioavailability? J Pharm Pharmacol 61:1303–1307

Nieto JE, Maher O, Stanley SD, Larson R, Snyder JR (2013) In vivo and in vitro evaluation of the effects of domperidone on the gastrointestinal tract of healthy horses. Am J Vet Res 74 (8):1103–1110

Niu XL, Liu L, Song ZX, Li Q, Wang ZH, Zhang JL, Li HH (2016) Prevalence of small intestinal bacterial overgrowth in Chinese patients with Parkinson's disease. J Neural Transm (Vienna) 123(12):1381–1386

Noorian AR, Rha J, Annerino DM, Bernhard D, Taylor GM, Greene JG (2012) Alpha-synuclein transgenic mice display age-related slowing of gastrointestinal motility associated with transgene expression in the vagal system. Neurobiol Dis 48(1):9–19

Nyholm D, Lennernäs H (2008) Irregular gastrointestinal drug absorption in Parkinson's disease. Expert Opin Drug Metab Toxicol 4(2):193–203

O'Brien MD, Phillips SF (1996) Colonic motility in health and disease. Gastroenterol Clin N Am 25 (1):147–162

Ohlsson B, Englund E (2019) Atrophic myenteric and submucosal neurons are observed in Parkinson's disease. Parkinsons Dis 2019:7935820

Pellegrini C, Fornai M, Colucci R, Tirotta E, Blandini F, Levandis G, Cerri S, Segnani C, Ippolito C, Bernardini N, Cseri K, Blandizzi C, Haskó G, Antonioli L (2016) Alteration of colonic excitatory tachykininergic motility and enteric inflammation following dopaminergic nigrostriatal neurodegeneration. J Neuroinflammation 13(1):146

Pellegrini C, Antonioli L, Colucci R, Tirotta E, Gentile D, Ippolito C, Segnani C, Levandis G, Cerri S, Blandini F, Barocelli E, Ballabeni V, Bernardini N, Blandizzi C, Fornai M (2017) Effects of L-DOPA/benserazide co-treatment on colonic excitatory cholinergic motility and enteric inflammation following dopaminergic nigrostriatal neurodegeneration. Neuropharmacology 123:22–33

Perez-Pardo P, Dodiya HB, Broersen LM, Douna H, Wijk N, Silva SL, Garssen J, Keshavarzian A, Kraneveld AD (2018) Gut-brain and brain-gut axis in Parkinson's disease models: effects of a uridine and fish oil diet. Nutr Neurosci 21(6):391–402

Petrov VA, Saltykova IV, Zhukova IA, Alifirova VM, Zhukova NG, Dorofeeva YB, Tyakht AV, Kovarsky BA, Alekseev DG, Kostryukova ES, Mironova YS, Izhboldina OP, Nikitina MA, Perevozchikova TV, Fait EA, Babenko VV, Vakhitova MT, Govorun VM, Sazonov AE (2017) Analysis of gut microbiota in patients with Parkinson's disease. Bull Exp Biol Med 162 (6):734–737

Pfeiffer RF (2011) Gastrointestinal dysfunction in Parkinson's disease. Parkinsonism Relat Disord 17(1):10–15

Pfeiffer RF (2018) Gastrointestinal dysfunction in Parkinson's disease. Curr Treat Options Neurol 20(12):54

Phan H, DeReese A, Day AJ, Carvalho M (2014) The dual role of domperidone in gastroparesis and lactation. Int J Pharm Compd 18(3):203–207

Pierantozzi M, Pietroiusti A, Sancesario G, Lunardi G, Fedele E, Giacomini P, Frasca S, Galante A, Marciani MG, Stanzione P (2001) Reduced L-dopa absorption and increased clinical fluctuations in Helicobacter pylori-infected Parkinson's disease patients. Neurol Sci 22:89–91

Poirier AA, Côté M, Bourque M, Morissette M, Di Paolo T, Soulet D (2016) Neuroprotective and immunomodulatory effects of raloxifene in the myenteric plexus of a mouse model of Parkinson's disease. Neurobiol Aging 48:61–71

Pouclet H, Lebouvier T, Coron E, des Varannes SB, Rouaud T, Roy M, Neunlist M, Derkinderen P (2012) A comparison between rectal and colonic biopsies to detect Lewy pathology in Parkinson's disease. Neurobiol Dis 45(1):305–309

Qian Y, Yang X, Xu S, Wu C, Song Y, Qin N, Chen SD, Xiao Q (2018) Alteration of the fecal microbiota in Chinese patients with Parkinson's disease. Brain Behav Immun 70:194–202

Qin XY, Wang ZG, Fei J, Liu FL, Cui DF, Chen SL (2003) Involvement of dopamine D3 and neuropeptide Y Y5 receptors in diabetic gastroparetic rats without response to erythromycin. Sheng Wu Hua Xue Yu Sheng Wu Wu Li Xue Bao (Shanghai) 35(9):811–815

Qin X, Li X, Xin Z, Li Z (2019) Gastrointestinal dysfunction in chinese patients with Parkinson's disease. Parkinsons Dis 2019:3897315

Rahne KE, Tagesson C, Nyholm D (2013) Motor fluctuations and helicobacter pylori in Parkinson's disease. J Neurol 260:2974–2980

Ramirez R, Stalcup P, Croft B, Darracq MA (2017) Haloperidol undermining gastroparesis symptoms (HUGS) in the emergency department. Am J Emerg Med 35(8):1118–1120

Rao SS, Camilleri M, Hasler WL, Maurer AH, Parkman HP, Saad R, Scott MS, Simren M, Soffer E, Szarka L (2011) Evaluation of gastrointestinal transit in clinical practice: position paper of the American and European Neurogastroenterology and motility societies. Neurogastroenterol Motil 23(1):8–23

Reddymasu SC, Soykan I, McCallum RW (2007) Domperidone: review of pharmacology and clinical applications in gastroenterology. Am J Gastroenterol 102(9):2036–2045

Robertson CS, Evans DF, Hicks F, Atkinson M (1988) A double-blind placebo-controlled trial of BRL 24924 on lower oesophageal sphincter pressure and gastro-oesophageal reflux in healthy volunteers. Aliment Pharmacol Ther 2(6):501–506

Rossi M, Merello M, Perez-Lloret S (2015) Management of constipation in Parkinson's disease. Expert Opin Pharmacother 16(4):547–557

Rota L, Pellegrini C, Benvenuti L, Antonioli L, Fornai M, Blandizzi C, Cattaneo A, Colla E (2019) Constipation, deficit in colon contractions and alpha-synuclein inclusions within the colon precede motor abnormalities and neurodegeneration in the central nervous system in a mouse model of alpha-synucleinopathy. Transl Neurodegener 8:5

Saavedra JM, Axelrod J (1976) Octopamine as a putative neurotransmitter. Adv Biochem Psychopharmacol 15:95–110

Saltzman MB, McCallum RW (1983) Diabetes and the stomach. Yale J Biol Med 56(3):179–187

Sampath C, Kalpana R, Ansah T, Charlton C, Hale A, Channon KM, Srinivasan S, Gangula PR (2019) Impairment of Nrf2- and nitrergic-mediated gastrointestinal motility in an MPTP mouse model of Parkinson's disease. Dig Dis Sci 64(12):3502–3517

Sánchez J, Costa G, Benedito S, Rivera L, García-Sacristán A, Orensanz LM (1990) Alpha 2-mediated effect of dopamine on the motility of the chicken esophagus. Life Sci 46(2):121–126

Sanders KM (2019) Spontaneous electrical activity and rhythmicity in gastrointestinal smooth muscles. Adv Exp Med Biol 1124:3–46

Sanders KM, Koh SD, Ro S, Ward SM (2012) Regulation of gastrointestinal motility--insights from smooth muscle biology. Nat Rev Gastroenterol Hepatol 9(11):633–645

Sarna SK (1991) Physiology and pathophysiology of colonic motor activity (1). Dig Dis Sci 36 (6):827–862

Scarpignato C (1997) Pharmacological stimulation of gastrointestinal motility: where we are and where are we going? Dig Dis 15(Suppl 1):112–136

Scheperjans F, Aho V, Pereira PA, Koskinen K, Paulin L, Pekkonen E, Haapaniemi E, Kaakkola S, Eerola-Rautio J, Pohja M, Kinnunen E, Murros K, Auvinen P (2015) Gut microbiota are related to Parkinson's disease and clinical phenotype. Mov Disord 30(3):350–358

Schuurkes JA, Van Nueten JM (1981) Is dopamine an inhibitory modulator of gastrointestinal motility? Scand J Gastroenterol Suppl 67:33–36

Schuurkes JA, Van Nueten JM (1982) Dose-dependent stimulation of antroduodenal coordination by domperidone via specific dopamine-receptors. Arch Int Pharmacodyn Ther 256(2):311–314

Schuurkes JA, Van Nueten JM (1984a) Control of gastroduodenal coordination: dopaminergic and cholinergic pathways. Scand J Gastroenterol Suppl 92:8–12

Schuurkes JA, Van Nueten JM (1984b) Domperidone improves myogenically transmitted antroduodenal coordination by blocking dopaminergic receptor sites. Scand J Gastroenterol Suppl 96:101–110

Schuurkes JA, Helsen LF, Van Nueten JM (1985) A comparative study on the effects of domperidone, metoclopramide, clebopride and trimebutine on the gastro-duodenal preparation of the guinea pig. Jpn J Pharmacol 39(2):123–130

Schwiertz A, Schäfer KH (2016) Short chain fatty acids and gut microbiota differ between patients with Parkinson's disease and age-matched controls. Parkinsonism Relat Disord 32:66–72

Serra J (2010) Levosulpiride in the management of functional dyspepsia and delayed gastric emptying. Gastroenterol Hepatol 33(8):586–590

Shannon KM, Keshavarzian A, Mutlu E, Dodiya HB, Daian D, Jaglin JA, Kordower JH (2012) Alpha-synuclein in colonic submucosa in early untreated Parkinson's disease. Mov Disord 27 (6):709–715

Shultz JM, Resnikoff H, Bondarenko V, Joers V, Mejia A, Simmons H, Emborg ME (2016) Neurotoxin-induced catecholaminergic loss in the colonic myenteric plexus of rhesus monkeys. J Alzheimers Dis Parkinsonism 6(6). pii: 279

Shuto K, Shiozaki Z, Kojima T, Tanaka M (1980a) Antagonism of KW-5338 (domperidone) against emesis and depression of intestinal motility induced by L-DOPA. J Pharmacobiodyn 3 (12):709–714

Shuto K, Shiozaki S, Kojima T, Tanaka M (1980b) Effects of KW-5338 (domperidone) on gastric motility. J Pharmacobiodyn 3(12):715–719

Sigala S, Missale G, Raddino R, Cestari R, Lojacono L, Missale C, Spano PF (1994) Opposing roles for D-1 and D-2 dopamine receptors in the regulation of lower esophageal sphincter motility in the rat. Life Sci 54(15):1035–1045

Singaram C, Ashraf W, Gaumnitz EA, Torbey C, Sengupta A, Pfeiffer R, Quigley EM (1995) Dopaminergic defect of enteric nervous system in Parkinson's disease patients with chronic constipation. Lancet 346(8979):861–864

Smout AJ (2008) Advances in esophageal motor disorders. Curr Opin Gastroenterol 24(4):485–489

Solnes LB, Sheikhbahaei S, Ziessman HA (2018) Nuclear scintigraphy in practice: gastrointestinal motility. AJR Am J Roentgenol 211(2):260–266

Song J, Zheng L, Zhang X, Feng X, Fan R, Sun L, Hong F, Zhang Y, Zhu J (2014) Upregulation of β1-adrenoceptors is involved in the formation of gastric dysmotility in the 6-hydroxydopamine rat model of Parkinson's disease. Transl Res 164(1):22–31

Soykan I, Sarosiek I, Shifflett J, Wooten GF, McCallum RW (1997a) Effect of chronic oral domperidone therapy on gastrointestinal symptoms and gastric emptying in patients with Parkinson's disease. Mov Disord 12(6):952–957

Soykan I, Sarosiek I, McCallum RW (1997b) The effect of chronic oral domperidone therapy on gastrointestinal symptoms, gastric emptying, and quality of life in patients with gastroparesis. Am J Gastroenterol 92(6):976–980

Stanghellini V, De Giorgio R, Barbara G, Cogliandro R, Tosetti C, De Ponti F, Corinaldesi R (2004) Delayed gastric emptying in functional dyspepsia. Curr Treat Options Gastroenterol 7 (4):259–264

Stellamor K, Hruby W (1983) Comparative study on the effect of the dopamine receptor blockers metoclopramide and domperidone on stomach emptying and the visualization of the small intestine in monocontrast. Wien Klin Wochenschr 95(18):649–652

Stocchi F, Torti M (2017) Constipation in Parkinson's disease. Int Rev Neurobiol 134:811–826

Sturm A, Holtmann G, Goebell H, Gerken G (1999) Prokinetics in patients with gastroparesis: a systematic analysis. Digestion 60(5):422–427

Sun MF, Zhu YL, Zhou ZL, Jia XB, Xu YD, Yang Q, Cui C, Shen YQ (2018) Neuroprotective effects of fecal microbiota transplantation on MPTP-induced Parkinson's disease mice: gut microbiota, glial reaction and TLR4/TNF-α signaling pathway. Brain Behav Immun 70:48–60

Sustić A, Zelić M, Protić A, Zupan Z, Simić O, Desa K (2005) Metoclopramide improves gastric but not gallbladder emptying in cardiac surgery patients with early intragastric enteral feeding: randomized controlled trial. Croat Med J 46(2):239–244

Szarka LA, Camilleri M (2012) Methods for the assessment of small-bowel and colonic transit. Semin Nucl Med 42(2):113–123

Takahashi T, Kurosawa S, Wiley JW, Owyang C (1991) Mechanism for the gastrokinetic action of domperidone. In vitro studies in guinea pigs. Gastroenterology 101(3):703–710

Takeuchi K, Nishiwaki H, Okabe S (1987) Role of local motility changes in the pathogenesis of duodenal ulcers induced by cysteamine in rats. Dig Dis Sci 32(3):295–304

Takeuchi K, Nishiwaki H, Okabe S (1988) Effects of dopamine on gastric mucosal lesions induced by ethanol in rats. Possible involvement of antigastric motor activity mediated with alpha 2-adrenoceptors. Dig Dis Sci 33(12):1560–1568

Tamhane MD, Thorat SP, Rege NN, Dahanukar SA (1997) Effect of oral administration of Terminalia chebula on gastric emptying: an experimental study. J Postgrad Med 43(1):12–13

Tan AH, Mahadeva S, Thalha AM, Gibson PR, Kiew CK, Yeat CM, Ng SW, Ang SP, Chow SK, Tan CT, Yong HS, Marras C, Fox SH, Lim SY (2014) Small intestinal bacterial overgrowth in Parkinson's disease. Parkinsonism Relat Disord 20(5):535–540

Tanaka Y, Kato T, Nishida H, Araki H, Murase M, Nagaki M, Moriwaki H, Inuzuka T (2009) Is there a difference in gastric emptying between Parkinson's disease patients under long-term L-dopa therapy with and without motor fluctuations? An analysis using the 13C-acetate breath test. J Neurol 256(12):1972–1976

Tasselli M, Chaumette T, Paillusson S, Monnet Y, Lafoux A, Huchet-Cadiou C, Aubert P, Hunot S, Derkinderen P, Neunlist M (2013) Effects of oral administration of rotenone on gastrointestinal functions in mice. Neurogastroenterol Motil 25(3):e183–e193

Taylor TN, Caudle WM, Shepherd KR, Noorian A, Jackson CR, Iuvone PM, Weinshenker D, Greene JG, Miller GW (2009) Nonmotor symptoms of Parkinson's disease revealed in an animal model with reduced monoamine storage capacity. J Neurosci 29(25):8103–8113

Tian YM, Chen X, Luo DZ, Zhang XH, Xue H, Zheng LF, Yang N, Wang XM, Zhu JX (2008) Alteration of dopaminergic markers in gastrointestinal tract of different rodent models of Parkinson's disease. Neuroscience 153:634–644

Tonini M, Cipollina L, Poluzzi E, Crema F, Corazza GR, De Ponti F (2004) Review article: clinical implications of enteric and central D2 receptor blockade by antidopaminergic gastrointestinal prokinetics. Aliment Pharmacol Ther 19(4):379–390

Torres ERS, Akinyeke T, Stagaman K, Duvoisin RM, Meshul CK, Sharpton TJ, Raber J (2018) Effects of sub-chronic MPTP exposure on behavioral and cognitive performance and the microbiome of wild-type and mGlu8 knockout female and male mice. Front Behav Neurosci 12:140

Toti L, Travagli RA (2014) Gastric dysregulation induced by microinjection of 6-OHDA in the substantia nigra pars compacta of rats is determined by alterations in the braingut axis. Am J Physiol Gastrointest Liver Physiol 307:G1013–G1023

Travagli RA, Anselmi L (2016) Vagal neurocircuitry and its influence on gastric motility. Nat Rev Gastroenterol Hepatol 13:389–401

Travagli RA, Gillis RA, Rossiter CD, Vicini S (1991) Glutamate and GABA-mediated synaptic currents in neurons of the rat dorsal motor nucleus of the vagus. Am J Phys 260:G531–G536

Ueki A, Otsuka M (2004) Life style risks of Parkinson's disease: association between decreased water intake and constipation. J Neurol 251(Suppl 7):VII18–VII23

Uemura N, Yagi H, Uemura MT, Hatanaka Y, Yamakado H, Takahashi R (2018) Inoculation of α-synuclein preformed fibrils into the mouse gastrointestinal tract induces Lewy body-like aggregates in the brainstem via the vagus nerve [published correction appears in Mol Neurodegener. 2019;14(1):31]. Mol Neurodegener 13(1):21

Uemura N, Yagi H, Uemura MT, Yamakado H, Takahashi R (2020) Limited spread of pathology within the brainstem of α-synuclein BAC transgenic mice inoculated with preformed fibrils into the gastrointestinal tract. Neurosci Lett 716:134651

Ulvestad A, Gerner T (1984) Effects of alpha- and beta-adrenergic agents and dopamine on gastric motor activity stimulated by acetylcholine or physostigmine. Scand J Gastroenterol Suppl 89:59–63

Ulvestad A, Gerner T (1985) Inhibitory effects of dopamine and isoprenaline on antral motor activity stimulated by acetylcholine or physostigmine in vitro. Scand J Gastroenterol 20 (2):243–246

Unger MM, Spiegel J, Dillmann KU, Grundmann D, Philippeit H, Bürmann J, Faßbender K, Schwiertz A, Schäfer KH (2016) Short chain fatty acids and gut microbiota differ between patients with Parkinson's disease and age-matched controls. Parkinsonism Relat Disord 32:66–72

Valenzuela JE (1976) Dopamine as a possible neurotransmitter in gastric relaxation. Gastroenterology 71(6):1019–1022

Valenzuela JE, Dooley CP (1984) Dopamine antagonists in the upper gastrointestinal tract. Scand J Gastroenterol Suppl 96:127–136

Valenzuela JE, Liu DP (1982) The effect of variations in intragastric pressure and gastric emptying of a saline meal in humans. Scand J Gastroenterol 17(2):293–296

van Kessel SP, Frye AK, El-Gendy AO, Castejon M, Keshavarzian A, van Dijk G, El Aidy S (2019) Gut bacterial tyrosine decarboxylases restrict levels of levodopa in the treatment of Parkinson's disease. Nat Commun 10(1):310

Van Nueten JM, Schuurkes JA (1984) Studies on the role of dopamine and dopamine blockers in gastroduodenal motility. Scand J Gastroenterol Suppl 96:89–99

Van Nueten JM, Ennis C, Helsen L, Laduron PM, Janssen PA (1978) Inhibition of dopamine receptors in the stomach: an explanation of the gastrokinetic properties of domperidone. Life Sci 23(5):453–457

Vegezzi G, Al Harraq Z, Levandis G, Cerri S, Blandini F, Gnudi G, Miduri F, Blandizzi C, Domenichini G, Bertoni S, Ballabeni V, Barocelli E (2014) Radiological analysis of gastrointestinal dysmotility in a model of central nervous dopaminergic degeneration: comparative study with conventional in vivo techniques in the rat. J Pharmacol Toxicol Methods 70 (2):163–169

Velasco M, Luchsinger A (1998) Dopamine: pharmacologic and therapeutic aspects. Am J Ther 5 (1):37–43

Vijayvargiya P, Camilleri M (2019) Use of prucalopride in adults with chronic idiopathic constipation. Expert Rev Clin Pharmacol 12(7):579–589

Vizcarra JA, Wilson-Perez HE, Fasano A, Espay AJ (2018) Small intestinal bacterial overgrowth in Parkinson's disease: tribulations of a trial. Parkinsonism Relat Disord 54:110–112

Wakabayashi K, Takahashi H, Takeda S, Ohama E, Ikuta F (1988) Parkinson's disease: the presence of Lewy bodies in Auerbach's and Meissner's plexuses. Acta Neuropathol 76 (3):217–221

Wakabayashi K, Takahashi H, Ohama E, Ikuta F (1990) Parkinson's disease: an immunohistochemical study of Lewy body-containing neurons in the enteric nervous system. Acta Neuropathol 79(6):581–583

Wang L, Fleming SM, Chesselet MF, Taché Y (2008) Abnormal colonic motility in mice overexpressing human wild-type alpha-synuclein. Neuroreport 19(8):873–876

Wang L, Magen I, Yuan PQ, Subramaniam SR, Richter F, Chesselet MF, Taché Y (2012) Mice overexpressing wild-type human alpha-synuclein display alterations in colonic myenteric ganglia and defecation. Neurogastroenterol Motil 24(9):e425–e436

Wang ZY, Lian H, Cai QQ, Song HY, Zhang XL, Zhou L, Zhang YM, Zheng LF, Zhu JX (2014) No direct projection is observed from the substantia nigra to the dorsal vagus complex in the rat. J Parkinsons Dis 4(3):375–383

Wang W, Song N, Jia F, Tang T, Bao W, Zuo C, Xie J, Jiang H (2018) Genomic DNA levels of mutant alpha-synuclein correlate with non-motor symptoms in an A53T Parkinson's disease mouse model. Neurochem Int 114:71–79

Wiley J, Owyang C (1987) Dopaminergic modulation of rectosigmoid motility: action of domperidone. J Pharmacol Exp Ther 242(2):548–551

Yang YL, Ran XR, Li Y, Zhou L, Zheng LF, Han Y, Cai QQ, Wang ZY, Zhu JX (2019) Expression of dopamine receptors in the lateral hypothalamic nucleus and their potential regulation of gastric motility in rats with lesions of bilateral substantia nigra. Front Neurosci 13:195

Yoshida N, Mizumoto A, Iwanaga Y, Itoh Z (1991) Effects of 5-hydroxytryptamine 3 receptor antagonists on gastrointestinal motor activity in conscious dogs. J Pharmacol Exp Ther 256 (1):272–278

Yoshikawa T, Yoshida N (2010) The possible involvement of dopamine D3 receptors in the regulation of gastric emptying in rats. Life Sci 87(19–22):638–642

Zar MA, Ebong O, Bateman DN (1982) Effect of metoclopramide in guinea-pig ileum longitudinal muscle: evidence against dopamine-mediation. Gut 23(1):66–70

Zhang JS, Ye LP, Zhang JL, Wang CY, Chen JY (2011) Intramuscular injection of metoclopramide decreases the gastric transit time and does not increase the complete examination rate of capsule endoscopy: a prospective randomized controlled trial. Hepato-Gastroenterology 58 (110–111):1618–1621

Zhang X, Guo H, Xu J, Li Y, Li L, Zhang X, Li X, Fan R, Zhang Y, Duan Z, Zhu J (2012) Dopamine receptor D1 mediates the inhibition of dopamine on the distal colonic motility. Transl Res 159(5):407–414

Zhang X, Li Y, Liu C, Fan R, Wang P, Zheng L, Hong F, Feng X, Zhang Y, Li L, Zhu J (2015) Alteration of enteric monoamines with monoamine receptors and colonic dysmotility in 6-hydroxydopamine-induced Parkinson's disease rats. Transl Res 166(2):152–162

Zhao S, Sha H, Li ZY, Ren CS (2012) Electrical bioimpedance gastric motility measurement based on an electrical-mechanical composite mechanism. World J Gastroenterol 18(25):3282–3287

Zheng LF, Wang ZY, Li XF, Song J, Hong F, Lian H, Wang Q, Feng XY, Tang YY, Zhang Y, Zhu JX (2011) Reduced expression of choline acetyltransferase in vagal motoneurons and gastric motor dysfunction in a 6-OHDA rat model of Parkinson's disease. Brain Res 1420:59–67

Zheng LF, Zhang Y, Chen CL, Song J, Fan RF, Cai QQ, Wang ZY, Zhu JX (2013) Alterations in TH- and ChAT-immunoreactive neurons in the DMV and gastric dysmotility in an LPS-induced PD rat model. Auton Neurosci 177(2):194–198

Zheng LF, Song J, Fan RF, Chen CL, Ren QZ, Zhang XL, Feng XY, Zhang Y, Li LS, Zhu JX (2014) The role of the vagal pathway and gastric dopamine in the gastroparesis of rats after a 6-hydroxydopamine microinjection in the substantia nigra. Acta Physiol (Oxf) 211(2):434–446

Zhong CB, Chen QQ, Haikal C, Li W, Svanbergsson A, Diepenbroek M, Li JY (2017) Age-dependent alpha-synuclein accumulation and phosphorylation in the enteric nervous system in a transgenic mouse model of Parkinson's disease. Neurosci Bull 33(5):483–492

Zieba D, Wegrzyn T, Cebrat E, Leroch Z (1991) Effect of dopamine on cecum motility in sheep. Pol Arch Weter 31(1-2):109–114

Zizzo MG, Mule F, Mastropaolo M, Serio R (2010) D1 receptors play a major role in the dopamine modulation of mouse ileum contractility. Pharmacol Res 61:371–378

Zizzo MG, Cavallaro G, Auteri M, Caldara G, Amodeo I, Mastropaolo M, Nuzzo D, Di Carlo M, Fumagalli M, Mosca F, Mule F, Serio R (2016) Postnatal development of the dopaminergic signaling involved in the modulation of intestinal motility in mice. Pediatr Res 80(3):440–447

Chapter 6
Dopamine in the Pancreas

Feng Hong, Guang-Wen Li, Sumei Liu, Yan Zhang, Xiao-Yan Feng, and Jin-Xia Zhu

Abstract Dopamine (DA) presents in the pancreas at a high level. Pancreatic endocrine or exocrine is able to synthesize DA, although DA may come from other sources such as the autonomic nervous system and blood circulation. Dopamine receptors (DARs), including D_1R, D_2R, D_3R, D_4R, and D_5R, are widely expressed in the pancreatic tissues. Pancreatic DA regulates insulin secretion and protects pancreas from inflammation. DA-related biomarkers such as the vesicular monoamine transporter 2 (VMAT2) and D_2R have become the research hot spots to look for the therapeutic targets and develop the pancreatic islets imaging probes used for monitoring β-cell mass morphology and physiology.

Keywords Pancreatic DA · Islet regulation · Pancreatic acinar protection · β-cell mass monitor · Pancreatic function

Abbreviations

6-OHDA	6-hydroxydopamine
COMT	Catechol-*O*-methyltransferase
D_1R	Dopamine D1 receptor
D_2R	Dopamine D2 receptor

F. Hong
Department of Physiology, School of Preclinical Medicine, Wannan Medical College, Wuhu, China

Department of Physiology and Pathophysiology, School of Basic Medical Science, Capital Medical University, Beijing, China

G.-W. Li · Y. Zhang · X.-Y. Feng · J.-X. Zhu (✉)
Department of Physiology and Pathophysiology, School of Basic Medical Science, Capital Medical University, Beijing, China
e-mail: fengxy@ccmu.edu.cn; zhu_jx@ccmu.edu.cn

S. Liu
Department of Biology, College of Science and Health, University of Wisconsin-La Crosse, La Crosse, WI, USA
e-mail: sliu@uwlax.edu

J.-X. Zhu (ed.), *Dopamine in the Gut*, https://doi.org/10.1007/978-981-33-6586-5_6

D_3R	Dopamine D3 receptor
D_4R	Dopamine D4 receptor
D_5R	Dopamine D5 receptor
DA	Dopamine
DARs	Dopamine receptors
DAT	DA transporter
DDC	DOPA decarboxylase
GLU	Glucagon
GSIS	Glucose-stimulated insulin secretion
i.p.	Intraperitoneal
IF	Immunofluorescence
INS	Insulin
IR	Immunoreactivity
LAT	L-amino acid transporter
L-DOPA	Levodopa
MAO	Monoamine oxidase
SST	Somatostatin
TH	Tyrosine hydroxylase
VMAT2	Vesicular monoamine transporter 2
WB	Western blot

6.1 Introduction

The pancreas consists of exocrine and endocrine parts. The exocrine pancreas produces digestive enzymes from pancreatic acini and secrets HCO_3^-/H_2O from pancreatic duct epithelia. The digestive enzymes and HCO_3^-/H_2O mix together to form pancreatic juice, which is then released to the lumen of the small intestine for food digestion. The endocrine pancreas, or islets of Langerhans, can synthesize and release several vital hormones, including insulin, glucagon, and somatostatin, which are involved in the regulation of blood glucose homeostasis. More and more studies suggest that the exocrine and endocrine pancreas are interdependent and interconnected in both structure and function. In the clinic, acute pancreatitis often leads to a high risk of diabetes mellitus (Das et al. 2014), while diabetic patients often manifest pancreatic exocrine dysfunction (Prasanna Kumar et al. 2018). Furthermore, the structural connection between endocrine and exocrine pancreas has been demonstrated (Bertelli et al. 2001).

Bromocriptine, a D_2R agonist, has been reported to reduce postprandial blood glucose levels and improve glucose tolerance in obese patients with type 2 diabetes (Barnett et al. 1980; Pijl et al. 2000). The rapid-release formulation of bromocriptine (Cycloset) has been approved by the United States Food and Drug Administration (FDA) for the treatment of type 2 diabetes in 2009 (Defronzo 2011). Although many drugs regulating insulin secretion have been developed, the studies related to the dopaminergic system in pancreas still attract special attention, and many unknown mechanisms need to be further investigated.

6.2 Sources of Pancreatic Dopamine

High levels of DA and levodopa (L-DOPA) are present in the pancreas tissue (Schafer et al. 2013; Mezey et al. 1996). The pancreatic DA may originate from autonomic nerves innervating pancreas (Mezey et al. 1996; Somvanshi et al. 2018, Fig. 6.1), peripheral circulating blood (Maffei et al. 2015), as well as pancreas itself. A study on the rats with chemical sympathectomy has demonstrated that about 60% of DA was still presented in the pancreas after intraperitoneal injection of 6-hydroxydopamine (6-OHDA), which suggests that part of pancreatic DA is by self-synthesis in the pancreatic tissue (Mezey et al. 1996).

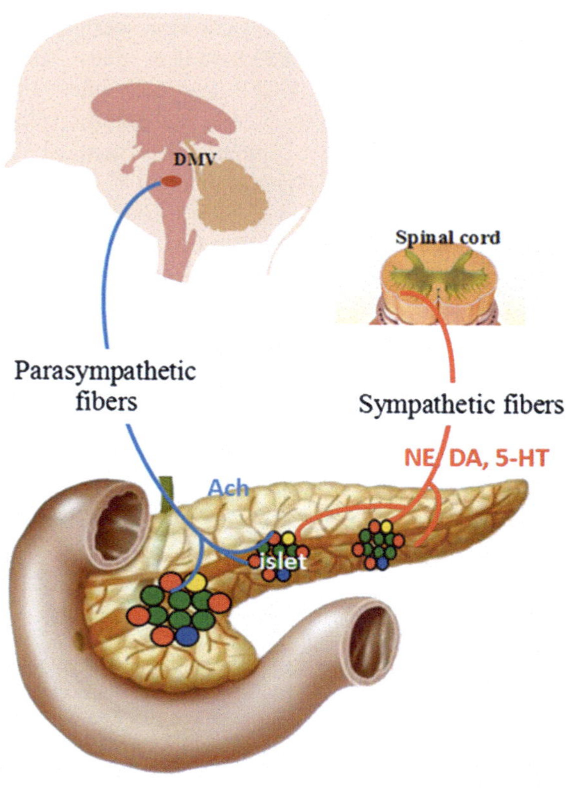

Fig. 6.1 DA released from the sympathetic fibers innervating the pancreas. The pancreas receives inputs nerves from the sympathetic never (red) and parasympathetic never (blue). The parasympathetic efferent fibers mainly originate from the dorsal motor nucleus of the vagus (DMV), whereas the sympathetic efferent fibers project from the spinal cord. The sympathetic fibers secrete dopamine (DA), norepinephrine (NE), and serotonin (5-hydroxytryptamine, 5-HT) into pancreatic tissue. *Ach* Acetylcholine

6.2.1 *Distribution of DA Synthetases, Metabolic Enzymes, and Transporters in the Pancreas*

A series of studies have reported the expression of DA-related biomarkers, including DA synthetases tyrosine hydroxylase (TH) and DOPA decarboxylase (DDC), DA metabolic enzymes catechol-*O*-methyltransferase (COMT), monoamine oxidase (MAO), dopamine transporter (DAT), L-amino acid transporter (LAT), and vesicular monoamine transporter 2 (VMAT2) in pancreas tissues. However, their cellular distribution in the pancreas remains controversial across different species.

6.2.1.1 Tyrosine Hydroxylase and DOPA Decarboxylase

Since TH is the rate-limiting enzyme of DA synthesis, mapping the cellular localization of TH in the pancreas is a critical step for understanding pancreatic DA synthesis. It has been declared the distribution of TH in pancreatic exocrine acinar cells by means of immunofluorescence in rat pancreas (Mezey et al. 1996) and in pancreatic tissue and primary isolated acinar cells of Balb/C mouse by Western blot (Oomori et al. 1994; Han et al. 2017; Fig. 6.2). Furthermore, the expression of TH in pancreas tissue and pancreatic acinar cells was increased in the pancreas of caerulein plus lipopolysaccharide (LPS) treatment mouse and pancreatic acinar cells stimulated by cholecystokinin (CCK) (Fig. 6.2). Interestingly, the distribution of TH in pancreatic islets, mainly on β cells, and some δ cells in mouse pancreas have also been reported by means of immunohistochemical staining (Persson-Sjögren et al. 2002) and in the islets of db/db mouse and MIN6 β cell lines by qPCR (Cataldo et al. 2016). Furthermore, TH has also been found to be expressed in sympathetic fibers innervating the pancreas in human (Schafer et al. 2013).

Expression of DDC mRNA has been reported in INS-1E cell lines and isolated islets of mouse, rat, and human (Cataldo et al. 2016; Rubí et al. 2005). The localization study by immunohistochemical staining suggests the distribution of DDC protein in the islets of human pancreas, which is mainly on the β cells and

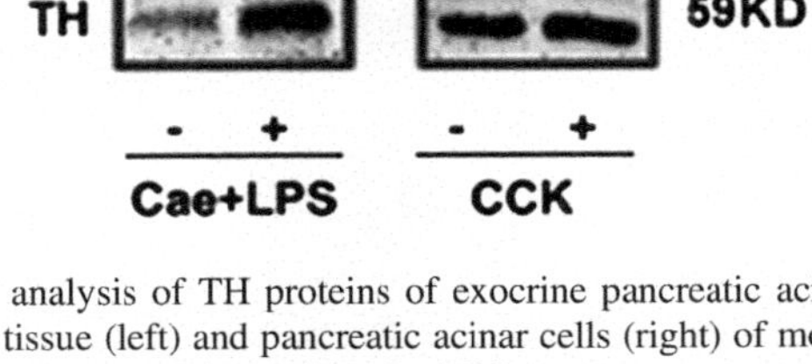

Fig. 6.2 Immunoblot analysis of TH proteins of exocrine pancreatic acinar cells of mouse. TH expressed in pancreas tissue (left) and pancreatic acinar cells (right) of mouse, and the expression levels were increased in pancreas of caerulein plus LPS treatment mouse and pancreatic acinar cells stimulated by CCK. *LPS* lipopolysaccharide, *CCK* cholecystokinin. (Han et al., Br J Pharmacol; 2017.174(24):4751–70)

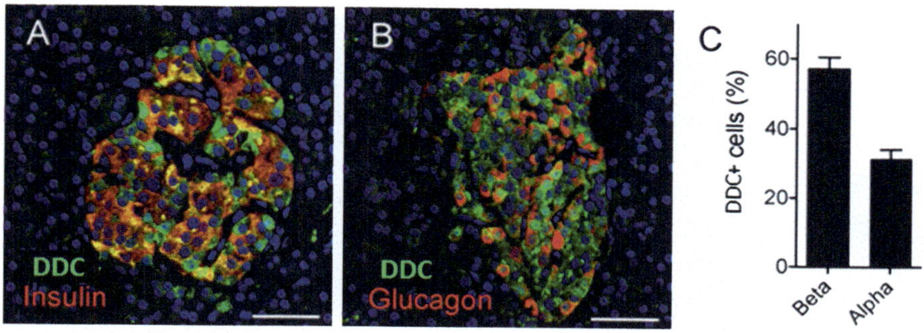

Fig. 6.3 Localization of DDC in α and β cells of human islets. (**a**) Confocal image of DDC (green) and insulin (red). (**b**) Confocal image of DDC (green) and glucagon (red). (**c**) Quantification of the number of DDC-positive cells that are β (insulin positive) or α (glucagon positive) cells. Scale bars = 50 μm. (Almaça et al., Cell Rep. 2016.20;17(12):3281–91)

Table 6.1 Distribution of TH and DDC in pancreas

Synthetases	Distribution	Species	Methods	References
TH	Sympathetic fibers	Human	IF	Schafer et al. (2013)
	Acini cells	Mouse	WB[a]	Somvanshi et al. (2018)
	Acini cells	Rat	WB[a]	Han et al. (2017)
	Acini cells	Rat	IF	Mezey et al. (1996)
	Islets β and δ cells	Fetal mouse	IHC	Persson et al. (2002)
	Islets	MIN-6 cell and mouse	RT-PCR	Cataldo et al. (2016)
DDC	Acini cells	Mouse	WB[a]	Han et al. (2017)
	Islets	MIN-6 cell and mouse	RT-PCR	Cataldo et al. (2016)
	Islets	Rat and human	RT-PCR	Rubí et al. (2005)
	Islets β and α cells	Human	IF	Almaça et al. (2016)

IF immunofluorescence, *WB* Western blot, *PCR* polymerase chain reaction
[a]The proteins from isolated pancreatic acini cells were utilized for WB

also in some α cells (Almaça et al. 2016, Fig. 6.3). Western blot analysis reveals no expression of DDC in exocrine acinar cells of mouse (Han et al. 2017). The results of these studies suggest that both TH and DDC are expressed in the pancreas, but their cellular localization in the pancreas remains controversial (Table 6.1).

6.2.1.2 Monoamine Oxidase and Catechol-*O*-Methyltransferase

In 1994, Karhunen et al. have localized COMT in pancreatic islet β cells and δ cells in rat by using immunohistochemical staining (Karhunen et al. 1994). The mRNA expression of COMT is also found in the mouse islets (Cataldo et al. 2016). Both isoforms of MAO, MAO-A and MAO-B, both are expressed in the pancreatic tissue of mouse and human (Ganic et al. 2015). Further localization studies have shown that MAO-A is mainly distributed in the exocrine pancreas, while relatively lower

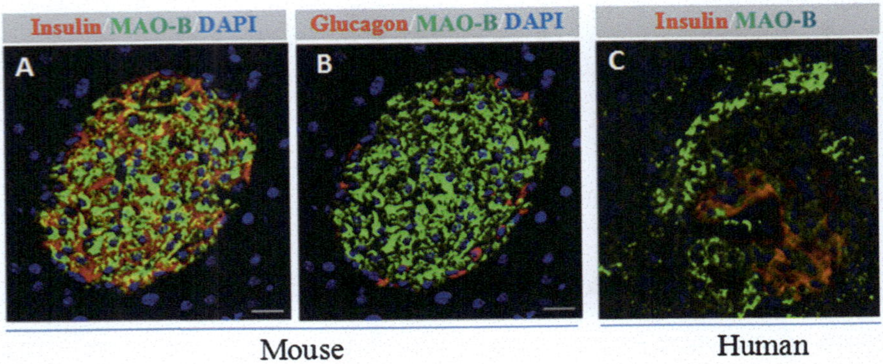

Fig. 6.4 Cellular localization of MAO-B in the islets of mouse and human. (**a**) Localization of MAO-B (green) and insulin (red) in mouse pancreatic islets. (**b**) Localization of MAO-B (green) and glucagon (red) in mouse pancreatic islets. (**c**) Localization of MAO-B (green) and insulin (red) in human pancreatic islets. Scale bar = 20 μm. (Ganic et al., Biochem Biophys Res Commun.2015. 25;468(4):629–35)

MAO-A immunoreactivity is observed in the α cells and β cells of pancreatic islets (Ganic et al. 2015). MAO-B is predominantly distributed in the β cells of mouse islets, while absent in β cells of human islets (Ganic et al. 2015; Cataldo et al. 2016, Fig. 6.4). Notably, reduction of MAO-B levels has been reported in the β cells of db/db diabetes mouse and whole pancreatic islets from human type 2 diabetic donors (Ganic et al. 2015).

6.2.1.3 Dopamine Transporter and Vesicular Monoamine Transporter

DAT always exists on the membrane of dopaminergic cells and transports DA from extracellular to intracellular side. Immunofluorescence in mouse pancreatic section illuminates the distribution of DAT in the whole pancreas, including endocrine cells, exocrine cells, excreting pancreatic ducts cells, and also some in the venous part of the vasculature (Mezey et al. 1996). The in situ hybridization histochemistry (Mezey et al.) and Western blot results (Ustione and Piston 2012, Fig. 6.5) demonstrate the expression of DAT in mouse islets. LAT is responsible for the uptake of amino acids including L-DOPA from the outside to the inside of cells. LAT includes two subtypes, LAT1 and LAT2. LAT1 is expressed in the pancreatic islets of rat (Fukushima et al. 2010), mouse (Zhou et al. 2014), and human (Nakada et al. 2014). Recently, Farino et al. have reported the expression of both LAT1 and LAT2 in mouse pancreas tissue by using qPCR assay (Farino et al. 2019).

VMAT is present on the vesicular membrane and responsible for transporting the synthesized DA from cytoplasm into vesicles, which prevents DA from being metabolized by cellular enzymes such as COMT (Wimalasena 2011). Two isoforms of VMAT, VMAT1 and VMAT2, have been characterized in rodent and human (Erickson and Eiden 1993; Erickson et al. 1996). VMAT1 is expressed in the

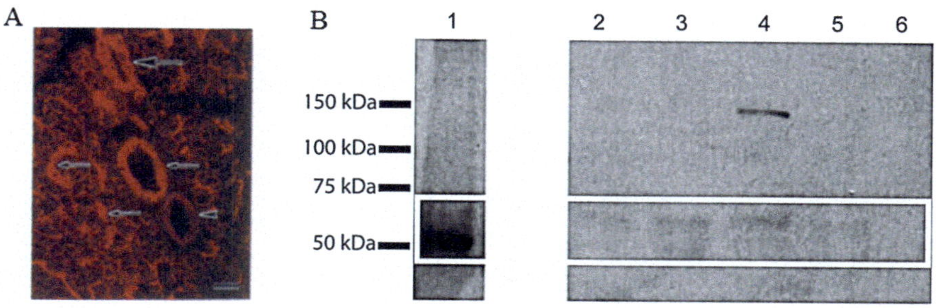

Fig. 6.5 Expression of DAT in pancreatic islets. Immunoblot for DAT. lane 1, mouse brain extract; lanes 2–5, pancreatic islets of lysate; lane 6, empty control. White box indicates the expression of the 50–62 kDa bands in the islet membrane fraction. (Ustione and Piston, Mol Endocrinol. 2012; 26 (11):1928–40)

enterochromaffin (EC) cells of the pancreatic exocrine system, while VMAT2 is predominantly present in the pancreatic islets in human pancreas (Anlauf et al. 2003). Notably, the VMAT2 in pancreas has attracted more attention, since it has become the hot target for the monitor of potential β-cell mass. The pancreatic uptake value of the VMAT2 selective tracers, [^{18}F]fluoropropyl-(+)-dihydrotetrabenazine ([^{18}F]FP-(+)-DTBZ or [^{18}F]AV133), correlated with insulin secretion from β cells, provides the possibility for noninvasive detection of type 2 diabetes (Cline et al. 2018). VMAT2-related probes have been also used to detect the secretory vesicles of β cells in human islets (Pecic et al. 2019). However, some studies do not support the distribution of VMAT2 in the islet β cells. Dr. Jinxia Zhu's lab has reported that VMAT2 is absent in β cells in pancreatic islets, but mainly distributed in α and pancreatic polypeptide (PP) cells of human, monkey, and rat (Hong et al. 2014, Fig. 6.6). Indeed, the expression of VMAT2 has no reduction when the β cells significantly destructed by the alloxan injection in type 1 diabetes rat model (Hong et al. 2014). Schafer et al. have also reported that the VMAT2 does not distribute in the islet β cells in rat and mouse, but expresses in human islet β cells. They have also reported VMAT2-positive nerve fibers in the pancreas of rodent (Schafer et al. 2013; Fig. 6.7).

All the above studies suggest that the pancreas contains all the enzymes synthesizing and degrading DA and transporters including DAT, LAT, and VMAT. Indeed, the differential results on the expression of these DA-related biomarkers in the endocrine or exocrine parts of the pancreas should be noticed, which might be attributed to the difference in species, specimens, experimental condition, etc. It is certain that with the emergence of new technologies such as flow cytometry and single-cell sequencing, the cellular localization of DA synthetases and metabolic enzymes in the pancreatic tissue in different species and different disease models would be clarified, which will be significant for further functional study and drug target exploration.

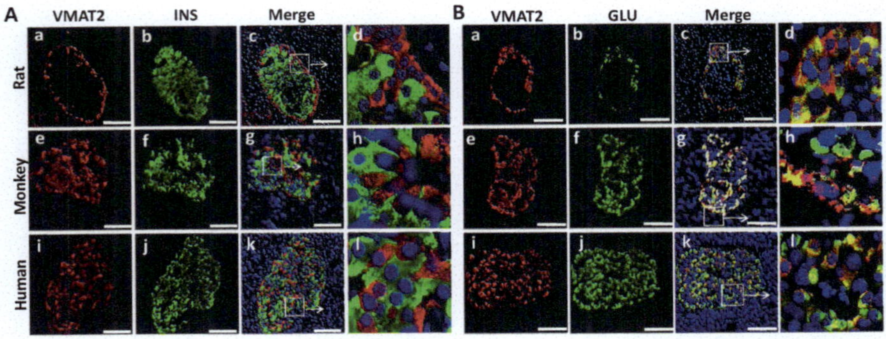

Fig. 6.6 Localization of VMAT2 immunoreactivity in the α cells, not β cells, of pancreatic islets in rat, monkey, and human. (**A**) Colocalization study of VMAT2 and insulin in rat, monkey, and human pancreas. (**B**) Colocalization study of VMAT2 and glucagon in rat, monkey, and human pancreas. VMAT2 was colocalized with glucagon in **B**, but not with insulin in **A**. Scale bars = 50 μm. (Hong et al., Transl Res. 2014. 163(2):171–82)

6.2.2 *Synthesis of Pancreatic Dopamine*

In addition to sympathetic nerve fibers which can release DA, there is no doubt that the pancreatic tissue itself can also produce DA. However, the synthetic process in pancreas is not fully understood. In 1977, a high level of DA (6.07–16.70 ng/mg) in the primary isolated islets of mouse and pancreatic tissue has been reported by using highly sensitive double-isotope method (Hansen and Hedeskov 1977). In 1996, the high content of DA (12.87 ± 5.51 ng/mg) has been detected in rat pancreatic tissue by liquid chromatography with electrochemical detection (Mezey et al. 1996). Meanwhile, the content of DA in the serum of mouse, rat, and human has been detected. (Maffei et al. 2015; Goldstein and Holmes 2008; Han et al. 2017). In addition, Mezey et al. have also reported a high level of L-DOPA in rat pancreatic tissue (Mezey et al. 1996).

Since Mezey et al. have reported that pancreas itself can generate DA (Mezey et al. 1996), Beek et al. have further detected a significant physiological uptake of [^{18}F]-labeled DOPA ([^{18}F]FDOPA) in human pancreas by means of PET/CT imaging technique (van Beek et al. 2017). The pancreatic DA level increases with the intake of tyrosine from circulating blood (Korner et al. 2019). In addition, by means of amperometry and voltammetry microelectrodes, Simpson et al. have demonstrated the secretion of DA along with insulin release after glucose stimulation in the isolated islet samples of human (Simpson et al. 2012).

Recently, by means of Western blot, Han et al. have detected the expressions of DA synthesis enzymes, TH and DDC, and metabolism enzymes, dopamine beta hydroxylase (DBH) and MAO, in the pancreatic tissue and isolated pancreatic acinar cells of mouse. The studies from both Mezey's and Han's lab suggest that the pancreatic acinar cells are able to synthesize DA by using circulating L-DOPA, but

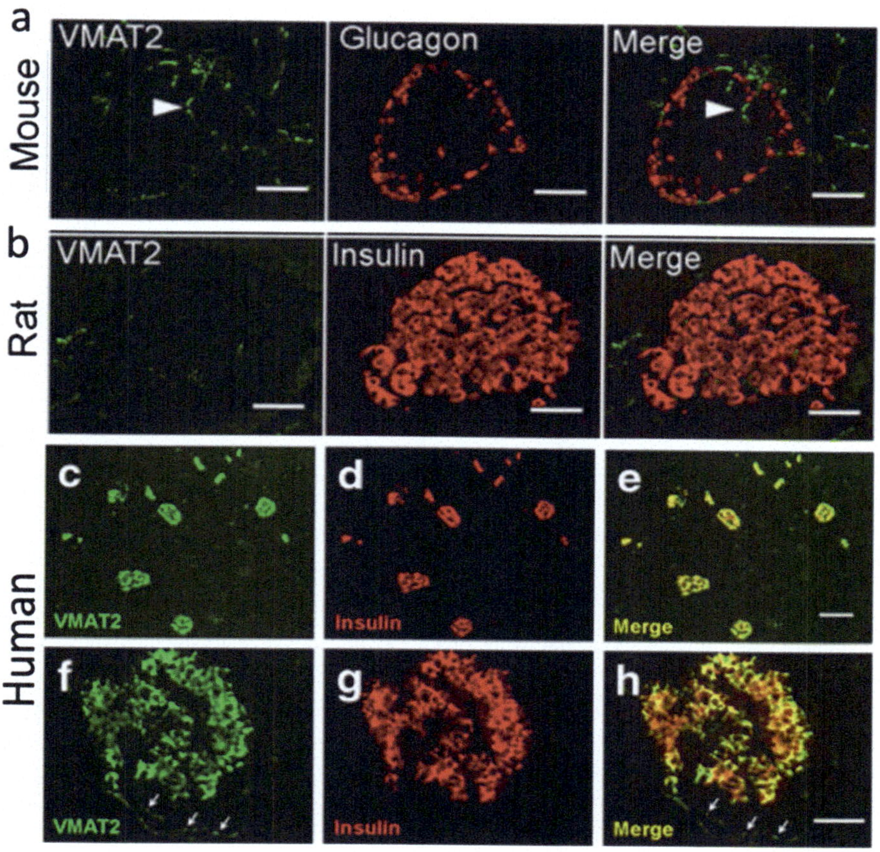

Fig. 6.7 VMAT2 immunoreactivity was absent in rodent pancreatic islets, while presented in human islets. (**a**) VMAT2 (green) was not localized with glucagon (red) in mouse. (**b**) VMAT2 (green) was not localized with insulin (red) in rat. (**c–h**) VMAT2 (green) and insulin (red) were colocalized in β cells of human islets. Arrowhead in (**a**) manifested the positive expression of VMAT2 on nerve fibers which innervated the islets. Scale bars = 100 μm. (Schafer et al., Diabetologia. 2013. 56(5):1047–56)

the cellular localization of DDC in pancreas are still unclear. Indeed, many studies illuminate the expression of DDC in endocrine islet of pancreas (see Sect. 6.2.1.1).

As early in 1977, Ericson et al. have reported that injection of radiolabeled L-DOPA through tail vein leads to the DA production in the pancreatic islets of mouse (Ericson et al. 1977). Later, in vitro study demonstrated the primary isolated pancreatic islets from mouse can utilize L-DOPA to synthesize DA, and the in vivo study further demonstrated that intraperitoneal injection of 50 mg/kg L-DOPA could also result in the production of DA in the isolated pancreatic islets from mouse (Ustione and Piston 2012). Ustione et al. proposed that pancreatic islets are potential sites for DA synthesis based on the localization of DDC in islet β cells, and

Table 6.2 Contents of DA and L-DOPA in serum and pancreatic tissues

Specimens	Sample	DA content	L-DOPA content	References
Human	Serum	15.32–91.91 ng/ml	8.9 nmol/ml	Goldstein and Holmes (2008)
		55.5 ng/ml	\	Iwen et al. (2017)
		18 ng/ml	\	Han et al. (2017)
Rat	Serum	\	3 nmol/L	Nissinen et al. (1992)
		0.3 ng/ml	\	Maffei et al. (2015)
		0.76 ng/ml	\	Kavelaars et al. (2005)
	Pancreatic tissue	12.87 ± 5.51 ng/mg	\	Mezey et al. (1996)
Mouse	Serum	0.04 ng/ml	\	Han et al. (2017)
	Pancreatic tissue	0.029 ng/mg	\	Nagler et al. (2018)
	Pancreatic tissue	6.07–16.70 ng/mg	\	Hansen and Hedeskov (1977)

Note: "\" means data not available

the application of L-DOPA resulted in the detection of DA in the supernatant of primary cultured islets (Ustione et al. 2013). Furthermore, it has also been demonstrated that the pancreatic islet β cells can store and import DA in insulin storage vesicles and then co-secrete DA with insulin in response to glucose stimulation in the isolated islets of both rat and human by using modified amperometry and voltammetry at microelectrodes detection system (Maffei et al. 2015; Simpson et al. 2012).

Based on the expression of DA-relative biomarkers in pancreatic tissues and the functional detection of DA production in pancreatic tissues after addition of L-DOPA, the pancreatic islets and exocrine may all participate in the synthesis of pancreatic DA, in which the detail needs further investigation. Table 6.2 summarizes the studies on the content of DA and L-DOPA in serum and pancreatic tissues.

6.3 Distribution of Dopamine Receptors in the Pancreas

DARs include D_1R, D_2R, D_3R, D_4R, and D_5R. DARs are widely present in the pancreatic endocrine and exocrine parts. More research focuses on the expression of D_2R in endocrine islets, but the results still remain contradictory. Some study declares that D_2R is localized in islet β cells (Rubí et al. 2005), while others illuminate that D_2R is absent in β cells and present in islet δ cells (Chen et al. 2014). It has also been reported that D_1R and D_3R are expressed on islet β cells (Chen et al. 2014; Simpson et al. 2012), while D_5R is expressed on α and PP cells

(Chen et al. 2014). Furthermore, it is also demonstrated that both D_1R and D_2R are expressed on the pancreatic acinar cells (Han et al. 2017).

6.3.1 Dopamine Receptors in the Endocrine Pancreas

By means of RT-PCR, Rubí et al. have reported the mRNA expressions of all five subtypes of DARs in the INS-1 cell line and cultured rat islets (Rubí et al. 2005, Fig. 6.8). Ustione and Piston has detected a protein band of D_3R at approximately 48 kDa in lysate from mouse pancreatic islets (Ustione and Piston 2012, Fig. 6.9). The Western blot studies have revealed the expression of both D_2R and D_3R in the islets of mouse, whereas only D_2R in the islets of human (Simpson et al. 2012). A study using cultured INS-1 cells and HIT-Y15 cells reveals the distribution of all D_1R, D_2R, and D_5R in both β cell lines, while only D_1R is located in the pancreatic islet β cells of rat (Li et al. 2018).

By means of Western blot, Dr. Jinxia Zhu's lab reveals higher protein levels of D_1R, D_2R, and D_5R but hardly detected D_3R and D_4R in the pancreatic islets of rat

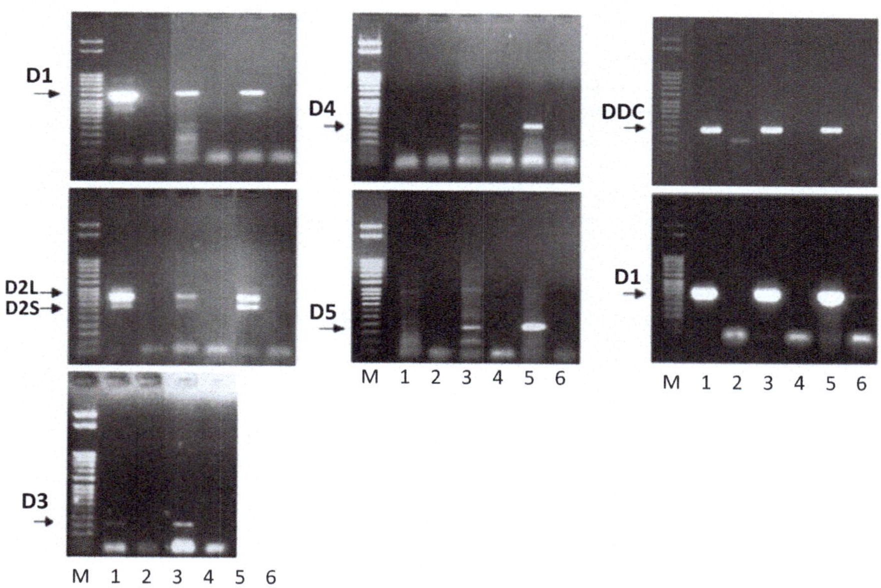

Fig. 6.8 RNA expressions of DARs in INS-1E cells and rat islets. Lane 1, INS-1E cDNAs; lane 2, INS-1E cDNAs without RT-PCR; lane 3, rat islet cDNAs; lane 4, rat islet cDNAs without RT-PCR; lane 5, rat brain cDNAs; lane 6, rat brain cDNAs without RT-PCR; M, molecular weight markers, 50 bp. (Rubí et al., J Biol Chem. 2005. 4; 280 (44):36824–32)

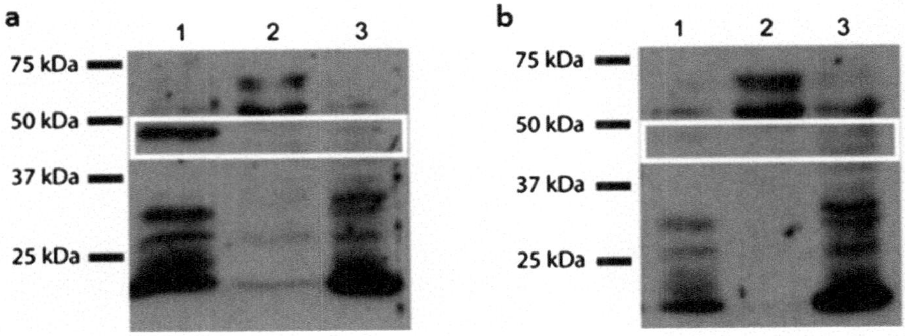

Fig. 6.9 Expression of D_3R in pancreatic islets. (**a**) Immunoblot for D_3R. (**b**) Negative control. Lane 1, islet membrane; lane 2, islet cytosolic; lane 3, islet nuclear. White box manifested the 48 kDa band of D_3R present in the islet membrane. (Ustione and Piston, Mol Endocrinol. 2012. 26 (11):1928–40)

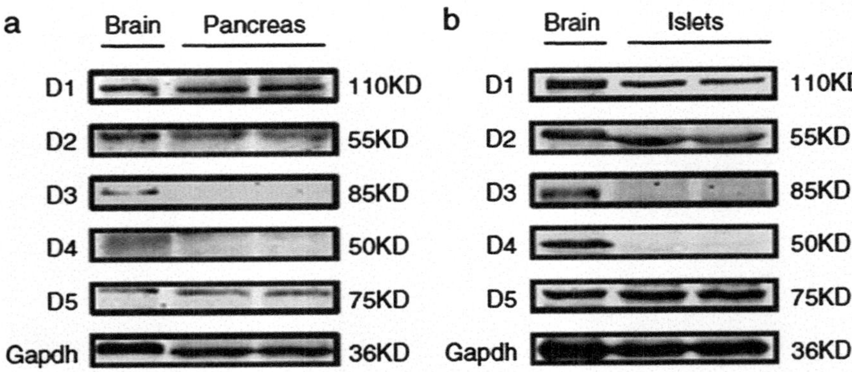

Fig. 6.10 Western blot results of DARs protein expression in rat pancreas and islets. a. Protein expression of five types of DARs (D_1R–D_5R) in rat pancreas. b. Protein expression of five types of DARs in rat-isolated islets. Brain: brain tissues from rat; Islets: islets from various normal rat. The rat brain was used as a positive control. Gapdh: D-glyceraldehyde-3-phosphate dehydrogenase. (Chen et al., Cell and Tissue Research, 2014. 357(3): p. 597–606)

(Chen et al. 2014, Fig. 6.10). By means of immunofluorescence, they have further observed the localization of D_1R in the β cells and D_5R in the α cell of rat pancreatic islets (Chen et al. 2014, Fig. 6.11). Meanwhile the distribution of D_2R in the islet δ cells has been observed in either paraffin-embedded or OCT-embedded frozen pancreatic tissues (Chen et al. 2014, Fig. 6.12). But the study in cultured β cell line, the INS-1 cells, and cultured isolated islets of rat and mouse (Fig. 6.13) revealed the localization of D_2R in islet β cells (Rubí et al. 2005). The reasons for the differential results of D_2R might be various, but there is no doubt that D_2R is expressed in islets and plays a critical role in mediating the regulation of DA on islet endocrine function.

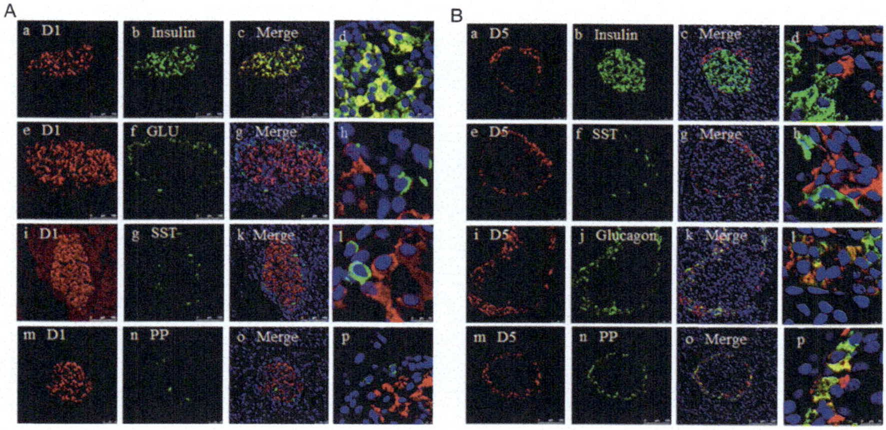

Fig. 6.11 Localization of D_1R and D_5R in pancreatic endocrine islets of rat pancreas. (**A**) D_1R was colocalized with insulin in islets β cells of rat. (**B**) D_5R was colocalized with glucagon in islets α cell of rat. Scale bars = 25 μm in d, h, l, p and scale bars = 100 μm in others

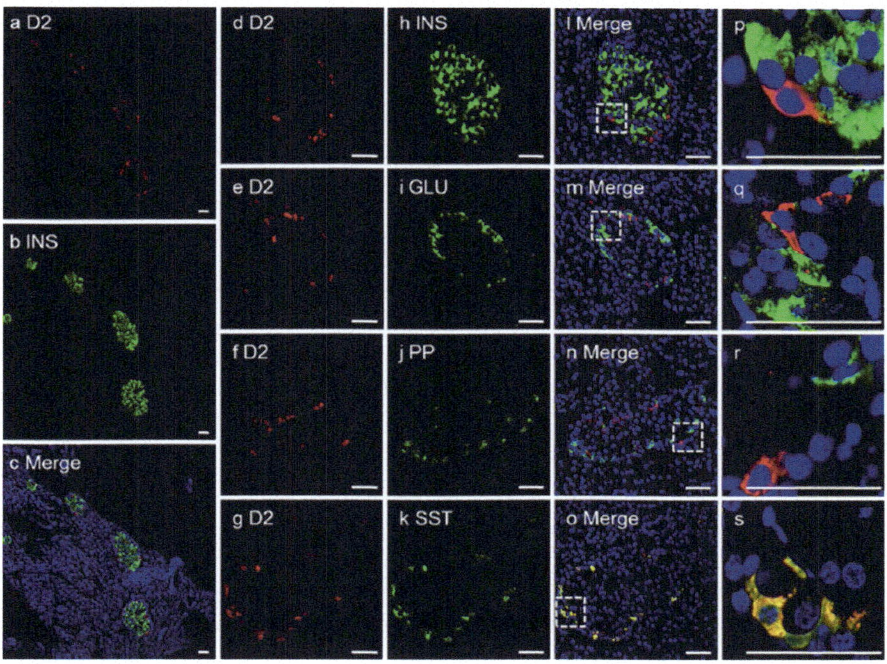

Fig. 6.12 Localization of D_2R in pancreatic tissue of rat. (**a–c**) Distribution of D_2R in pancreas in low-magnification images. (**d–s**) Colocalization study of D_2R and INS, GLU, PP, and SST, respectively, in pancreas in higher-magnification images. Red manifests the localization of D_2R and green represents the related islet cells, including INS secretion β cells. Scale bars = 100 μm. (Chen et al., Cell and Tissue Research, 2014. 357(3): p. 597–606)

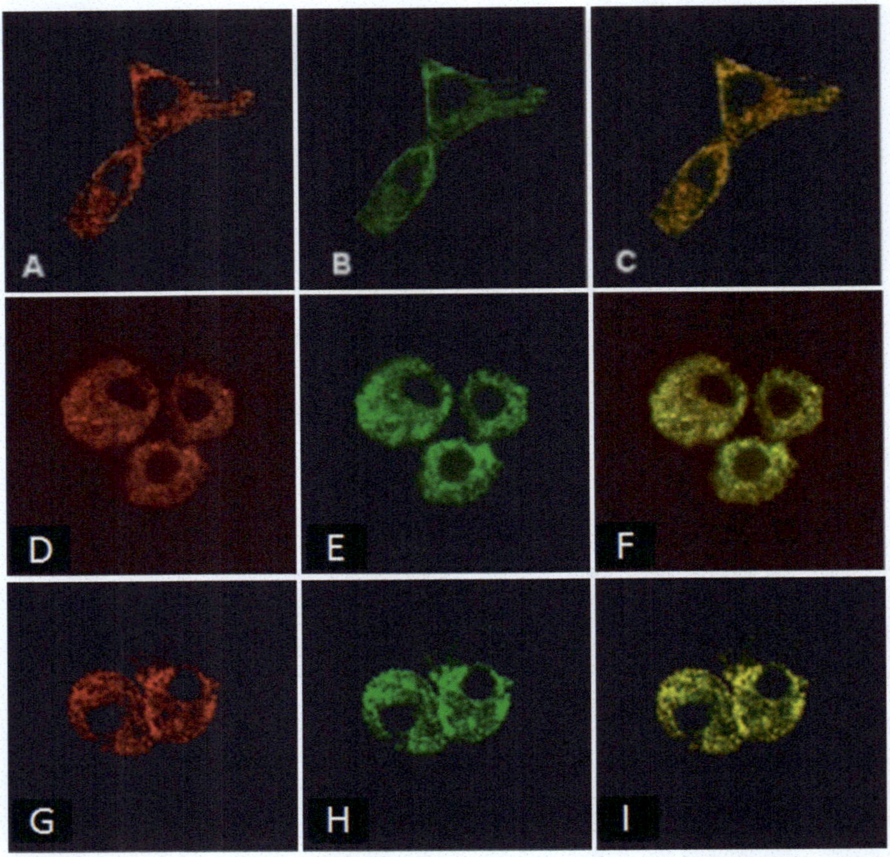

Fig. 6.13 Expression of D_2R in INS-1 cells, cultured isolated islets of mouse and rat. (**a–c**) Localization of D_2R in INS-1 cells. (**d–f**) Localization of D_2R in cultured islets of rat. (**g–i**) Localization of D_2R in cultured islets of mouse. Anti-insulin (red staining) and polyclonal anti-D_2R (green staining) are colocalized in islets INS-1 cells, cultured isolated islets of mouse and rat. (Rubí et al., J Biol Chem; 2005. 4; 280 (44):36824–32)

A D_2/D_3R radioactive ligand, fallypride, has been evaluated in the measurement of islet cell mass. This ligand can be labeled with either Fluorine-18 (110 min half-life) (Mukherjee et al. 1995) or Carbon-11 (20 min half-life) (Mukherjee et al. 2004). Garcia et al. have demonstrated that the ratio of [^{18}F] fallypride uptake in the pancreas to reference tissue (erector spinae muscle) is 5.5, while the affinity of [^{18}F] fallypride to endocrine islets is four times higher than exocrine tissue (Garcia et al. 2011). In addition, binding of [^{18}F] fallypride to pancreas is reduced in streptozotocin-treated rat, in which the islet β cells are chemically destroyed (Garcia et al. 2011). This study indicates that D_2R and/or D_3R are present in the pancreatic islet β cells. However, it is also reported that D_2R is distributed in the pancreatic δ cells, not β cells of rat islets (Chen et al. 2014). Whether the tracer targeting D_2/D_3R is suitable for detecting pancreatic islets β cells needs further demonstration.

6.3.2 *Dopamine Receptors in the Exocrine Pancreas*

Early study has demonstrated the expression of DARs in pancreatic acinar cells of dog by using [^{3}H]DA-binding experiment (Vayssette et al. 1986), which is mainly binding to D_1R (Iwatsuki et al. 1995). Recently, both D_1R and D_2R are reported to be expressed in the pancreatic acinar cells through Western blot in the pancreatic tissue and the isolated pancreatic acinar cells of mouse, and the gene level of D_2R is three times of D_1R expression in the mouse pancreas detected by gene chips (Han et al. 2017). The protein expression of D_2R protein has also been reported in the human pancreatic ductal adenocarcinoma by means of tissue microarray analysis (Jandaghi et al. 2016). Until now, no morphological research has examined the cellular localization of DARs in the pancreatic exocrine of any species, including rodents and human.

In summary, all five subtypes of DARs are expressed in the pancreatic tissue, including endocrine islets and exocrine acinar cells (Table 6.3), and contributed to the regulation of pancreatic endocrine and exocrine function. The expression of DARs in the pancreatic islets is definite, but the cellular localization of each DAR subtype in the endocrine cells of islets is still controversial. Since the expression of DARs in the acinar cells of pancreas has only been detected by Western blot at the protein levels, further morphological studies are needed to illustrate the cellular distribution of DARs in the exocrine part of the pancreas. In other tissues, DARs are often expressed in neuronal cells, immune cells, etc. One report has demonstrated the protein expression of D_1R and D_2R in M_1-phenotype bone marrow-derived

Table 6.3 The localization of DARs in the pancreas endocrine and exocrine parts

DARs	Distribution	Methods	Species	References
D_1R	β cells	IF, WB, PCR	Human, rat	Chen et al. (2014)
	Acini cells	WB	Dog	Iwatsuki et al. (1995)
			Mouse	Han et al. (2017)
D_2R	β cells	IF, RT-PCR, IHC	Human, mouse	Rubí et al. (2005)
			INS-1E cells	Simpson et al. (2012)
	δ cells	IF, WB	Human, rat	Chen et al. (2014)
	Acini cells	WB	Mouse	Han et al. (2017)
D_3R	β cells	RT-PCR	INS-1E cells	Rubí et al. (2005), Simpson et al. (2012)
D_5R	α and PP cells	IF, WB	Rat	Chen et al. (2014)

Note: *IF* immunofluorescence, *WB* Western blot, *RT-PCR* reverse transcription polymerase chain reaction, *PP* pancreatic polypeptide macrophages by immunoblot analysis (Han et al. 2020). However, the cellular distribution of DARs in pancreatic neuronal cells and pancreatic immune cells needs to be further explored, which will improve the network diagram of the roles of the dopamine system in pancreas.

6.4 Dopamine and the Regulation of Pancreatic Function

After the discovery of DA acting as a paracrine or autocrine factor to regulate metabolic function (Uehara et al. 2007; Ter Horst et al. 2018), a significant amount of works demonstrate that DA can directly decrease insulin secretion from freshly isolated rat pancreatic islets and cultured pancreatic β cells (an INS-1 cell-line) (Rubí et al. 2005; Shankar et al. 2006; Ustione et al. 2013) via an autocrine manner (Korner et al. 2019; Simpson et al. 2012; Farino et al. 2019). New findings about DA regulating pancreatic function keep springing up, and a critical advance more recently is that DA protects pancreatic acinar cells from inflammation in rat with acute pancreatitis (Han et al. 2017, 2020). There is a growing consensus that DARs play a critical role in the protection and regulation of pancreatic acinar cells.

6.4.1 Dopamine and Pancreatic Endocrine Function

DA participates in the regulation of insulin secretion from the pancreatic islets. Meanwhile, norepinephrine (NE) and 5-hydroxytryptamine (5-HT) also regulate pancreatic endocrine function.

DA and Pancreatic Islets

The insulin secretion from islet β cells is simultaneously regulated by blood glucose, arginine, and sympathetic, parasympathetic, and sensory nerves. NE directly affects K^+-ATP channel by activating adrenergic α_2 receptors on β cells to inhibit insulin secretion. Acetylcholine (ACh) binds to cholinergic M receptors on β cells to stimulate insulin secretion via PLC-IP_3-Ca^{2+} pathway. Neuropeptide Y (NPY) directly affects vesicle release and inhibits insulin secretion through a Ca^{2+}-independent pathway by activating the Y1 receptor (Schwetz et al. 2013). As typical G-protein-coupled receptors, the DARs can also regulate islet β cell endocrine secretions by affecting intracellular cAMP or Ca^{2+} levels (Simpson et al. 2012).

DA regulates pancreatic islet endocrine function. An earlier study has reported that long-term L-DOPA therapy in patients with Parkinson's disease (PD) can increase blood glucose level (Rosati et al. 1976). A following study has reported that injection of L-DOPA (i.p.) causes the accumulation of DA in the pancreas and decrease of insulin in the serum of mouse (Ericson et al. 1977). Similar result has also been reported that the D_2R agonist bromocriptine (i.p.) can reduce arginine-induced insulin secretion including both first and second phases during the

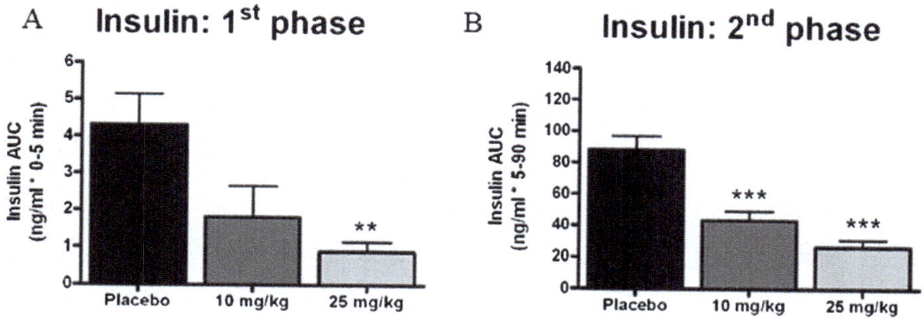

Fig. 6.14 Effects of acute injection of bromocriptine on arginine-induced insulin secretion in mouse subjected to a hyperglycemic clamp. (**a**) The first phase of insulin secretion was calculated from 0 to 5 min in the clamp. (**b**) The second phase of insulin secretion were calculated from 5 to 90 min. (Weenen et al., Biochem Pharmacol. 2010. 15; 79(12):1827–36)

hyperglycemic clamp test (de Leeuw van Weenen et al. 2010, Fig. 6.14). In an in vitro study, incubation of islets isolated from rat, mouse, and human with DA at 10 μM or quinpirole at 5 μM significantly inhibited glucose-stimulated insulin secretion (GSIS) (Rubí et al. 2005). D_2R knockout in INS-1 cell line by using interfering RNA (RNAi) results in an increased insulin secretion (Wu et al. 2008). Further experiments in human islets isolated from donors have also demonstrated the inhibition of DA on insulin secretion from β cells via an autocrine mode (Simpson et al. 2012). Studies in type 2 diabetic mouse model and diabetic human patients suggest a reduced expression of MAO-B, which may result in an elevation of DA level in the pancreas and inhibition of insulin release (Ganic et al. 2015).

It should be emphasized that in 2009, the United States FDA approved the rapid release form of bromocriptine for the treatment of type 2 diabetes. The mechanism of action of bromocriptine may be through the regulation of systemic metabolism through central hypothalamus, rather than directly acting on peripheral islet cells. Meanwhile, some studies have reported a stimulatory effect of DA on insulin secretion. For instance, the insulin release from β cells is transiently stimulated by islet DA accumulation in isolated rat islets (Ahrén and Lundquist 1985), and the impaired insulin response to glucose and elevated fasting glucose level happen in D_2R knockout mouse (García-Tornadú et al. 2010). The above studies suggest that pancreatic DA regulates insulin secretion from islets, but further study is needed to ultimately clarify the effect of DA on insulin secretion in mouse, rat, and human.

Being similar to the differential effects of DA on the insulin secretion, the mechanisms underlying the DA effect on insulin secretion also remain controversial. It has been reported that pancreatic DA exerts an inhibitory effect on insulin secretion through reducing intracellular cAMP level (Kozuka et al. 2015) or inhibiting intracellular Ca^{2+} concentration by directly acting on the D_2R on islet β cells (Liu et al. 2020). Rubí et al. have reported that activation of D2-like receptors significantly decreases intracellular cAMP levels and reduces the glucose-induced membrane depolarization of INS-1E cells (Rubí et al. 2005). DA can also inhibit

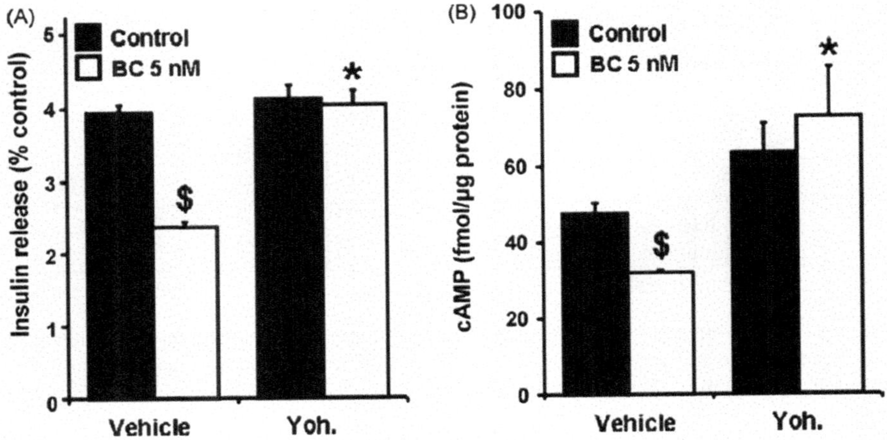

Fig. 6.15 Effects of the ⟨2-adrenergic receptor antagonist yohimbine on bromocriptine induced decrease in intracellular cAMP levels and insulin secretion in INS-1E cells. (**a**) Effects of bromocriptine (BC, 5 nM) on glucose-induced insulin secretion in vehicle- and yohimbine-treated INS-1E cells. (**b**) Effects of BC (5 nM) on intracellular cAMP levels in vehicle- and yohimbine-treated INS-1E cells. (Weenen et al., Biochem Pharmacol. 2010. 15; 79(12):1827–36)

insulin secretion by activating the D_3R on mouse β cells, and this effect is mainly mediated by activating the $G_{\beta\gamma}$ subunit, which subsequently inhibits the L-type calcium channels directly (Ivanina et al. 2000) or activates the potassium channels indirectly (Kuzhikandathil et al. 1998). Recently, Farino et al. have reported that the inhibitory effect of DA on GSIS in vivo was blocked by specific knockout of D_2R or D_3R in β cells of mouse (Farino et al. 2019). Surprisingly Weenen et al. have found that the D_2R antagonist domperidone is unable to block bromocriptine-induced inhibition of GSIS, while ⟨2-adrenergic receptor antagonist yohimbine can effectively block the inhibitory effect of bromocriptine on insulin secretion (de Leeuw van Weenen et al. 2010, Fig. 6.15), indicating the complexity of DA regulation on insulin secretion.

It is unclear about the reasons of the different results of DA in the pancreatic insulin release reported from different laboratories role. The following factors may partially explain the discrepancy, such as the studies are performed in vivo vs. in vitro, in cell lines vs. primary isolated islets, in mouse, rat, monkey, or human, the dosages of DA, the kinds of agonists or antagonists of DARs, or the research methods utilized in the studies, etc. An in vitro study in INS-1 cells has reported that DA significantly increases GSIS at lower concentration (10^{-8} M) and inhibits GSIS at higher concentration (10^{-7} M to 10^{-4} M) (Shankar et al. 2006, Fig. 6.16). The stimulatory effect of DA at lower concentration may be due to activation of D_1R on β cells. The distribution of D_1R in islets β cells suggests that DA might directly stimulate insulin secretion through the activation of D_1R on pancreatic β cells (Chen et al. 2014). The effects of D_1R activation on insulin secretion and glucose metabolism are poorly studied, and one report suggests that

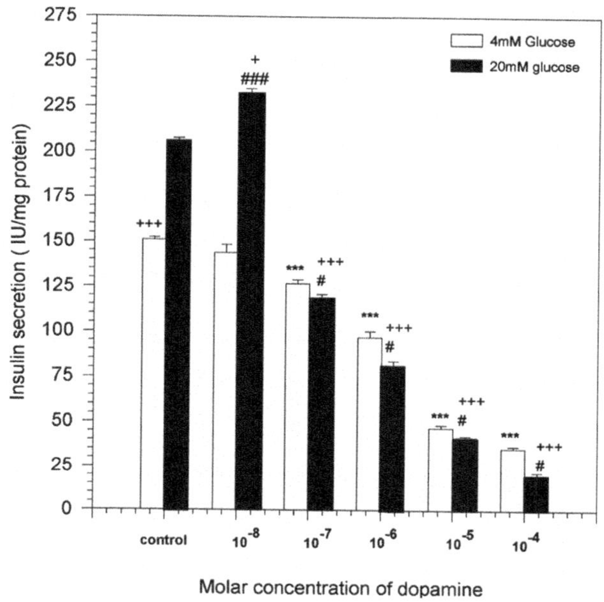

Fig. 6.16 Effect of dopamine on glucose-induced insulin secretion from pancreatic islets in vitro. DA at10^{-8} M significantly increased glucose-induced insulin secretion. +$P < 0.05$ when compared to 20 mM glucose in control, while DA at 10^{-7} M, 10^{-6} M, 10^{-5} M, 10^{-4} M all inhibited glucose-induced insulin secretion, ++ +$P < 0.001$when compared to 20 mM glucose in control. (Shankar et al., *IUBMB Life*. 2006; 58(3):157–163)

blocking D_1-like receptors with the antagonist SCH23390 has a preventative effect on diabetes mellitus in non-obese diabetic mouse (Hashimoto et al. 2009, Fig. 6.17). It might be possible that blocking the D_1-like receptors (the D_1R and D_5R) would directly inhibit the metabolism of insulin and glucagon by acting on β and α cells, respectively. The opposite (stimulatory and inhibitory) effects of DA at different concentrations have also been observed in the cardiovascular and renal systems (Goldberg et al. 1972) and gut motility (Song et al. 2014), which might be mediated by activating different DARs, D_1R or/and D_2R. Besides, DA may affect the secretion of insulin, glucagon, and somatostatin through binding to the D_1R, D_5R, and D_2R respectively, and the elevated serum insulin could be an integrated action of the whole endocrine islets.

The secretion of glucagon and somatostatin from pancreatic islets is also regulated by DA. Early report has demonstrated the direct stimulation by DA on glucagon secretion via a β-adrenergic mechanism (Matsumoto et al. 1984) and inhibition by DA on somatostatin secretion predominantly through dopaminergic receptors (Itoh et al. 1982). The D_2R is specifically distributed in the somatostatin (SST)-releasing δ cells, and D_5R is expressed in glucagon-secreting α cells (Chen et al. 2014). Activating the D_2R would induce SST release from δ cells. It has been reported that SST can inhibit insulin secretion by binding to SST receptor 5 ($SSTR_5$) on the β cells and inhibit glucagon release by binding to $SSTR_2$ on the α cells (Kailey et al. 2012). It is speculated that in D_2R knockout mouse, the lack of D_2R may lead to an increase in SST secretion, which can further inhibit insulin secretion through $SSTR_5$ (Tirone et al. 2003) and glucagon release via $SSTR_2$.

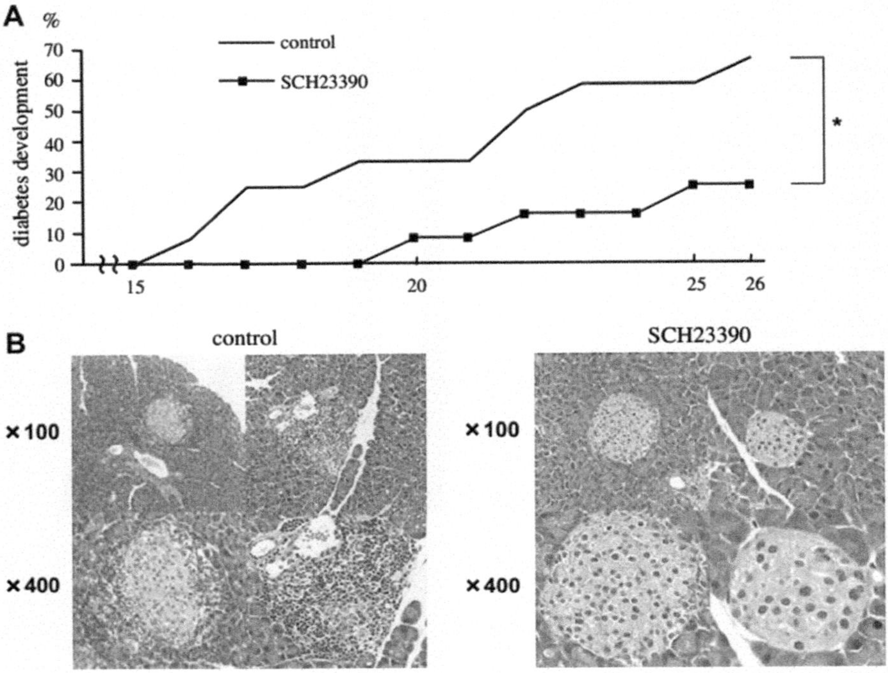

Fig. 6.17 The preventive effect of D_1R antagonist SCH23390 on NOD mouse. A The non-obese diabetic (NOD) mouse was orally administered with 0.3 mg/kg SCH23390 twice a week from week 6 to 26. The NOD mouse in the control group received 50 μl PBS. Treatment with SCH23390 prevented the development of diabetes in NOD mouse. B Histological examination of pancreatic tissue from control and SCH 23390-treated NOD mouse. The mouse without urinary sugar on week 26 was used for histological analysis in both control and SCH23390 groups. (Hashimoto et al., Biochemical and Biophysical Research Communications, 2009; 383(4): p. 460–63 * $p<0.05$)

Table 6.4 Effects of DARs activation/inhibition on insulin secretion

Receptors	Effects	References
D_2R	Promotes insulin secretion	García-Tornadú et al. (2010)
	Low content promotes insulin secretion High content inhibits insulin secretion	Shankar et al. (2006)
	Inhibits insulin secretion	Rubí et al. (2005), Simpson et al. (2012), de Leeuw van Weenen et al. (2010)
D_3R	Inhibits insulin secretion	Ustione et al. (2012), Farino et al. (2019)
D_2R and $SSTR_2$		Chen et al. (2014)

In conclusion, although the mechanisms underlying the effects of DA on the pancreatic endocrine function remain uncertain, more and more evidence suggests the important role of DARs in the pancreatic islet function (Table 6.4). More studies are required to explain the differential roles of DA on insulin secretion in different species, including biological models (mouse, rat). In addition, the effects of DA on other islet hormones, such as glucagon and somatostatin, are also need to be explored.

NE and Pancreatic Islets

NE in the pancreas is mainly released from sympathetic fibers which innervate pancreas. At least nine isoforms of adrenergic receptors exist, which can be classified into two classes, α and β. The β_1, β_2, and β_3 are coupled to the stimulatory G proteins (Gs) to produce excitatory effects, α1 is mainly coupled to the Gq, whereas the α_{2A}, α_{2B}, and α_{2C} are coupled to the inhibitory G proteins (Gi) to produce inhibitory role (S Jensen et al. 2011; Straub and Sharp 2012). Several isoforms of adrenergic receptors including α_{2A}, α_{2C}, and β_2 are reported in the pancreas of mouse and INS-1 cell line (Kelly et al. 2018). Some studies have demonstrated that NE inhibits insulin secretion via activating α_2 receptors (Lacey et al. 1996; Urano et al. 2004; Kelly et al. 2014). While an in vitro study has demonstrated that blocking the β receptor with propranolol significantly inhibits insulin secretion through the cAMP pathway (Sjoholm 1991), which suggests the stimulatory effect of NE on insulin secretion. Later reports have demonstrated that different doses of NE manifest differential effects on insulin secretion in rat, with higher concentrations of NE inhibit insulin secretion via activating α_2 receptors, while lower concentrations NE stimulate insulin secretion via binding to β_2 receptor on the β cells (Das et al. 2006). The mechanisms of the inhibition of NE on insulin secretion could be due to ATP-sensitive K^+ channels (K_{ATP}) activation, L-type Ca^{2+} channel inhibition, or adenylyl cyclase (AC) inhibition (Keahey et al. 1989; Zhao et al. 2010). Due to the complexity of adrenergic receptors in the pancreas, NE may manifest different effects on islet function and involve multi-mechanism. Furthermore, it has been reported that NE at 10^{-4} M inhibits the stimulation of DA at 10^{-8} M on GSIS, but enhances the inhibitory effect of DA at 10^{-4} M on GSIS (Shankar et al. 2006, Fig. 6.18).

5-HT and Pancreatic Islets

5-HT is also present in the pancreatic islets of mouse, rat, and human (Bennet et al. 2015; Gylfe 1978). Two kinds of enzymes, including tryptophan hydroxylase (TPH) and DDC, are needed for 5-HT biosynthesis. According to reports, both *TPH1* and *TPH2* genes are present in the islets of rodent and human (Walther et al. 2003), which suggests the possibility of islets utilizing tryptophan to synthesize 5-HT. But studies in the isolated islets from mouse suggest that the β cells utilize 5-hydroxytryptophan (5-HTP), but not tryptophan, to synthesize 5-HT and release 5-HT together with insulin (Gylfe 1978). There is no doubt that 5-HT plays critical regulatory roles on the pancreatic endocrine function, but the signaling pathways still remain unraveled due to complexity of its receptors. 5-HT receptors have at least 15 subtypes, which can be divide into seven subfamilies, many of them are expressed in the islets of rodent (Cataldo Bascuñan et al. 2019) and human (Bennet et al. 2015). Fourteen different 5-HT receptors are expressed in human islets and have different effects on insulin secretion, which is similar to the complexity of DA

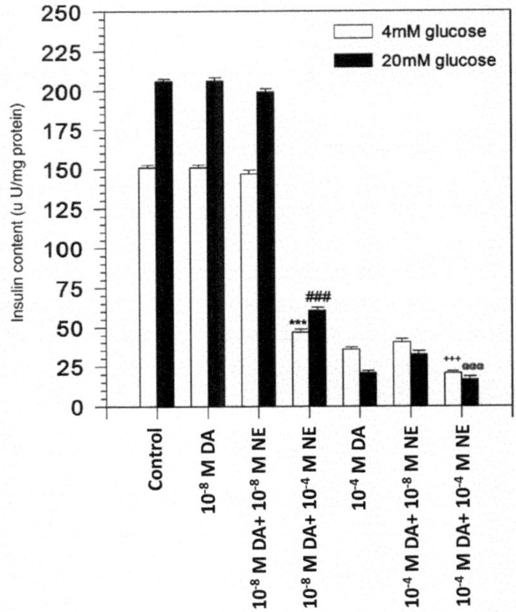

Fig. 6.18 Effect of NE on the role of DA in glucose-induced insulin secretion from pancreatic islets in vitro. $^{***}P < 0.001$ when compared to 4 mM glucose $+10^{-8}$ M DA; $^{\#\#\#}P < 0.001$ when compared to 20 mM glucose$+10^{-8}$ M DA; $^{+++}P < 0.001$ when compared to 4 mM glucose $+10^{-4}$ M DA; $^{@@@}P < 0.001$ when compared to 20 mM glucose $+10^{-4}$ M DA. (Shankar et al., *IUBMB Life*. 2006; 58 (3):157–163)

regulation. Recently, Dr. Zhu's lab has reported the localization of 5-HT_4 receptor in β cells of rat islets, and activation of 5-HT4 receptor promotes insulin secretion from islets (Chen et al. 2016).

6.4.2 Dopamine and Pancreatic Exocrine Function

DA exerts protective effect on experimental acute pancreatitis by reducing inflammatory cytokines and microvascular permeability in the pancreas (Karanjia et al. 1991, 1994; Zhang et al. 2017). Pancreatic D_2R knockout mouse displays a more severe acute pancreatitis, compared with normal control mouse (Han et al. 2017). Recently, Han et al. have reported that DA activates D_2R on acinar cells, which then exerts anti-inflammatory effects through the PP_2A-dependent Akt-NFκB pathway in rat models of acute pancreatitis (Han et al. 2017). Since DA is widely involved in the body's anti-inflammatory of systemic and periphery inflammation (Yan et al. 2015), the regulation of DA and DARs on the pancreatic exocrine would be necessary, although the study of Kaya et al. shows that DA at low dose fails to protect the pancreas in an acute pancreatitis rat model induced by Na-taurocholic acid 5% (Kaya et al. 2005). Myeloid-specific D_2R expressed on M_1 macrophage could control pancreatic injury and systemic inflammation by inhibiting M_1 macrophage polarization (Han et al. 2020). It is still unclear whether DA regulates the function of pancreatic duct or not.

6.4.3 Progress of Dopamine in Parkinson's Disease, Diabetes, and Pancreatitis

DA plays an important role in the development of Parkinson's disease (PD) and diabetes and closely links those two diseases. The 50 ~ 80% of PD patients show impaired glucose tolerance (Sandyk 1993; Camargo Maluf et al. 2019). Treatment with L-DOPA or bromocriptine can impair the glucose tolerance in PD patients. One year after chronic L-DOPA treatment, PD patients usually manifest impaired glucose tolerance, with a delayed insulin response (Sirtori et al. 1972). Oral glucose tolerance experiment has demonstrated that L-DOPA alone can reduce insulin secretion in PD patients (Rosati et al. 1976). Diabetes can significantly increase the risk of PD (Yang et al. 2017). Compared with the non-diabetic PD patients, the PD patients with diabetes manifest more severe pathological features, lower levels of DA in the striatum (Elbassuoni and Ahmed 2019), and worse performance in attention and working memory (Bohnen et al. 2014). Given these findings, the key problem that needs to be solved is how to maintain blood glucose stability during treatment for the PD patients with diabetes. Studies in animal models have also demonstrated the relationship among DA, PD, and diabetes. In a rat model of nigrostriatal DA depletion, insulin signaling in the basal ganglia is impaired and blood glucose is slightly reduced (Morris et al. 2008). In unilateral striatum-injected 6-OHDA middle-aged (16 months) rat, the serum insulin level and the brain's insulin resistance are increased at 6 weeks after operation, but glucose tolerance is unaffected (Morris et al. 2011). DARs including D_1R and D_2R are present in the paraventricular nucleus (PVN) of the hypothalamus and coexistence with vasopressin and oxytocin (Ran et al. 2019). In the PD rat induced by central bilateral SN administration of 6-OHDA for 6 weeks, an attenuated glucose tolerance has been observed. Decreased D_2R in hypothalamus evidenced by the upregulated expression of corticotropin releasing hormone in the periventricular nucleus of the hypothalamus, the increased content of adrenocorticotropic hormone in the pituitary gland, and the increased content of corticosterone in the blood, suggesting the involvement of DA in the hypothalamus–pituitary–adrenal axis (Zhou et al. 2019). In the periphery, DA inhibits insulin secretion (Simpson et al. 2012; Farino et al. 2019; Rubí et al. 2005) and affects glucose homeostasis, while enhancing glycolysis can slow down the neurodegenerative changes in PD patients (Cai et al. 2019).

DA participates in anti-inflammatory of systemic and periphery inflammation by regulating IL1β and IL18 levels (Yan et al. 2015). Expressions of DA synthetase and metabolic enzyme were significantly increased in acute pancreatitis (AP). Simultaneously, the expressions of D_1R and D_2R are decreased in AP (Han et al. 2017). Furthermore, DA is able to reduce the inflammation reaction in CCK-stimulated pancreatic acinar cells, which is mediated by inhibiting the NF-κB pathway (Han et al. 2017). Clinically, the expression of D_2R is significantly increased in chronic pancreatitis patients (Jandaghi et al. 2016). Several studies have suggested the interaction between dopaminergic system and the immune system. For example, follicular helper T cells can produce DA to promote synapse formation in the germinal center, which in turn promotes B lymphocytes to produce antibodies (Papa et al. 2017). D1-like receptors are expressed on the macrophages, and activation of D1-like receptors is able to inhibit oxidative modified low-density lipoprotein-induced macrophage proliferation through the PI_3K/Akt and MAPK/ERK signaling pathways (Yao et al. 2016). The upregulated expression of D_2R has been reported in the pancreatic ductal adenocarcinoma (Jandaghi et al. 2016) and neuroendocrine tumors (Grossrubatscher et al. 2008), and the D_2R inhibitors, pimozide and haloperidol, are able to inhibit the migration and proliferation of pancreatic cancer cells (Jandaghi et al. 2016). These studies suggest that the pancreatic DA plays important roles in the regulation of immune system function and the occurrence and development of tumor, but further studies are needed.

6.5 Perspectives

Pancreatic tissue produces DA, and high level of DA exists in the pancreatic tissue and pancreatic juice. The alteration of pancreatic DA system is closely related to the pancreatic islet dysfunction, diabetes, pancreatitis, pancreatic cancer etc., although many details are not clear, and the underlying mechanisms remain to be explored. As a bio-active substances or bio-target, the research of DA system in the pathogenesis, diagnosis, and treatment of the pancreatic diseases needs to attract scientists' attention, and more drugs related to the DA system would be discovered in the near future. Meanwhile, the role of the DA system in the endocrine microenvironment of the pancreatic islets, and the regulation of pancreatic cells development will also be research hot spots.

References

Ahrén B, Lundquist I (1985) Effects of L-dopa-induced dopamine accumulation on 45Ca2+ efflux and insulin secretion in isolated rat islets. Pharmacology 30(2):71–82

Almaça J, Molina J, Menegaz D et al (2016) Human β cells produce and release serotonin to inhibit glucagon secretion from alpha cells. Cell Rep 17(12):3281–3291

Anlauf M, Eissele R, Schäfer MK et al (2003) Expression of the two isoforms of the vesicular monoamine transporter (VMAT1 and VMAT2) in the endocrine pancreas and pancreatic endocrine tumors. J Histochem Cytochem 51(8):1027–1040

Barnett AH, Chapman C, Gailer K, Hayter CJ (1980) Effect of bromocriptine on maturity onset diabetes. Postgrad Med J 56(651):11–14

Bennet H, Balhuizen A, Medina A et al (2015) Altered serotonin (5-HT) 1D and 2A receptor expression may contribute to defective insulin and glucagon secretion in human type 2 diabetes. Peptides 71:113–120

Bertelli E, Regoli M, Orazioli D, Bendayan M (2001) Association between islets of Langerhans and pancreatic ductal system in adult rat. Where endocrine and exocrine meet together? Diabetologia 44(5):575–584

Bohnen NI, Kotagal V, Müller ML, Koeppe RA, Scott PJ, Albin RL, Frey KA, Petrou M (2014) Diabetes mellitus is independently associated with more severe cognitive impairment in Parkinson disease. Parkinsonism Relat Disord 20(12):1394–1398

Cai R, Zhang Y, Simmering JE, Schultz JL, Li Y, Fernandez-Carasa I, Consiglio A, Raya A, Polgreen PM, Narayanan NS, Yuan Y, Chen Z, Su W, Han Y, Zhao C, Gao L, Ji X, Welsh MJ, Liu L (2019) Enhancing glycolysis attenuates Parkinson's disease progression in models and clinical databases. J Clin Invest 129(10):4539–4549

Camargo Maluf F, Feder D, Alves de Siqueira Carvalho A (2019) Analysis of the relationship between type ii diabetes mellitus and Parkinson's disease: a systematic review. Parkinsons Dis 2019:4951379

Cataldo Bascuñan LR, Lyons C, Bennet H, Artner I, Fex M (2019) Serotonergic regulation of insulin secretion. Acta Physiol (Oxf) 225(1):e13101

Cataldo LR, Mizgier ML, Busso D, Olmos P, Galgani JE, Valenzuela R, Mezzano D, Aranda E, Cortes VA, Santos JL (2016) Serotonin- and dopamine-related gene expression in db/db mouse islets and in MIN6 β-cells treated with palmitate and oleate. J Diabetes Res 2016:3793781

Chen Y, Hong F, Chen H, Fan RF, Zhang XL, Zhang Y, Zhu JX (2014) Distinctive expression and cellular distribution of dopamine receptors in the pancreatic islets of rats. Cell Tissue Res 357 (3):597–606

Chen H, Hong F, Chen Y, Li J, Yao YS, Zhang Y, Zheng LF, Zhu JX (2016) Activation of islet 5-HT$_4$ receptor regulates glycemic control through promoting insulin secretion. Eur J Pharmacol 89:354–361

Cline GW, McCarthy TJ, Carson RE, Calle RA (2018) Clinical and scientific value in the pursuit of quantification of beta cells in the pancreas by PET imaging. Diabetologia 61(12):2671–2673

Das VA, Robinson R, Paulose CS (2006) Enhanced β-adrenergic receptors in the brain and pancreas during pancreatic regeneration in weanling rats. Mol Cell Biochem 289(1–2):11–19

Das SL, Singh PP, Phillips AR, Murphy R, Windsor JA, Petrov MS (2014) Newly diagnosed diabetes mellitus after acute pancreatitis: a systematic review and meta-analysis. Gut 63 (5):818–831

de Leeuw van Weenen JE, Parlevliet ET, Maechler P, Havekes LM, Romijn JA, Ouwens DM, Pijl H, Guigas B (2010) The dopamine receptor D2 agonist bromocriptine inhibits glucose-stimulated insulin secretion by direct activation of the alpha2-adrenergic receptors in beta cells. Biochem Pharmacol 79(12):1827–1836

Defronzo RA (2011) Bromocriptine: a sympatholytic, d2-dopamine agonist for the treatment of type 2. Diabetes 34(6):1442

Elbassuoni EA, Ahmed RF (2019) Mechanism of the neuroprotective effect of GLP-1 in a rat model of Parkinson's with pre-existing diabetes. Neurochem Int 131:104583

Erickson JD, Eiden LE (1993) Functional identification and molecular cloning of a human brain vesicle monoamine transporter. J Neurochem 61:2314–2317

Erickson JD, Schafer MK, Bonner TI et al (1996) Distinct pharmacological properties and distribution in neurons and endocrine cells of two isoforms of the human vesicular monoamine transporter. Proc Natl Acad Sci U S A 93:5166–5171

Ericson LE, Hakanson R, Lundquist I (1977) Accumulation of dopamine in mouse pancreatic B-cells following injection of L-DOPA. Localization to secretory granules and inhibition of insulin secretion. Diabetologia 13:117–124

Farino ZJ, Morgenstern TJ, Maffei A et al (2019) New roles for dopamine D2 and D3 receptors in pancreatic β cell insulin secretion. Mol Psychiatry 25(9):2070–2085

Fukushima D, Doi H, Fukushima K et al (2010) Glutamate exocrine dynamics augmented by plasma glutamine and the distribution of amino acid transporters of the rat pancreas. J Physiol Pharmacol 61:265–271

Ganic E, Johansson JK, Bennet H, Fex M, Artner I (2015) Islet-specific monoamine oxidase a and B expression depends on MafA transcriptional activity and is compromised in type 2 diabetes. Biochem Biophys Res Commun 468(4):629–635

Garcia A, Mirbolooki MR, Constantinescu C et al (2011) 18F-Fallypride PET of pancreatic islets: in vitro and in vivo rodent studies. J Nucl Med 52(7):1125–1132

García-Tornadú I, Ornstein AM, Chamson-Reig A et al (2010) Disruption of the dopamine d2 receptor impairs insulin secretion and causes glucose intolerance. Endocrinology 151 (4):1441–1450

Goldstein DS, Holmes C (2008) Neuronal source of plasma dopamine. Clin Chem 54 (11):1864–1871

Goldberg D (1972) The detection of psychiatric illness by questionnaire. Maudsley monograph, No. 21. Oxford University Press, London

Grossrubatscher E, Veronese S, Ciaramella PD, Pugliese R, Boniardi M, De Carlis L, Torre M, Ravini M, Gambacorta M, Loli P (2008) High expression of dopamine receptor subtype 2 in a large series of neuroendocrine tumors. Cancer Biol Ther 7(12):1970–1978

Gylfe E (1978) Association between 5-hydroxytryptamine release and insulin secretion. J Endocrinol 78(2):239–248

Han X, Li B, Ye X, Mulatibieke T, Wu J, Dai J, Wu D, Ni J, Zhang R, Xue J, Wan R, Wang X, Hu G (2017) Dopamine D2 receptor signalling controls inflammation in acute pancreatitis via a PP2A-dependent Akt/NF-kappaB signalling pathway. Br J Pharmacol 174(24):4751–4770

Han X, Ni J, Wu Z et al (2020) Myeloid-specific dopamine D2 receptor signalling controls inflammation in acute pancreatitis via inhibiting M1 macrophage. Br J Pharmacol 177 (13):2991–3008

Hansen SE, Hedeskov CJ (1977) Simultaneous determination of the content of serotonin, dopamine, noradrenaline and adrenaline in pancreatic islets isolated from fed and starved mouse. Acta Endocrinol 86:820–832

Hashimoto K, Inoue T, Higashi T, Takei S, Awata T, Katayama S, Takagi R, Okada H, Matsushita S (2009) Dopamine D1-like receptor antagonist, SCH23390, exhibits a preventive effect on diabetes mellitus that occurs naturally in NOD mouse. Biochem Biophys Res Commun 383 (4):460–463

Hong F, Liu L, Fan RF et al (2014) New perspectives of vesicular monoamine transporter 2 chemical characteristics in mammals and its constant expression in type 1 diabetes rat models. Transl Res 163(2):171–182

Itoh M, Furman BL, Gerich JE (1982) Dopaminergic suppression of pancreatic somatostatin secretion. Acta Endocrinol 101(1):56–61

Ivanina T, Blumenstein Y, Shistik E, Barzilai R, Dascal N (2000) Modulation of L-type Ca2+ channels by $G_{\beta\gamma}$ and calmodulin via interactions with N and C termini of 1C. J Biol Chem 275:39846–39854

Iwatsuki K, Horiuchi A, Ren LM, Chiba S (1995) D1 dopamine receptors mediate dopamine-induced pancreatic exocrine secretion in anesthetized dogs. Hypertens Res 18(Suppl 1):S173–S174

Iwen KA, Backhaus J, Cassens M, Waltl M, Hedesan OC, Merkel M, Heeren J, Sina C, Rademacher L, Windjäger A, Haug AR, Kiefer FW, Lehnert H, Schmid SM (2017) Cold-induced brown adipose tissue activity alters plasma fatty acids and improves glucose metabolism in men. J Clin Endocrinol Metabol 102(11):4226–4234

Jandaghi P, Najafabadi HS, Bauer AS et al (2016) Expression of DRD2 is increased in human pancreatic ductal adenocarcinoma and inhibitors slow tumor growth in mouse. Gastroenterology 151(6):1218–1231

Jensen BC, O'Connell TD, Simpson PC (2011) Alpha-1-adrenergic receptors: targets for agonist drugs to treat heart failure. J Mol Cell Cardiol 51(4):518–528

Kailey B, van de Bunt M, Cheley S, Johnson PR, MacDonald PE, Gloyn AL, Rorsman P, Braun M (2012) SSTR2 is the functionally dominant somatostatin receptor in human pancreatic β- and α-cells. Am J Physiol Endocrinol Metab 303(9):E1107–E1116

Karanjia ND, Widdison AL, Lutrin FJ, Chang YB, Reber HA (1991) The antiinflammatory effect of dopamine in alcoholic hemorrhagic pancreatitis in cats. Studies on the receptors and mechanisms of action. Gastroenterology 101:1635–1641

Karanjia ND, Widdison AL, Lutrin FJ, Reber HA (1994) Dopamine in models of alcoholic acute pancreatitis. Gut 35:547–551

Karhunen T, Tilgmann C, Ulmanen I, Julkunen I, Panula P (1994) Distribution of catechol-O-methyltransferase enzyme in rat tissues. J Histochem Cytochem 42(8):1079–1090

Kaya E, Dervisoglu A, Eroglu C, Polat C, Sunbul M, Ozkan K (2005) Acute pancreatitis caused by leptospirosis: report of two cases. World J Gastroenterol 11(28):4447–4449

Kavelaars A, Cobelens PM, Teunis MA, Heijnen CJ (2005) Changes in innate and acquired immune responses in mouse with targeted deletion of the dopamine transporter gene. J Neuroimmunol 161(1–2):162–168

Keahey HH, Boyd AE, Kunze DL (1989) Catecholamine modulation of calcium currents in clonal pancreatic beta-cells. Am J Phys 257(6 Pt 1):C1171–C1176

Kelly AC, Steyn LV, Kitzmann JP, Anderson MJ, Mueller KR, Hart NJ, Lynch RM, Papas KK, Limesand SW (2014) Function and expression of sulfonylurea, adrenergic, and glucagon-like peptide 1 receptors in isolated porcine islets. Xenotransplantation 21(4):385–391

Kelly AC, Camacho LE, Pendarvis K et al (2018) Adrenergic receptor stimulation suppresses oxidative metabolism in isolated rat islets and Min6 cells. Mol Cell Endocrinol 473:136–145

Korner J, Cline GW, Slifstein M, Barba P, Rayat GR, Febres G, Leibel RL, Maffei A, Harris PE (2019) A role for foregut tyrosine metabolism in glucose tolerance. Mol Metab 23:37–50

Kozuka C, Sunagawa S, Ueda R et al (2015) A novel insulinotropic mechanism of whole grain-derived γ-oryzanol via the suppression of local dopamine D2 receptor signalling in mouse islet. Br J Pharmacol 172(18):4519–4534

Kuzhikandathil EV, Yu W, Oxford GS (1998) Human dopamine D3 and D2L receptors couple to inward rectifier potassium channels in mammalian cell lines. Mol Cell Neurosci 12:390–402

Lacey RJ, Chan SL, Cable HC, James RF, Perrett CW, Scarpello JH, Morgan NG (1996) Expression of alpha 2- and β-adrenoceptor subtypes in human islets of langerhans. J Endocrinol 148:531–543

Li J, Li GW, Zhang Y, Zhu JX (2018) Divergent distribution of dopamine receptors in tumor β-cell lines and. In Situ 39(6):877–822

Liu M, Ren L, Zhong X, Ding Y, Liu T, Liu Z, Yang X, Cui L, Yang L, Fan Y, Liu Y, Zhang Y (2020) D2-like receptors mediate dopamine-inhibited insulin secretion via ion channels in rat pancreatic β-cells. Front Endocrinol (Lausanne) 11:152

Maffei A, Segal AM, Alvarez-Perez JC, Garcia-Ocaña A, Harris PE (2015) Anti-incretin, anti-proliferative action of dopamine on β cells. Mol Endocrinol 29(4):542–557

Matsumoto P, Brown A, Dunbar JC (1984) Biogenic amine regulation of glucagon secretion. Acta Diabetol Lat 21(4):333–340

Mezey E, Eisenhofer G, Harta G, Hansson S, Gould L, Hunyady B, Hoffman BJ (1996) A novel nonneuronal catecholaminergic system: exocrine pancreas synthesizes and releases dopamine. Proc Natl Acad Sci U S A 93(19):10377–10382

Morris JK, Seim NB, Bomhoff GL, Geiger PC, Stanford JA (2011) Effects of unilateral nigrostriatal dopamine depletion on peripheral glucose tolerance and insulin signaling in middle aged rats. Neurosci Lett 504(3):219–222

Morris JK, Zhang H, Gupte AA, Bomhoff GL, Stanford JA, Geiger PC (2008) Measures of striatal insulin resistance in a 6-hydroxydopamine model of Parkinson's disease. Brain Res 1240:185–195

Mukherjee J, Yang ZY, Das MK, Brown T (1995) Fluorinated benzamide neuroleptics: III. Development of (S)-N-[(1-allyl-2-pyrrolidinyl) methyl]-5-(3- [18F]fluoropropyl) -2,3-dimethoxybenzamide as an improved dopamine D-2 receptor tracer. Nucl Med Biol 22:283–296

Mukherjee J, Shi B, Christian BT, Chattopadhyay S, Narayanan TK (2004) ^{11}C-fallypride: radiosynthesis and preliminary evaluation of novel dopamine D2/D3 receptor PET radiotracer in nonhuman primate brain. Bioorg Med Chem 12:95–102

Nagler J, Schriever SC, De Angelis M, Pfluger PT, Schramm KW (2018) Comprehensive analysis of nine monoamines and metabolites in small amounts of peripheral murine (C57Bl/6 J) tissues. Biomed Chromatogr 32(4). https://doi.org/10.1002/bmc.4151

Nakada N, Mikami T, Hana K et al (2014) Unique and selective expression of L-amino acid transporter 1 in human tissue as well as being an aspect of oncofetal protein. Histol Histopathol 29(2):217–227

Nissinen E, Lindén IB, Schultz E, Pohto P (1992) Biochemical and pharmacological properties of a peripherally acting catechol-O-methyltransferase inhibitor entacapone. Naunyn Schmiedeberg's Arch Pharmacol 346(3):262–266

Oomori Y, Iuchi H, Ishikawa K, Satoh Y, Ono K (1994) Immunocytochemical study of tyrosine hydroxylase and dopamine β-hydroxylase immunoreactivities in the rat pancreas. Histochemistry 101(5):313–323

Papa I, Saliba D, Ponzoni M, Bustamante S, Canete PF, Gonzalez-Figueroa P, McNamara HA, Valvo S, Grimbaldeston M, Sweet RA, Vohra H, Cockburn IA, Meyer-Hermann M, Dustin ML, Doglioni C, Vinuesa CG (2017) TFH-derived dopamine accelerates productive synapses in germinal centres. Nature 547(7663):318–323

Pecic S, Milosavic N, Rayat G, Maffei A, Harris PE (2019) A novel optical tracer for VMAT2 applied to live cell measurements of vesicle maturation in cultured human β-cells. Sci Rep 9 (1):5403

Persson-Sjögren S, Forsgren S, Täljedal IB (2002) Tyrosine hydroxylase in mouse pancreatic islet cells, in situ and after syngeneic transplantation to kidney. Histol Histopathol 17(1):113–121

Pijl H, Ohashi S, Matsuda M, Miyazaki Y, Mahankali A, Kumar V, Pipek R, Iozzo P, Lancaster JL, Cincotta AH, DeFronzo RA (2000) Bromocriptine: a novel approach to the treatment of type 2 diabetes. Diabetes Care 23(8):1154–1161

Prasanna Kumar HR, Gowdappa HB, Hosmani T, Urs T (2018) Exocrine dysfunction correlates with endocrinal impairment of pancreas in type 2 diabetes mellitus. Indian J Endocrinol Metab 22(1):121–125

Ran X, Yang Y, Meng Y, Li Y, Zhou L, Wang Z, Zhu J (2019) Distribution of D1 and D2 receptor-immunoreactive neurons in the paraventricular nucleus of the hypothalamus in the rat. J Chem Neuroanat 98:97–103

Rosati G, Maioli M, Aiello I, Farris A, Agnetti V (1976) Effects of long-term L-dopa therapy on carbohydrate metabolism in patients with Parkinson's disease. Eur Neurol 14(3):229–239

Rubí B, Ljubicic S, Pournourmohammadi S, Carobbio S, Armanet M, Bartley C, Maechler P (2005) Dopamine D2-like receptors are expressed in pancreatic β cells and mediate inhibition of insulin secretion. J Biol Chem 280(44):36824–36832

Sandyk R (1993) The relationship between diabetes mellitus and Parkinson's disease. Int J Neurosci 69:125–130

Schafer MK, Hartwig NR, Kalmbach N, Klietz M, Anlauf M, Eiden LE, Weihe E (2013) Species-specific vesicular monoamine transporter 2 (VMAT2) expression in mammalian pancreatic β cells: implications for optimising radioligand-based human β cell mass (BCM) imaging in animal models. Diabetologia 56(5):1047–1056

Schwetz TA, Ustione A, Piston DW (2013) Neuropeptide Y and somatostatin inhibit insulin secretion through different mechanisms. Am J Physiol Endocrinol Metab 304(2):E211–E221

Shankar E, Santhosh KT, Paulose CS (2006) Dopaminergic regulation of glucose-induced insulin secretion through dopamine D2 receptors in the pancreatic islets in vitro. IUBMB Life 58 (3):157–163

Simpson N, Maffei A, Freeby M, Burroughs S, Freyberg Z, Javitch J, Leibel RL, Harris PE (2012) Dopamine-mediated autocrine inhibitory circuit regulating human insulin secretion in vitro. Mol Endocrinol 26(10):1757–1772

Sirtori CR, Bolme P, Azarnoff DL (1972) Metabolic responses to acute and chronic L-dopa administration in patients with parkinsonism. N Engl J Med 287:729–733

Sjoholm A (1991) A-adrenergic inhibition of rat pancreatic b-cell replication and insulin secretion is mediated through a pertussis toxin-sensitive G-protein regulating islet cAMP content. Biochem Biophys Res Commun 180(1):152–155

Song J, Zheng LF, Zhang XL, Feng XY, Fan RF, Sun L, Hong F, Zhang Y, Zhu JX (2014) Upregulation of β1-adrenoceptors is involved in the formation of gastric dysmotility in the 6-hydroxydopamine rat model of Parkinson's disease. Transl Res 164(1):22–31

Somvanshi RK, Jhajj A, Heer M, Kumar U (2018) Characterization of somatostatin receptors and associated signaling pathways in pancreas of R6/2 transgenic mouse. Biochim Biophys Acta Mol basis Dis 1864(2):359–373

Straub SG, Sharp GW (2012) Evolving insights regarding mechanisms for the inhibition of insulin release by norepinephrine and heterotrimeric G proteins. Am J Physiol Cell Physiol 302(12): C1687–C1698

Ter Horst KW, Lammers NM, Trinko R, Opland DM, Figee M, Ackermans MT, Booij J, van den Munckhof P, Schuurman PR, Fliers E, Denys D, DiLeone RJ, la Fleur SE, Serlie MJ (2018) Striatal dopamine regulates systemic glucose metabolism in humans and mouse. Sci Transl Med 10(442):eaar3752

Tirone TA, Norman MA, Moldovan S, DeMayo FJ, Wang XP, Brunicardi FC (2003) Pancreatic somatostatin inhibits insulin secretion via SSTR-5 in the isolated perfused mouse pancreas model. Pancreas 26(3):e67–e73

Uehara T, Sumiyoshi T, Itoh H, Kurachi M (2007) Dopamine D1 and D2 receptors regulate extracellular lactate and glucose concentrations in the nucleus accumbens. Brain Res 1133 (1):193–199

Urano Y, Sakurai T, Ueda H, Ogasawara J, Sakurai T, Takei M, Izawa T (2004) Desensitization of the inhibitory effect of norepinephrine on insulin secretion from pancreatic islets of exercise-trained rats. Metabolism 53:1424–1432

Ustione A, Piston DW (2012) Dopamine synthesis and D3 receptor activation in pancreatic β-cells regulates insulin secretion and intracellular Ca^{2+} oscillations. Mol Endocrinol 26 (11):1928–1940

Ustione A, Piston DW, Harris PE (2013) Mini review: dopaminergic regulation of insulin secretion from the pancreatic islet. Mol Endocrinol 27(8):1198–1207

van Beek AP, de Heide LJ, van Ginkel B, Slart RH, van der Horst-Schrivers AN, Boellaard R, de Vries EG, Noordzij W, Glaudemans AW (2017) Pancreatic uptake by 18F-FDOPA PET/CT in patients with hypoglycemia after gastric bypass surgery compared with controls with or without carbidopa pretreatment. Clin Nucl Med 42(3):163–168

Vayssette J, Vaysse N, Ribet A (1986) Dopamine receptors in pancreatic acinar cells from dog. Eur J Pharmacol 122(3):321–328

Walther DJ, Peter JU, Bashammakh S et al (2003) Synthesis of serotonin by a second tryptophan hydroxylase isoform. Science 299(5603):76

Wimalasena K (2011) Vesicular monoamine transporters: structure-function, pharmacology and medicinal chemistry. Med Res Rev 31(4):483–519

Wu W, Shang J, Feng Y et al (2008) Identification of glucose-dependant insulin secretion targets in pancreatic cells by combining defined-mechanism compound library screening and siRNA gene silencing. J Biomol Screen 13:128–134

Yan Y, Jiang W, Liu L, Wang X, Ding C, Tian Z, Zhou R (2015) Dopamine controls systemic inflammation through inhibition of NLRP3 inflammasome. Cell 160(1–2):62–73

Yang Y-W, Hsieh T-F, Li C-I, Liu C-S, Lin W-Y, Chiang J-H, Li T-C, Lin C-C (2017) Increased risk of Parkinson disease with diabetes mellitus in a population-based study. Medicine 96(3): e5921

Yao Y, Yang D, Han Y, Wang W, Wang N, Yang J, Zeng C (2016) Dopamine D1-like receptors suppress the proliferation of macrophages induced by Ox-LDL. Cell Physiol Biochem 38 (1):415–426

Zhang S, Peng XG, Liu CC, Liu H, Lu Y (2017) Low-dose dopamine reduces inlfammatory factors of acute pancreatitis in rats. Hepatobiliary Pancreat Dis Int 6:646–649

Zhao Y, Fang Q, Straub SG, Lindau M, Sharp GW (2010) Noradrenaline inhibits exocytosis via the G protein βγ subunit and refilling of the readily releasable granule pool via the α(i1/2) subunit. J Physiol 588(Pt 18):3485–3498

Zhou Y, Leonie FW, Holmseth S, Guo C, Urs VB, Li Y, Lehre A-C, Knut PL, Niels CD (2014) Proteome analysis and conditional deletion of the EAAT2 glutamate transporter provide evidence against a role of EAAT2 in pancreatic insulin secretion in mouse. J Biol Chem 289(3):1329–1344

Zhou L, Ran XR, Hong F, Li GW, Zhu JX (2019) Downregulated dopamine receptor 2 and upregulated corticotrophin releasing hormone in the paraventricular nucleus are correlated with decreased glucose tolerance in rats with bilateral substantia nigra lesions. Front Neurosci 13:751

The manufacturer's authorised representative in the EU is Springer Nature Customer Service Centre GmbH, Europaplatz 3, 69115 Heidelberg, Germany. If you have any concerns regarding our products, please contact ProductSafety@springernature.com

Printed and bound by CPI Group (UK) Ltd, Croydon, CR0 4YY
30/09/2026
02225240-0002